# EVOLUTION OF IONIZING RADIATION RESEARCH

Edited by **Mitsuru Nenoi**

**Evolution of Ionizing Radiation Research**
http://dx.doi.org/10.5772/59330
Edited by Mitsuru Nenoi

**Contributors**

Şeyda Çolak, Ernesto Lamanna, Cataldo Bianco, Giulia Marvaso, Lidia Strigari, Marco D'Andrea, Gonzalo Martínez-Barrera, Liliana Ivette A?vila-Co?doba, Miguel Martínez-López, Eduardo Sadot Herrera-Sosa, Enrique Vigueras-Santiago, Carlos Eduardo Barrera-Díaz, Fernando Ureña-Nuñez, Nelly González-Rivas, Bing Wang, Kaoru Tanaka, Takanori Katsube, Kouichi Maruyama, Yasuharu Ninomiya, Mitsuru Nenoi, Noriyuki Bob Ouchi, José Díaz, Antonio Bensussen, Suk Lee, Chul Yong Kim, Yuan Jie Cao, Haitham Sghaier, Cherif Ben Hamda, Alia Benkahla, Maher Gtari, Ameur Cherif, Kaïs Ghedira, Slimane Ben Miled, Houria Ouled-Haddar, Benjamin Hofner, María Del Carmen Montero-Calasanz, Marcia Silva, Semih Yılmaz, Abdurrahman Ayvaz, Marius Treutwein, Petra M. Härtl, Christian Gröger, Zaira Katsilieri, Barbara Dobler, Thiago Mastrangelo, Valter Arthur, Andre Machi

**Published by InTech**

Janeza Trdine 9, 51000 Rijeka, Croatia



**Publishing Process Manager** Ivona Lovric

**Technical Editor** InTech DTP team

**Cover** InTech Design team

First published September, 2015

Printed in Croatia

Additional hard copies can be obtained from orders@intechopen.com

Evolution of Ionizing Radiation Research, Edited by Mitsuru Nenoi

p. cm.

ISBN 978-953-51-2167-1

For EU product safety concerns:
IN TECH d.o.o., Prolaz Marije Krucifikse Kozulić 3, 51000 Rijeka, Croatia,
info@intechopen.com or visit our website at intechopen.com.

Section 1

# Biological Process

# Meet the editor

Mitsuru Nenoi graduated from Kyoto University, Graduate School of Sciences, Japan, in 1983. He started his career as a scientist at the National Institute of Radiological Sciences (NIRS), Japan. He received a PhD from Kyoto University in 1992 for the study on induced accumulation of polyubiquitin gene transcripts after exposure to ultraviolet light and treatment with 12-*O*-tetradecanoylphorbol 13-acetate. In 1991, he stayed at the University of Cincinnati College of Medicine, USA, as a visiting scientist and was involved in the study of mechanisms for transcriptional regulation of small heat shock genes of Drosophila. He is now the director of the Radiation Risk Reduction Research Program, Research Center for Radiation Protection of NIRS. His research interest is radiation biology, especially on the mechanism for gene regulation after exposure to ionizing radiation.

# Contents

# Preface

Since the publication of *Current Topics in Ionizing Radiation Research* in March 2012, industrial and medical applications of radiation have been augmented and scientific insight into mechanisms for radiation action notably progressed. In addition, the public concern about radiation risk has also grown extensively. Today the importance of risk communication among stakeholders involved in radiation-related issues is emphasized much more than any time in the past. Thus, the circumstances of radiation research have drastically changed, and the demand for a novel approach to radiation-related issues is increasing. It is thought that the publication of the book *Evolution of Ionizing Radiation Research* at this time would have enormous impacts on the society. The editor believes that technical experts would find a variety of new ideas and hints in this book that would be helpful to them to tackle ionizing radiation.

In the section "Biological Process," the dynamic feature of p53-Mdm2 feedback loop in various conditions is discussed with a mathematical model by Antonio Bensussen and Jose Diaz. The classical quantitative radiation biology is overviewed by Noriyuki B. Ouchi. Bing Wang describes a laborious but beneficial research where moderation of harmful effects of ionizing radiation by modulating dietary habit is investigated. Cherif Ben Hamda introduces a web resource "RadioP," which was created to gather information about radio-resistant prokaryotes based on the published literature. In the section "Medical Uses," Suk Lee describes evaluation of radiotherapy treatment plan in view of optimization of radiation therapy treatment process in contemporary clinical use, which is thought to be an important aspect of medical utilization of ionizing radiation. Ernesto Lamanna overviews a variety of radiation used in radiotherapy and outlines new approaches and tools for quality control of therapeutic beams. Moreover, a commissioning procedure and a quality assurance program for linac twins with flattening filter free option are introduced by Marius Treutwein. In the Detection and Measurement section, the basic knowledge on radiation detectors and principles of radiation measurement are compactly summarized by Marcia Dutra R. Silva. In the section "Industrial Application section," radiation entomology is introduced by Valter Arthur, with an emphasis on its utilization in insect disinfestation and sterilization by ionizing radiations. Abdurrahman Ayvaz and Semih Yilmaz also described disinfestation of pests and insects by use of ionizing radiation. Finally, an industrial utilization of $\gamma$ radiation in recycling waste materials in concrete is described by Gonzalo Martinez-Berrera.

The editor wishes that the publication of *Evolution of Ionizing Radiation Research* could extensively drive research activities in the related areas similarly as the preceding book *Current Topics in Ionizing Radiation Research.*

**Dr. Mitsuru Nenoi**
Research Center for Radiation Protection,
National Institute of Radiological Sciences,
Chiba, Japan

Chapter 1

# Dynamics of p53 and Cancer

Antonio Bensussen and José Díaz

Additional information is available at the end of the chapter

http://dx.doi.org/10.5772/60916

**Abstract**

Cancer is a multifactorial disease in which cell types lost their capability to regulate growth, proliferation, and cell death pathways, causing the uncontrolled proliferation of tumor cells. Cell death pathway is supported by the operation of the p53–Mdm2-negative feedback loop that has a central role to prevent the development of tumor cells. Under severe DNA damage, this loop takes the control of the apoptotic pathway and activates Bax, which, in turn, activates the caspase cascade to produce the death of the injured cell. However, events like Mdm2 overexpression or the suppression of caspase-9 gene can block the transmission of the death signal to the caspase cascade allowing the survival of the mutated cell. In this chapter, a mathematical model that explores the effect of Mdm2 overexpression and the suppression of caspase-9 on the control of death by the p53–Mdm2 loop is presented. From the model, two strategies for tumor cell survival are indentified, showing how mutations that affect the death pathway allow the survival of transformed cells. The model suggests that the combination of different simultaneous treatments against these mutations can be a suitable strategy against cancer.

**Keywords:** Cancer, p53–Mdm2 feedback loop, Mdm2 overexpression, suppression of caspase-9 gene

## 1. Introduction

Cancer has become one of the most important chronic diseases around the world [1]. A great number of studies on cancer have been focused on the identification for the genetic factors responsible of the predisposition for it. All these works have reported that a wide range of genes is implicated, as well as a great number of environmental factors that increase cancer risk [1]. From a dynamical point of view, cancer is a complex robust system, in which many of the feedback motives that assure the homeostasis of cells have been altered, allowing a perturbed state of the cell in which it growths and proliferates at an increased rate.

The complexity of cancer lies on the high number of punctual DNA mutations (~105 per cell), which are grouped in a small number of signaling circuits, which seems to be the same recurring pathways [2]. In malignant tumors, mutations comprise oncogenes and tumor suppression genes like Mdm2, which takes part in the negative p53–Mdm2 feedback loop that controls the entrance of the cell to apoptosis [3]. Mdm2 is an E3 ubiquitin ligase that maintains low p53 levels in normal cells. However, under stress of genotoxic signals, Mdm2 inhibition on p53 is released, and several different responses are activated, including cell arrest and apoptosis.

The importance of the p53–Mdm2 loop in the tumor suppression is reflected in the fact that almost 50% of malignant tumors present a mutation in the p53 gene [4]. In cells without p53 mutation, the tumor suppression function of the p53 transcription factor can be stopped by overexpression of Mdm2, blocking the entrance of the cell to apoptosis. Another mechanism to override p53 tumor suppression function is the uncoupling of the caspase cascade from the intrinsic apoptosis pathway; mutations in caspase-3 and caspase-9 can be responsible for this uncoupling.

This chapter explores the dynamics of the p53 tumor suppression system under normal and stress conditions in order to understand how overexpression of Mdm2, and mutations in the components of the caspases cascade, alters the form in which p53 controls the entrance of the cell to apoptosis. The dynamic features of the ordinary and altered operation conditions of p53–Mdm2 feedback loop are explored with a mathematical model that links it with the caspase cascade. The biological implications of the results obtained from the model are discussed in deep to understand how the deregulation of the intrinsic apoptotic pathway leads the modified cells to escape from death.

## 2. Molecular biology of the p53 pathway

### 2.1. The p53 pathway

Activation of the p53 pathway starts under the command of a complex feed forward structure implied in DNA damage detection, which is formed by proteins like the histone acetyl transferase Tip60 and the MRN complex. These proteins induce the activation of the ATM kinase [5–7]. In the next step, ATM phosphorylates and activates histone H2AX and other proteins with a BRCT domain (such as Nbs1, 53BP1, and MDC1) [7, 8]. These proteins will bind to the broken segment of DNA in order to stop the spread of damage [8]. Similarly, ATM activates Chk2 kinase, partially responsible for promoting cell cycle arrest [8, 9]. Chk2 also promotes the activity of the transcription factor E2F-1 [9], which regulates the expression of Chk2 and proapoptotic proteins ASPP [10] (Figure 1).

The active nuclear form of ATM assembles a complex with NEMO, which is ubiquitinated and transported to the cytoplasm where induces the release of the transcription factor NF-κB [11], a factor that enters the nucleus and promotes the expression of various genes [12, 13], including BCL-XL [14, 15]. Simultaneously, in the nucleus, active ATM phosphorylates the p53 store, and Mdm2 becomes less effective to recognize p53 delaying its degradation. In addition, ATM phosphorylates Mdm2 and induces its self-ubiquitination and degradation in the proteosome, allowing the increase of p53 nuclear concentration over time [16].

On the other hand, active Chk2 phosphorylates p53 to stabilize its structure [8]. At the end of this initial cascade of phosphorylation, p53 can interact either with accessory proteins ASPP1, ASPP2, or with MUC1 [17, 18] (Figure 1). In response to genotoxic stimuli, the transmembrane glycoprotein Musin-1 (MUC1) is cleaved, and the cytoplasmic segment is targeted to the nucleus, allowing it to bind to phosphorylated p53, inducing cell cycle arrest [19–24]. However, if there is severe DNA damage, p53 binds mainly to ASPP proteins, enhancing the proapoptotic function of p53 [25]. Thus, according to the intensity of the stimulus, p53 can be oriented either to transcribe proapoptotic genes or antiapoptotic genes, depending on the accessory proteins to which it is attached.

When the DNA damage signal is weak, nuclear concentration of MUC1 increases and enhances its binding to p53. After that, p53/MUC1 interacts with p300/CBP-associated factor (PCAF) to form a new complex [18, 26]. PCAF acetylates p53 on lysine 320, allowing it to recognize specifically the promoters of genes related with DNA damage repair and cell cycle arrest, as p21WAF1/cip1 [18]. MUC1 blocks Bax dimerization and the Bax-mediated release of cytochrome *c* [27]. MUC1 is capable of enhancing the expression of ARF, resulting in Mdm2 repression [28]. When p53/MUC1 complex starts its nuclear functions, the deacetylase SIRT1 opposes the PCAF-dependent acetylation of p53 by removing the acetyl group, blocking its transcriptional activity when the cell does not have enough energy to express proteins [29–31]. However, SIRT1 is not the only control point of p53. As it was discussed above, active ATM, along with NEMO, turns on NF-κB, which promotes the expression of antiapoptotic genes like BCL-XL [15].

NF-κB competes with p53 for the cofactors acetyltransferases p300, CBP, and PCAF, which are required for the binding of p53 to DNA [32, 33]. The effect of such interaction leads to the indirect repression of the transcriptional activity of p53 because the concentration of the cofactors remains relatively constant [11, 32, 34] (Figure 1).

Once nuclear p53 is activated, it starts the transcription of sensor genes like the phosphatase Wip1 and Mdm2 (in its p90 isoform) in order to activate the negative feedback control of the pathway. Wip1 function is to dephosphorylate Chk2, ATM, activated p53, and phosphorylated Mdm2. This action leads to inactivation of all of these proteins, except Mdm2 that is activated [25, 33, 35–38]. The dephosphorylated form of Mdm2, together with activated Wip1 and new synthesized Mdm2, inhibits p53 and reduces its nuclear concentration, avoiding its transcriptional activity [39]. Because of Wip1 function, the signal generated by ATM and Chk2 blinks while DNA damage is not repaired since ATM activator proteins remain linked to broken DNA [33, 40, 41]. When DNA damage is repaired, the nuclear topology is restored and ATM activating proteins stops the signal. In this case, Wip1 definitely inactivates ATM and its

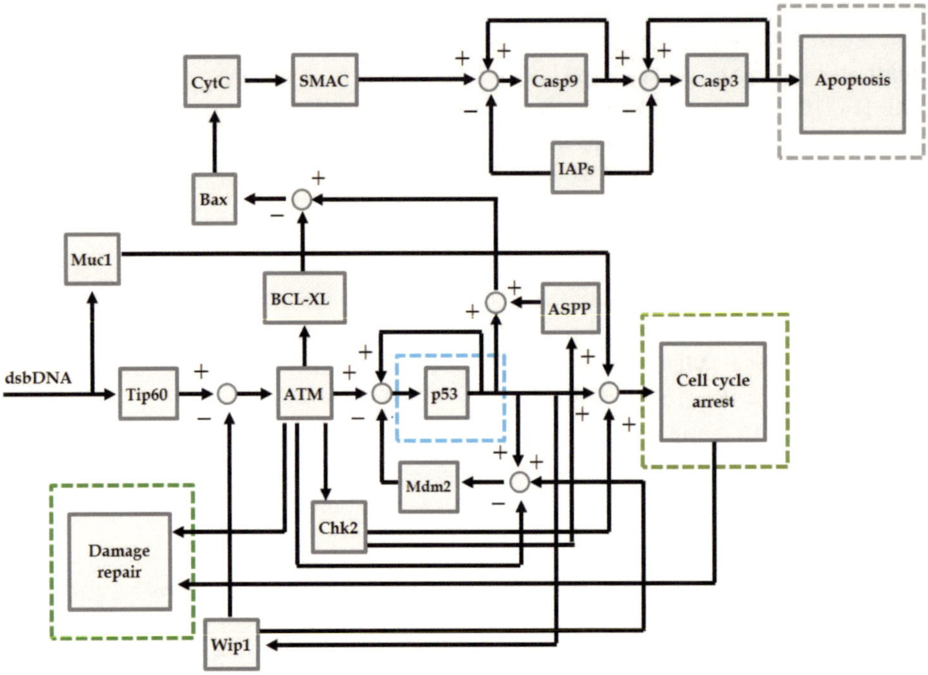


**Figure 1.** The p53 pathway.

effector molecules like Chk2. This encourages Mdm2 (p90 isoform) to suppress efficiently p53 activity, allowing the cell to return to its normal state. The levels of p90Mdm2 and Wip1 also return to their basal value [38] (Figure 1). If the signal of DNA damage is either too strong or persistent, and the cell has not enough energy to repair the damage, p53 binds to ASPP proteins and forms a complex with the histone acetyltransferase p300/CBP to acetylate p53 on lysine 373 [26, 42].

Another regulatory interaction is held by proteins p90 and Tip60. The interaction of p90 with p53 allows Tip60 to transfer an acetyl group to Lysine 120 of p53 [43]. Together, these chemical modifications permit the selective recognition of the promoter sites of proapoptotic genes like PUMA, Bak, and Bax by p53 [43, 44]. It is noteworthy that the formation of the p53/p300 complex is reversible, and p300 is separated from the complex by the deacetylation of p53 [30]. The enzyme responsible for this step is the deacetylase SIRT1 [30, 45]. This enzyme uses NAD +, from the cellular metabolism, as a cofactor; typically, SIRT1 activity can delay the apoptotic intrinsic pathway when there is no sufficient metabolic energy and whether DNA damage is not enough to activate apoptosis directly; nonetheless, this enzymatic regulation has no importance when DNA damage is severe [30].

In the cytoplasm, p53 forms a complex with the antiapoptotic protein BCL-XL in order to create a cytoplasmic reserve of p53 [16, 46]; in parallel to this, the excess of free BAX is neutralized in the cytoplasm by forming a complex with the antiapoptotic protein BCL2 [3]. Once the cell has chosen to die, p53 enhances the expression of proapoptotic genes like Bax, PUMA, NOXA,

BID, p53AIP, DR5, caspase-6, PERP, and FAS [47]. As a result, PUMA cytosolic concentration increases and enables its interaction with the p53/BCL-XL complex, replacing p53; in turn, free cytoplasmic p53 interacts with complex BCL2/BAX in order to form a new complex BCL2/p53, releasing BAX into the cytoplasm [46–48]. Free BAX monomers will interact with other Bak and Bax monomers to form the mitochondrial apoptosis-induced channel (MAC), and those complexes will perforate mitochondrial outer membrane, allowing the release of cytochrome *c*, SMAC/DIABLO, and Omi/HTRA [47, 49], triggering the commitment step in the apoptotic intrinsic pathway. Free cytochrome *c* will interact with apoptotic protease activating factor 1 (Apaf-1) and procaspase-9 to form the apoptosome. At the same time, cytosolic SMACs block the repression of inhibitor of apoptosis proteins (IAPs) on the initiator and executer caspases, ensuring apoptosis effectiveness [50–57]. Then the apoptosome cleaves the inactive procaspase-9 to active initiator caspase-9, and then it will activate executer caspase-3 [58]. Once this protein is activated, it starts a positive feedback loop for its self-activation as well for the activation of initiator caspases [59]. Caspase-3 starts the degradation of the cytoskeleton and other important cellular components, triggering the exposure of phosphotidylserine on the outer leaflet of the apoptotic cells, allowing the noninflammatory phagocytic recognition of these cells [47, 60] (Figure 1).

## 3. Nonlinear dynamics

### 3.1. Cells are complex networks

Cells are formed by several types of molecules (nodes) that are linked by the interactions between them forming a complex and hierarchical network. This network consists of a series of subnetworks that controls cellular processes like signaling, metabolism, apoptosis, transcription, translation, and DNA repairing and modification. The links between nodes can be of several types, including protein–protein, protein–DNA (transcription factors), small molecule–DNA (acetylation, methylation, etc.), small molecule–protein (posttranslation modifications), and metabolic [12]. Furthermore, these interactions are organized in clusters, with specific motives such as feedforward structures, single-input modules, or feedback modules [12]. This modular architecture of the cell's network is regulated by global regulators or "hubs" such as p53, mTOR, or AMPK [61].

### 3.2. The control principles of biology

A remarkable property of the cellular network is that all of its connections respond to specific and well-defined stimuli; in other words, such interactions are controllable. In this form, the modular structure of cells allows them to control several processes simultaneously, depending of the received inputs. Control is exerted by specific regulatory structures or motives like the negative feedback loop.

In order to understand why the negative feedback loop is so important for the control of the cellular processes, the block diagram of Figure 2A shows the functioning of this motive when a reference signal (or input) $r$ is transformed into an output signal $y$. From this diagram,

$$y = rG + rG\Delta + \delta - yG - yG\Delta \tag{1}$$

where $G$ corresponds to a system, $\Delta$ is the intrinsic noise proper of the system, and G and $\delta$ is the extrinsic noise due to external factors.

Working Equation (1) algebraically, we obtain

$$y = \frac{rG(1+\Delta)}{1+G(1+\Delta)} + \frac{\delta}{1+G(1+\Delta)} \tag{2}$$

That is equivalent to

$$y = \frac{r}{1/G(1+\Delta)+1} + \frac{\delta}{1+G(1+\Delta)} \tag{3}$$

According to Equation (3), if the regulation of the system $G$ tends to be stronger (i.e., $G \to \infty$), then the input signal will be totally transformed into the output, regardless the presence of intrinsic and extrinsic noise. In other words,

$$y = r \tag{4}$$

In this form, negative feedback loops are able to neutralize interferences due to extrinsic and intrinsic noise. This characteristic is known as *robustness*, and it confers to the system the capability to function in a changing environment like the intracellular medium [62, 63], which explains why negative feedback motives are so common in cell biology. In regard to robustness, it was thought that *redundancy* was its obligated synonym because if the cell loses a determined connection, and there are redundant interactions, then the cell can compensate the absence of such a connection. However, one system can be redundant but not robust depending on the type of connections that are redundant. At the molecular level, redundant interactions contribute to enhance the regulation of a biological process, which is a sine qua non condition to reject all effects of intrinsic and extrinsic noises [62].

Besides redundant connections, the cell also has contradictory interactions in which a biological molecule simultaneously activates and represses one process. These motives are better known as feedforward loops, and sometimes they are called *incoherent* feedforward motives [64]. In order to understand why the cell needs to activate and to repress one process at the same time, from Figure 2B, the block diagram can be written as follows:

$$y = -rCAP - rC\Delta + rAP + r\Delta + yCAP + yC\Delta + \delta \tag{5}$$

where $P$ is the process to be controlled (e.g., transcription or translation), $A$ is the molecule that physically modifies and executes the process (e.g., RNA polymerase), and $C$ is a biological controller unit (e.g., a signaling circuit).

Working this algebraic expression, we obtain

$$y = \frac{-rC(AP+\Delta)}{1-C(AP+\Delta)} + \frac{r(AP+\Delta)}{1-C(AP+\Delta)} + \frac{\delta}{1-C(AP+\Delta)} \tag{6}$$

That is equivalent to

$$y = \frac{-r}{\left(C(AP+\Delta)\right)^{-1}-1} + \frac{r(AP+\Delta)}{1-C(AP+\Delta)} + \frac{\delta}{1-C(AP+\Delta)} \tag{7}$$

Once again, if the control of the system (Equation 7) tends to be stronger (i.e., $C \to \infty$), then we obtain Equation (4), which means that this structure also neutralizes the effects of intrinsic and extrinsic noises. Thus, this apparently incoherent motive has robustness as well as negative feedback loops. Nevertheless, in order to exert their full robustness properties, these motives must be faster than the negative feedback ones during the regulation of a process, and their robustness is given by the fact that they can prevent a disturbance before it occurs [65]. For such reasons, feedforward motives are useful to control complex systems like positive feedback loops or nonlinear processes like splicing.

However, none of these cellular motives could be effective without biological controllers. In fact, such devices are specifically designed for operating a determined control motive such as negative feedback loops, feedforward loops, or open loops (i.e., without feedback, Figure 2C) [66].

However, there is a general structure for assembling controllers (Figure 3A), which consists in coupling a comparator and a control action module (the core of controller) with an amplifier [66]. The comparator is a device that collects and contrasts the reference signal (input) with the signals produced by either a sensor (device that measures the output of the system) or a timer (in open-loop systems). The difference between these signals is known as the error signal ($e(t)$), and it is used by the core of the controller to activate its inner mechanism in order to produce a control signal. Then the amplifier augments the potency of the control signal for regulating the entire system [66].

There are many types of controller cores, but the simplest are on–off cores (all or nothing control). Mathematically, this controller core is described by the following equation:

$$u(t) = \begin{cases} 1, & e(t) \geq \alpha \\ 0, & e(t) < \alpha \end{cases} \tag{8}$$

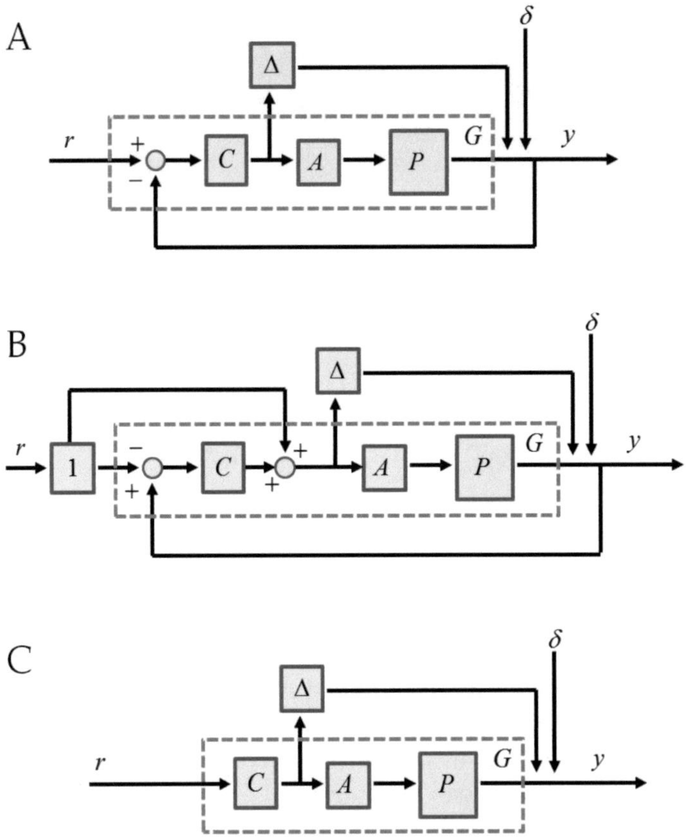


**Figure 2.** Control motives. (A) Negative feedback loops with the presence of intrinsic (Δ) and extrinsic noises (δ). (B) Feedforward motives with intrinsic and extrinsic noises. (C) Open-loop motives. In all cases, C is the controller, A is the effector, P is the plant or process to be controlled, and G represents the whole system.

It means that the controller core will produce a control signal only if the magnitude of error signal is higher than a threshold value ($\alpha$) [66]. In turn, this signal will be interpreted by the system as an order for turning on its functioning.

However, if the error signal magnitude is below the threshold value, the controller core will turn off the system. It is important to remark that such controller cores could be problematic when the error signal has an irregular behavior as shown in Figure 3B. To deal with that error signals, there is a special variant of on–off cores that includes a memory range (Figure 3C). This type of core is known as on–off with hysteresis [66] and generically is described by the following equation:

$$u(t) = \begin{cases} 1, & e(t) \geq \alpha \\ 0, & e(t) < -\alpha \end{cases} \tag{9}$$

http://dx.doi.org/10.5772/60916

It means that the core will produce a control signal only when the magnitude of the error signal is higher than a threshold value ($\alpha$), and it will stop when the error signal is below another threshold value (- $\alpha$). With this modification, the controller core does not change unexpectedly because of fluctuations in the error signal [66]. In cell biology, we can find this controller core almost everywhere. For example, in molecular switches (Figure 3D) at transcriptional level, one gene cannot be fully transcribed unless there is enough concentration of its inducer [63, 67].

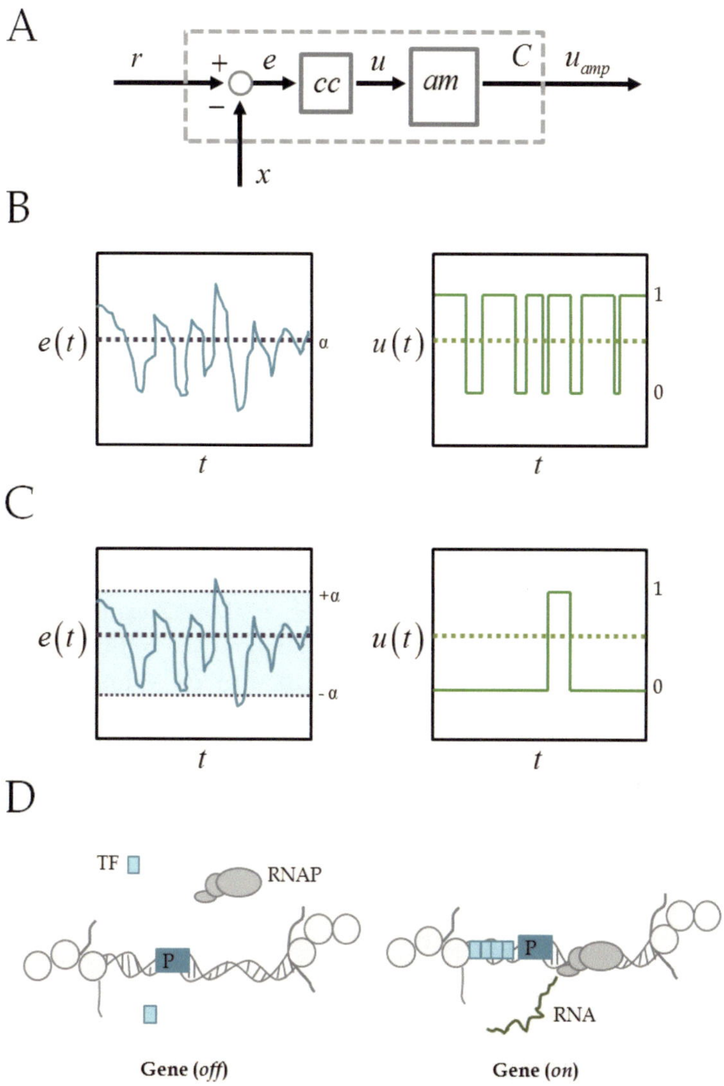


**Figure 3.** Biological controllers. (A) The canonical structure of a controller. (B) Temporal response of a core controller *all or nothing*. (C) Temporal response of a core controller *all or nothing with hysteresis*. (D) Biological equivalence to a core controller *all or nothing with hysteresis*.

### 3.3. Control principles on the p53 pathway

An example of control principles applied to biology can be observed during the evaluation of the DNA damage (Figure 1). This process is performed by some specialized kinases such as ataxia-telangiectasia mutated (ATM) and ataxia telangiectasia and Rad3 related (ATR), as well as some DNA binding proteins like the MRN complex (Mre11, Rad50, and Nbs1) [8]. If these proteins do not detect any DNA damage, the cell cycle will continue normally [68]. But if so, they will activate a feedforward signaling structure in order to induce the cell cycle arrest through effector molecules such as Chk2 and p21Waf1/cip1 [69]. Furthermore, when DNA damage is severe, the cell will turn on apoptosis through p53 transcriptional activity [68].

The p53 transcription factor also regulates the expression of genes responsible for the cell cycle arrest, repair of DNA damage, senescence, apoptosis, and other signaling pathways [70, 71]. The expression of p53 gene is activated by many inputs from transcriptional factors that include the p50 subunit of NF-κB, C/EPEβ-2, Ets-1, Pitx1, p73, and p53 itself [72–75]. Cytoplasmic p53 concentration is regulated in a negative feedback motive that starts when a fraction of p53 interacts with the anti apoptotic protein BCL-XL to generate a heterodimer [46, 76], and the remaining fraction of p53 is marked in its carboxyl-terminus with a nuclear import signal (NLS1) to be subsequently transported into the nucleus, where it interacts with its natural repressor: the ubiquitin ligase Mdm2 (Hdm2 in humans).

On the other hand, the negative regulator Mdm2 is an example of a biological sensor because its expression is a consequence of the p53 transcriptional activity, which allows the formation of a negative feedback loop. However, many studies suggest that p53–Mdm2 interactions are more complicated than it was thought; mainly because the Mdm2 gene has two promoters (P1 and P2) that can generate at least two isoforms of Mdm2 [35]. The first one isoform (p90Mdm2) is responsible of the p53 inhibition, but the second one (p76Mdm2) promotes the translation of p53 mRNA [35, 77]. Under ordinary conditions, p76Mdm2 expression is greater than p90Mdm2 [78], generating basal levels of p53 throughout the cell cycle. However, under cellular stress conditions, p53 induces the expression of p90Mdm2 in detriment of the p76Mdm2 isoform [77]. Remarkably, it was reported that other transcriptional factors can activate the Mdm2 gene in order to increase the regulation of p53, which is a clear example of the enhancement of biological robustness due to redundant interactions [78].

### 3.4. Nonlinear dynamical systems

All previous concepts are useful for understanding how the cell coordinates its molecular processes. However, it is important to introduce some concepts from dynamical systems prior to analyze how cells make operative decisions.

Most of the cellular processes can be represented in mathematical terms with nonlinear differential equations, which together with a set of initial conditions define a dynamical system. In this form, if the state of a system at time $t$ is determined by the set of variables $\{x_1(t), x_2(t), ..., x_n(t)\}$, then the respective dynamical system is the set of nonlinear differential equations $x_j(t)=f_j(x_1(t), x_2(t), ..., x_j(t), ..., x_n(t))$, $j=1, 2, ..., n$, subject to the initial conditions $\{x_{1o}(0), x_{2o}(0), ..., x_{no}(0)\}$. The set of variables $\{x_1(t), x_2(t), ..., x_n(t)\}$ defines the phase space of the dynamical system in which the motion of the system occurs. As time goes on, the point $\mathbf{x}(t)=\langle x_1(t),\ x_2(t),\ \ldots,\ x_n(t)\rangle$ moves along the phase space and defines a curve o trajectory of the system for each initial condition. Thus, the objective is to analyze the dynamical system in order to know the complete set of trajectories of the system in the phase plane (the phase portrait), and to discern the behavior of the trajectories in the neighborhood of the equilibrium o fixed points of the dynamical system. Equilibrium or fixed points are the points of the phase plane in which all the derivates vanish. As an example, in Figure 4, the phase portrait of the two-dimensional dynamical system $\dot{x}=\sin(y),\ \dot{y}=x-x^3$ is shown:

**Figure 4.** Portrait phase of the system $\dot{x}=\sin(y),\ \dot{y}=x-x^3$.

Nonlinear dynamical systems cannot be analytically solved; however, there are mathematical tools that can be used to know the qualitative behavior of the system. One of these tools is the linearization around a fixed point.

Let $\mathbf{x}^o=(x_1^o,\ x_2^o)$ be a fixed point of the two-dimensional dynamical system:

$$\begin{aligned}\dot{x}_1 &= f_1(x_1,x_2)\\ \dot{x}_2 &= f_2(x_1,x_2)\end{aligned} \tag{10}$$

If fluctuations $\delta x_1$ and $\delta x_2$ perturb the fixed point $\mathbf{x}^o$, the system is displaced to a new state $\delta\mathbf{x}^o=(x_1^o+\delta x_1,\ x_2^o+\delta x_2)$, and the trajectory that emerges from this point of the phase plane can either bring closer to the original equilibrium or go far away from it. If the trajectory tends to the fixed point in an asymptotic form, then $\mathbf{x}^o$ is stable, and it is an attractor. If the trajectory moves away from the fixed point, then $\mathbf{x}^o$ is unstable, and it is a repeller.

The dynamical behavior of the fluctuations determines the stability of the system, and it is essential to establish a form to analyze the evolution of the fluctuation along time. In order to achieve this goal, it is necessary to assume that the behavior of the fluctuations in the neighborhood of the fixed point is linear, i.e., the nonlinear terms can be neglected.

From Equation (10),

$$\begin{aligned} f_1\left(x_1^o+\delta x_1, x_2^o+\delta x_2\right) &\approx f_1\left(x_1^o, x_2^o\right)+\left.\frac{\partial f_1}{\partial x_1}\right|_{\left(x_1^o, x_2^o\right)} \delta x_1+\left.\frac{\partial f_1}{\partial x_2}\right|_{\left(x_1^o, x_2^o\right)} \delta x_2+O\left(\delta x_1^2, \delta x_2^2, \delta x_1 \delta x_2\right) \\ f_2\left(x_1^o+\delta x_1, x_2^o+\delta x_2\right) &\approx f_2\left(x_1^o, x_2^o\right)+\left.\frac{\partial f_2}{\partial x_1}\right|_{\left(x_1^o, x_2^o\right)} \delta x_1+\left.\frac{\partial f_2}{\partial x_2}\right|_{\left(x_1^o, x_2^o\right)} \delta x_2+O\left(\delta x_1^2, \delta x_2^2, \delta x_1 \delta x_2\right) \end{aligned} \tag{11}$$

Neglecting the nonlinear terms and taking into a consideration that the derivates vanish in the fixed point, Equation (11) can be written as

$$\begin{aligned} \delta \dot{x}_1 &= \left.\frac{\partial f_1}{\partial x_1}\right|_{\left(x_1^o, x_2^o\right)} \delta x_1+\left.\frac{\partial f_1}{\partial x_2}\right|_{\left(x_1^o, x_2^o\right)} \delta x_2 \\ \delta \dot{x}_2 &= \left.\frac{\partial f_2}{\partial x_1}\right|_{\left(x_1^o, x_2^o\right)} \delta x_1+\left.\frac{\partial f_2}{\partial x_2}\right|_{\left(x_1^o, x_2^o\right)} \delta x_2 \end{aligned} \tag{12}$$

which can be written in a matrix form as

$$\begin{bmatrix} \delta \dot{x}_1 \\ \delta \dot{x}_2 \end{bmatrix}=\begin{bmatrix} \frac{\partial f_1}{\partial x_1} & \frac{\partial f_1}{\partial x_2} \\ \frac{\partial f_2}{\partial x_1} & \frac{\partial f_2}{\partial x_2} \end{bmatrix}_{\left(x_1^o, x_2^o\right)} \begin{bmatrix} \delta x_1 \\ \delta x_2 \end{bmatrix} \tag{13}$$

Equation (13) represents the dynamics of the fluctuations around the fixed point when the nonlinear terms are neglected. The stability of the fixed point is then determined by the matrix $\mathbf{J}(\mathbf{x}^o)=\begin{bmatrix} \frac{\partial f_1}{\partial x_1} & \frac{\partial f_1}{\partial x_2} \\ \frac{\partial f_2}{\partial x_1} & \frac{\partial f_2}{\partial x_2} \end{bmatrix}_{\mathbf{x}^o}$, which is the Jacobian of the dynamical system. In two-dimensional dynamical systems, the evolution of the trajectories in the phase plane is settled on by the roots of the characteristic equation of the Jacobian: $\lambda^2+tr(\mathbf{J})\lambda+\det(\mathbf{J})=0$, where $tr(\mathbf{J})$ is the trace of the matrix and $det(\mathbf{J})$ its determinant. This characteristic equation has two roots, known as eigenvalues, that can be real, imaginary, or complex numbers. In all cases, if both $\lambda$s are negative real numbers or complex numbers with a negative real part, the fixed point is stable, and it is an attractor. If at least one of the roots is a positive real number or has a positive real part, the system is unstable. Figure 5 summarizes the properties of eigenvalues for any system. In the general case of an $n$-dimensional dynamical system, there are $n$ eigenvalues, and if all of them are negative real numbers or are complex numbers with negative real part, the fixed point is an attractor. Otherwise, if a least one eigenvalue is a positive real number or a complex number with a positive real part, the fixed point is unstable.

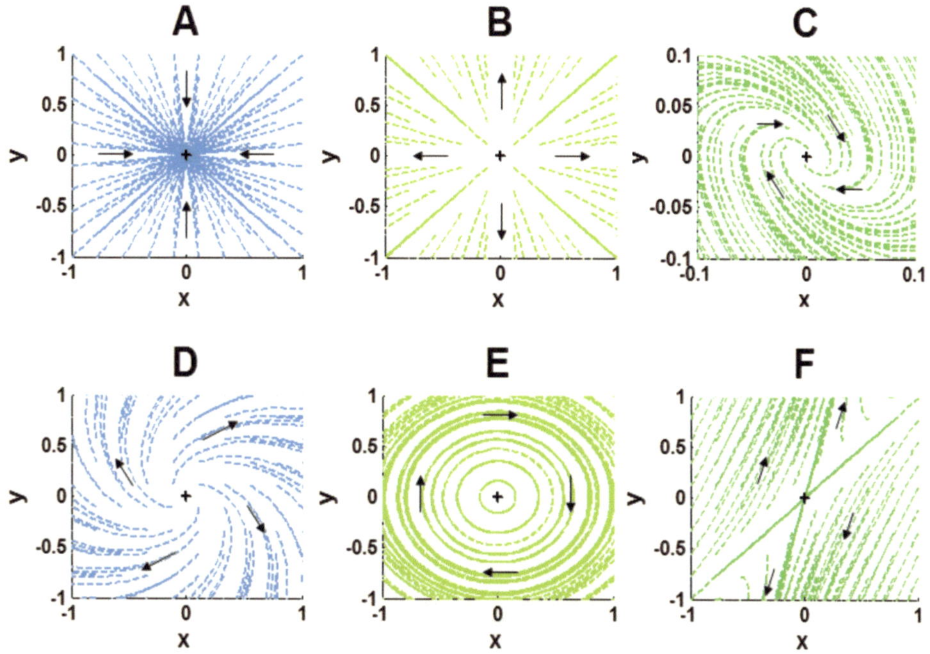


**Figure 5.** Portrait phases for different eigenvalues. (A) The portrait phase of one system is described by a stable node (sink) when $Re(\lambda_i) < 0$, $i = 1,2$, and there is no imaginary part. (B) On the other hand, if $Re(\lambda_i) > 0$, $i = 1, 2$, the portrait phase will show an unstable node (source). (C) Similarly, when $Re(\lambda_i) < 0$, $i = 1, 2$, and there is an imaginary part, the system presents a stable spiral (sink). (D) In case of $Re(\lambda_i) > 0$, $i = 1, 2$, and if there is an imaginary part, the system will present an unstable spiral (source). (E) When $Re(\lambda_i) = 0$, $i = 1, 2$, and there is an imaginary part, the portrait phase of the system will show a center (it is stable, but not asymptotically stable like a sink). (F) If the system has real eigenvalues with different signs, for instance, $\lambda_1 > 0$ and $\lambda_2 < 0$, the system will present a saddle point. In all cases, the origin is the equilibrium and is marked with a black cross, and the arrows show the direction of trajectories.

If $Re(\lambda_{1,2}) \neq 0$, the fixed points are hyperbolic points, which are robust points that cannot be altered by the small nonlinear terms. Fixed points like stable nodes and stable spirals are hyperbolic points. The Hartman–Grobman theorem affirms that the phase portrait in the neighborhood of a hyperbolic fixed point of a nonlinear system is topologically equivalent to the phase portrait of the linearization, i.e., the phase portrait of the nonlinear system is an invertible deformation of the phase portrait of the linearized system. A phase portrait whose topology cannot be altered by small nonlinear perturbations is structurally stable.

The phase portrait of a dynamical system can be modified as some parameter of the system varies, giving rise to qualitative changes in its structure. Such dynamical transitions are known as bifurcations. They generally occur in a one-dimensional subspace, and the remaining dimensions of the phase plane are affected as a consequence of the trajectories that can be attracted or repelled from this subspace. Bifurcations can lead to the rich qualitative behavior characteristic of nonlinear systems that includes bistability, biological switches, circadian rhythms, bursting, and traveling waves, among others.

Considering the imaginary plane, we can roughly classify bifurcations into two cases: (1) the eigenvalues of the Jacobian matrix are both real and bifurcations occur along the real axis as certain parameter changes. This kind of bifurcation comprises the saddle-node bifurcation, the transcritical bifurcation, and subcritical and supercritical pitchfork bifurcation. (2) The eigenvalues of the Jacobian matrix are complex conjugated. Bifurcations occur crossing the imaginary axis as certain parameter changes. This kind of bifurcation comprises the supercritical and subcritical Hopf bifurcation.

The formation of the complex by fibroblast growth factor (FGF) and its receptor and the subsequent homodimerization of the complexes can be taken as an example of a dynamical system. The biochemical reaction is

$$\begin{aligned} &\mathrm{FGF} + \mathrm{FGFR} \underset{k_{-1}}{\overset{k_1}{\rightleftharpoons}} \mathrm{FGF} - \mathrm{FGFR} \\ &2(\mathrm{FGF} - \mathrm{FGFR}) \underset{k_{-2}}{\overset{k_2}{\rightleftharpoons}} (\mathrm{FGF} - \mathrm{FGFR})_2 \end{aligned} \tag{14}$$

The corresponding dynamical system is

$$\begin{aligned} &\frac{dc}{dt} = k_1(\mathrm{FGF_T} - c)(\mathrm{FGFR_T} - c) - k_{-1}c, \\ &\frac{dc_2}{dt} = k_2 c^2 - k_{-2}c_2, \\ &c(0) = c_o \quad c_2(0) = c_{2o}. \end{aligned} \tag{15}$$

where $FGF_T$ is the total amount of the agonist in the medium, $FGFR_T$ is the total amount of FGF receptor in the cell's membrane, $c$ is the amount of the complex agonist–receptor FGF-FGFR, $c_2$ is the amount of the homodimer complex $(FGF\text{-}FGFR)_2$, and $k_1$, $k_{-1}$, $k_2$, and $k_{-2}$ are the rate constants. For the following particular values of parameters and constants, $FGF_T = 1\ \mu M$, $FGFR_T = 0.5\ \mu M$, $k_1 = 1\ \mu M^{-1}s^{-1}$, $k_{-1} = 0.004\ s^{-1}$, $k_2 = 2\ \mu M^{-1}s^{-1}$, and $k_{-2} = 1.8\ s^{-1}$; the fixed points are A(0.4652, 0.24) and B(1.0747, 1.282). The linearized system is

$$\begin{bmatrix} \delta \dot{c} \\ \delta \dot{c}_2 \end{bmatrix} = \begin{bmatrix} 2c - 1.54 & 0 \\ 4c & -1.8 \end{bmatrix} \begin{bmatrix} \delta c \\ \delta c_2 \end{bmatrix} \tag{16}$$

and the Jacobian of the linearized system is

$$J(\mathbf{x}^{\circ}) = \begin{bmatrix} 2c - 1.54 & 0 \\ 4c & -1.8 \end{bmatrix}_{\mathbf{x}^{\circ}} \tag{17}$$

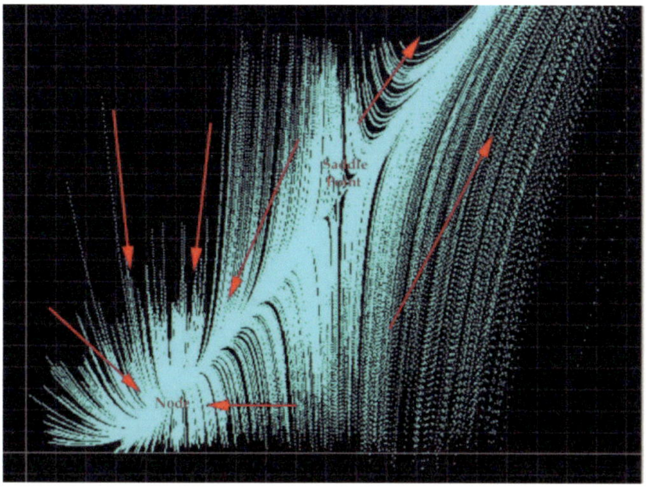

**Figure 6.** Portrait phase of the FGF model.

Evaluating the Jacobian (Equation 17) in each fixed point, the eigenvalues for the fixed point A are both real negative, and for the case of point B, one eigenvalue is real negative while the other is real positive. Thus, the fixed point A is a stable node, while the fixed point B is a saddle point. In Figure 6, the phase portrait of the dynamical system is shown, in which the saddle point is an energetic barrier that separates the dynamics of the system around the node from the dynamics of the outer region of the phase space, which is filled of trajectories that continually diverge. Both fixed points are hyperbolic and the phase portrait is structurally stable.

# 4. Modeling apoptosis

## 4.1. Mathematical modeling of the p53 pathway

p53 has a central role in the control of apoptosis and in the induction of death in cells after chemotherapy or radiotherapy. One of the main characteristics of the control of apoptosis by p53 is the existence of oscillations in the p53–Mdm2 negative feedback loop whose frequency and duration determine the entrance of the cell to apoptosis. These oscillations arise through a supercritical Hopf bifurcation in response to DNA damage signals [37, 79].

Many attempts have been done to understand how the negative p53–Mdm2 feedback loop sustains the observed p53–Mdm2 oscillations. These models have addressed that oscillations can arise in systems in which there is a sustained input signal and a single negative feedback loop with a delay term. However, models of this type are unable to produce multiple oscillations and are necessary to introduce a positive feedback in the core of equations to reproduce a wide range of oscillatory responses experimentally observed [80]. Nevertheless, the dynamical features of this kind of models are not robust against changes in the value of the parameters, and new attempts have been done to correct this problem by introducing additional feedback motives in the model that improve robustness but cannot explain the observed variability in the p53 and Mdm2 oscillations in cell populations [81]. Stochastic models have been proposed in order to overcome these limitations of deterministic models. In this kind of modeling, random variables like noise or DNA damage are taken into consideration in the core of the equations giving rise to a series of oscillations with variable amplitude in concordance with the experimental observations [80].

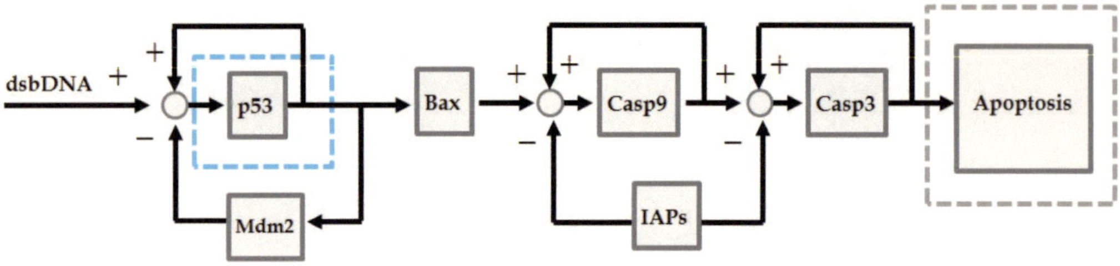


**Figure 7.** Minimal model of the p53 pathway.

Proctor and Douglas [81] proposed two stochastic mechanistic models of the p53–Mdm2 feedback loop. The first model explores the inhibitory effect of p14ARF on Mdm2, which leads to the stabilization of p53; the second model explores the effect of the phosphorylation of both Mmd2 and p53 by ATM. From both models, oscillations p53 and Mdm2 with high variability are obtained; however, the ATM model exhibits more variability. Both models are robust for a wide range of parameter values, indicating that stochasticity could be a basic component of the cell's response to DNA severe damage. Liu et al. [82] proposed a complete stochastic model of p53 and apoptosis with three-coupled modules. The first module represents the p53–Mdm2 feedback loop dynamics; the second module represents cytochrome *c* release from the mitochondria, and the third module represents caspase activation. Results from this model indicate that the major determinants of cell fate are the strength of p53 transcriptional activity together with its coupling to mitochondrial opening. This model indicates that p53 oscillations are a necessary but not sufficient condition for the onset of apoptosis. In this point, the Bax molecules located at the surface of the outer mitochondrial membrane have a central role in driving apoptosis, indicating that the possible role of the p53–Mdm2 system is to control the cytoplasmic levels of activated Bax.

Bistability is a broad dynamical feature of nonlinear biochemical systems, in which a saddle-point separates two stable nodes. The transition between nodes requires energy to overcome the high energetic barrier between them, leading to a saddle-node bifurcation. Generally, once the transition from one steady point to the other occurs, return to the original state needs a different pathway with different energy requirements, forming a hysteresis loop. Molecular switches are the common examples of this type of dynamics [83].

Bagci et al. [60] studied the role of Bax, Bcl2 synthesis, and degradation rates in mitochondria-dependent apoptosis, finding the existence of a saddle-node bifurcation in the caspase-3-Bax degradation rate diagram. This bifurcation drives the system either to a stable point with a high concentration of caspase-3 (cell death) or to a stable point with a low concentration of caspase-3 (cell survival). Out of this region, the systems dynamics is settled on by the existence of a single stable fix point that determines cell surviving independently of the concentration of caspase-3. An important conclusion from this work is that an abnormal state, like cancer, arises in cells when Bax degradation rate is above a threshold value, giving rise to a stable cell survival dynamics, i.e., cells do not die.

In order to explore the dynamics of the caspase cascade when coupled to the p53 pathway by Bax switch, a model of ordinary differential equations (ODEs) is proposed (Figure 7). However, this model takes into account only the most representative nodes of the pathway (Section 2.1): p53, Mdm2 and its mRNA, Bax, IAPs, active initiator caspase-9, and the active executioner caspase-3 (Figure 7B). Besides its simplicity, this model reproduces the oscillatory dynamics of the p53–Mdm2 and its effects on the activation of caspase-3. In this model of the p53 pathway, the concentration of p53 is given by

$$\frac{d[\text{p53}]}{dt} = \alpha_{\text{act}}[\text{p53}] + k_1[\text{p53}][\text{Mdm2}] - \delta_1[\text{p53}] \tag{18}$$

where $\alpha_{\text{act}}$ is the rate of p53 synthesis, $k_1$ is the affinity of Mdm2 to p53, and $\delta_1$ is the p53 rate of decay. In a similar form, the concentration of the Mdm2 mRNA is given by

$$\frac{d[\text{mRNA}]}{dt} = \alpha_1[\text{p53}] - \delta_2[\text{mRNA}] \tag{19}$$

where $\alpha_1$ is the rate of mRNA synthesis due to p53 activity and $\delta_2$ is the mRNA rate of decay. Continuing with the model, the concentration of Mdm2 is given by the following equation:

$$\frac{d[\text{Mdm2}]}{dt} = \alpha_2[\text{mRNA}] + k_1[\text{p53}][\text{Mdm2}] - \delta_3[\text{Mdm2}] \tag{20}$$

In this equation, $\alpha_2$ is the rate of Mdm2 synthesis, and $\delta_3$ is the Mdm2 rate of decay. The next equation represents the fraction of free Bax protein:

$$\frac{d[B^*]}{dt} = k_2([B]_T - [B^*])[\text{p53}] - \delta_4[B^*] \tag{21}$$

where $k_2$ is the rate of Bax release due to p53 activity, $[B^*]_T$ is the total concentration of Bax protein and $\delta_4$ is its rate of decay. The following equation represents the decrease of IAPs activity due to free Bax:

$$\frac{d[I]}{dt} = -k_3[B^*][I] \tag{22}$$

Finally, we have the equation for the activation of initiator caspase-9:

$$\frac{d[C9^*]}{dt} = k_4([C9]_T - [C9^*])([B^*] + [C3^*]) - (\delta_5 + k_5[I])[C9^*] \quad (23)$$

Moreover, the equation for the activation of executioner caspase-3 is as follows:

$$\frac{d[C3^*]}{dt} = k_6([C3]_T - [C3^*])([C9^*] + [C3^*]) - (\delta_6 + k_5[I])[C3^*] \quad (24)$$

where $k_4$ and $k_6$ are the rates of activation, $\delta_5$ and $\delta_6$ are the rates of caspases decay, and $k_5$ is the rate of repression due to IAPs activity. All parameters of the model are reported in Table 1. The initial concentrations (μM) of the variables are

$$[p53]_0 = 0.01 \qquad [I]_0 = 1 \qquad [mRNA]_0 = [Mdm2]_0 = [B^*]_0 = [C9^*]_0 = [C3^*]_0 = 0$$

| Parameter | Value | Biological meaning |
|---|---|---|
| $\alpha$ act | 0.15 $s^{-1}$ Low damage<br>0.2 $s^{-1}$ Medium damage<br>0.3 $s^{-1}$ Strong damage | Rate of p53 synthesis |
| $k_1$ | 1 $\mu M^{-1} s^{-1}$ | Rate of p53 degradation due to Mdm2 repression |
| $\delta_1$ | 0.01 $s^{-1}$ | Basal degradation of p53 |
| $\alpha$ 1 | 1 $s^{-1}$ | Rate of Mdm2 RNA synthesis due to p53 |
| $\delta$ 2 | 0.1 $s^{-1}$ | Basal degradation of Mdm2 RNA |
| $\alpha$ 2 | 0.1 $s^{-1}$ | Rate of synthesis of Mdm2 |
| $\delta$ 3 | 0.5 $s^{-1}$ | Basal degradation of Mdm2 |
| $k_2$ | 0.1 $\mu M^{-1} s^{-1}$ | Rate of Bax releasing due to p53 activity |
| $\delta$ 3 | 0.1 $s^{-1}$ | Basal degradation of Bax |
| $k_3$ | 0.02 $s^{-1}$ | IAPs inhibition due to Bax activity |
| $k_4$ | 0.01 $\mu M^{-1} s^{-1}$ | Rate of procaspase-9 activation |
| $\delta$ 5 | 0.1 $s^{-1}$ | Basal degradation of caspase-9 |
| $k_5$ | 10 $\mu M^{-1} s^{-1}$ | Inhibition of caspase-9 due to IAPs activity |
| $k_6$ | 0.1 $\mu M^{-1} s^{-1}$ | Rate of procaspase-3 activation |
| $\delta$ 6 | 0.1 $s^{-1}$ | Basal degradation of caspase-3 |
| **[B]**$_T$ | 1 μM | Total concentration of Bax |
| **[C9]**$_T$ | 1 μM | Total concentration of caspase-9 |
| **[C3]**$_T$ | 1 μM | Total concentration of caspase-3 |

Note: all parameters were estimated for this work.

**Table 1.** Model parameters

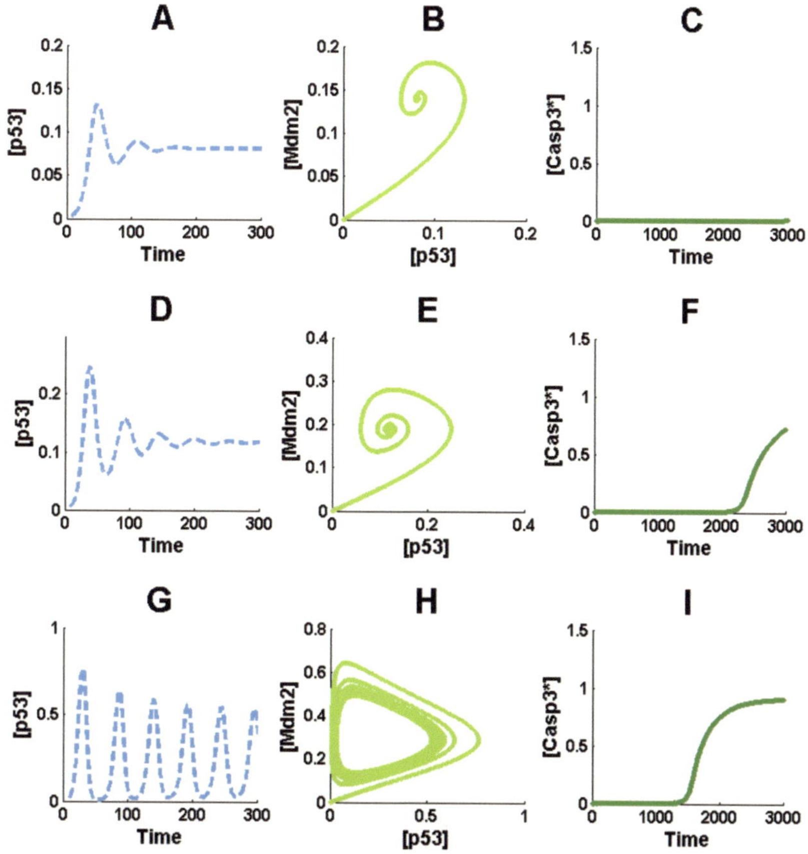


**Figure 8.** Dynamics of a simplistic model of p53 pathway and apoptosis. When the cell receives a soft damage on DNA, a few damped oscillations on p53 concentration (A) appear, which correspond to a stable spiral (B), and there is no increase in active caspase-3 concentration (C). When the damage on DNA is stronger, p53 exhibits stronger damped oscillations (D), and after a certain time, apoptosis starts (F). Finally, when DNA damage is severe, p53 oscillations are not damped (G), and a limit cycle is settled on (H), leading to a faster start of apoptosis (I). For all simulations, the time scale is in minutes, and concentrations are in micromolar units.

## 4.2. Dynamical features of the p53 pathway

From the above model, it is possible to visualize the main molecular events in the p53 pathway, as well as the activation of apoptosis. In the model, soft damage in DNA is interpreted as an increase of p53 synthesis. In this scenario, the p53 concentration presents damped oscillations (Figure 8A), which are represented by a stable spiral in the p53-Mdm3 phase plane (Figure 8B). With this level of DNA damage, the system does not activate apoptosis because the caspases-3 concentration remains zero (Figure 8C). When DNA damage reaches a medium level, the p53 concentration exhibits more pronounced damped oscillations, i.e., the stable spiral has a higher initial amplitude and a larger relaxation time (Figure 8D). However, if the DNA damage persists for long time, the active caspase-3 concentration will augment until it starts apoptosis irreversibly (Figure 8F). Moreover, when DNA damage is severe, the p53 pathway undergoes a supercrital Hopf bifurcation (reviewed in [83]), which means that the stable spiral (Figure 8E) becomes a stable limit cycle through an unstable spiral (Figure 8G). During the bifurcation process, the caspase-3 activation is faster, which implies that the cell has no option but to die (Figure 8I). Thus, p53 concentration acts as a biological timer and allows the cell to response in a specific way for certain perturbations, such as DNA damage intensity and duration. In addition, the activation of caspase-3 is governed by a control action of an "all or nothing" scheme, which avoids the random activation of apoptosis.

### 4.3. Over expression of Mdm2

In the nature, there are many factors that can permanently perturb the p53 network functioning, such as genetic alterations like deletions or constitutive overexpression due to polymorphisms [84]. A classical example of this point is the overexpression of Mdm2, which is present in more than forty different types of cancer, taking a central role in cancer development [85].

In order to know how the overexpression of Mdm2 perturbs apoptosis, Equation (20) of the model (see Section 5.1) can be modified as follows:

$$\frac{d[\mathrm{Mdm2}]}{dt} = \alpha_2[\mathrm{mRNA}] + \varepsilon_{onco} + k_1[\mathrm{p53}][\mathrm{Mdm2}] - \delta_3[\mathrm{Mdm2}] \tag{25}$$

where $\varepsilon_{onco}$ is a parameter and represents the external factors that induce the Mdm2 overexpression.

From a dynamical point of view and according to the model results, Mdm2 overexpression disrupts the basal dynamics of the p53-pathway and of apoptosis. From Figures 9A to 9C, clearly the normal response of the system to strong DNA damage is a series of oscillations of p53 concentration, i.e., a limit cycle dynamics that induces the activation of caspases-3 if the oscillations last for time enough. However, Mdm2 overexpression suppresses the p53 oscillations (Figures 9D to 9F) and avoids cascapse-3 activation (see Section 2.1), allowing the survival of cells with strong DNA damage. This could be one of the possible dynamical forms in which cells with intense DNA damage loss the control of apoptosis and survive, leading to a malignant tumor.

### 4.4. Suppression of caspase-9

Caspase-9 is a key protease that is constitutively expressed in a variety of fetal and adult tissues. Previous studies on cancer risk [58] show that polymorphism of caspase-9 is strongly correlated with lung cancer development in light smokers, reflecting an intense gene-environment interaction at this level of the apoptotic pathway. The loss of this caspase interrupts apoptosis, leading to abnormal proliferation of cells with severe DNA damage.

In the model of the p53 pathway, the missing of caspase-9 gene is modeled, assuming that

$$\frac{d[C9^*]}{dt} = 0 \tag{26}$$

Results from the minimal model presented in Section 5.1, modified with Equation (26), show that deletions on the caspase-9 gene suppress the activation of the effector caspase-3, uncoupling the p53 dynamics from the caspases activation module. Although the limit cycle in the p53–Mdm2 phase plane remains, the absence of caspase-9 avoids the activation of caspase-3 and apoptosis even if the DNA damage is severe.

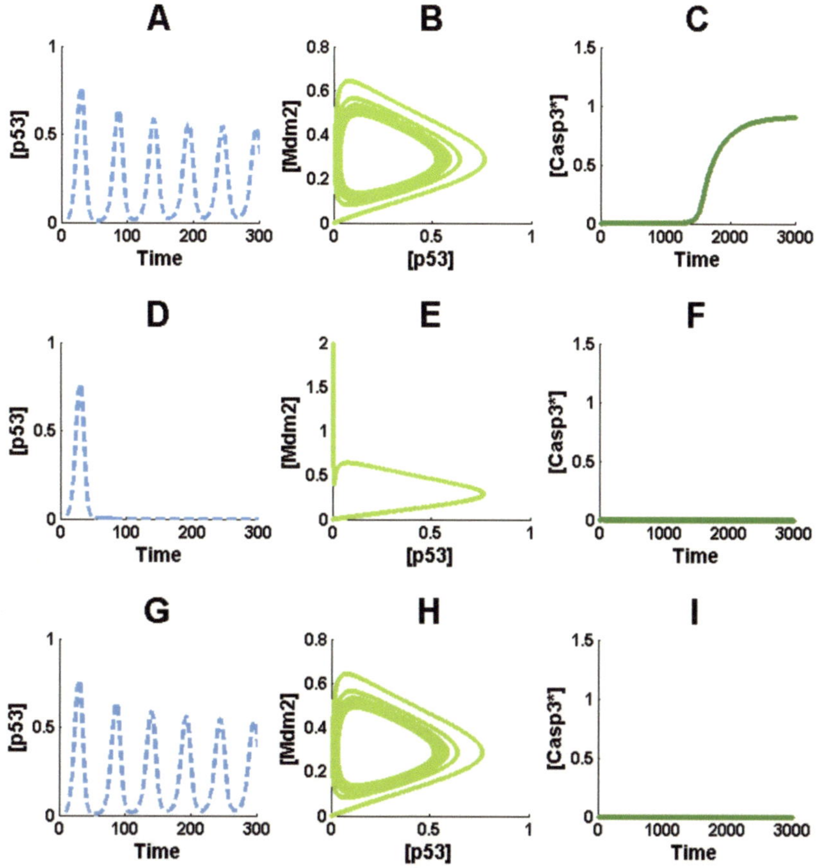


**Figure 9.** Oncogenic perturbations on the p53 and apoptosis pathways. Normally, when the cell receives severe DNA damage, the p53 concentration oscillates (A) following a limit cycle dynamics (B), which eventually starts apoptosis (C). However, overexpression of Mdm2 abrogates the p53 oscillations (D), dampens the normal dynamics (E), and avoids the caspase-3 activation (F). On the other hand, deletions on the caspace-9 gene simply abrogate the activation of effector caspase-3 (I). For all simulations, the time scale is in minutes and concentrations are in micromolar units.

### 4.5. Survival of malignant cells

These two examples of genetic perturbations show that there are two possible dynamical mechanisms implicated in the deregulation of apoptosis in cancerous cells. Such mechanisms are (1) uncoupling between the p53–Mdm2 limit cycle (pacemaker) and the caspase cascade and (2) suppression of p53 activity by overexpression of a negative regulator (like Mdm2) that leads to the not activation of caspase-3.

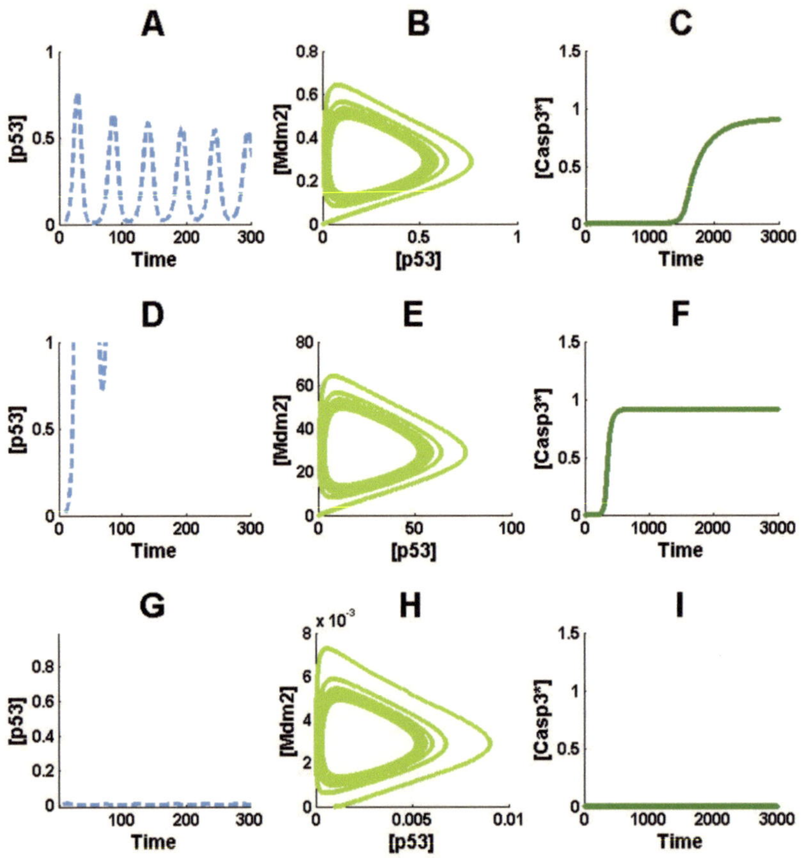


**Figure 10.** Variation of the rate of p53 degradation due to Mdm2 repression ($k_1$). Compared with the normal cell response to severe DNA damage (A, B, and C), the effect of decreasing in two orders the rate of p53 degradation due to Mdm2 $k_1$ augments the magnitude of p53 oscillations (D), preserving its limit cycle (E) and starts quicker apoptosis (F). On the other hand, the effect of increasing $k_1$ in two orders reduces the magnitude of p53 oscillations (G), avoiding caspase activation (I) but preserves the limit cycle (H).

In fact, p53 inhibition decreases miR-34a and BTG3 levels, releasing the inhibition on the activator E2fs and promoting the proliferation of abnormal cells [4]. Thus, a perturbed functioning of the network may give rise to the malignant transformation of damaged cells.

Besides genetic predisposition, there are other factors that contribute in transforming defective cells, such as viral and bacterial infections, metabolic disorders, and exposure to toxic substances. These exogenous agents modify the global behavior of the p53 network by targeting strategic interactions like (1) p53 repression due to Mdm2 (represented by $k_1$), (2) the rate of Mdm2 degradation (represented in the model by $\delta_3$), (3) the rate of Bax releasing (represented

http://dx.doi.org/10.5772/60916

by $k_2$), and (4) the rate of caspase-9 activation (represented by $k_3$). In order to observe the effect of alterations on these interactions, the corresponding parameters were varied in two orders of magnitude, and then the stability of the net was studied.

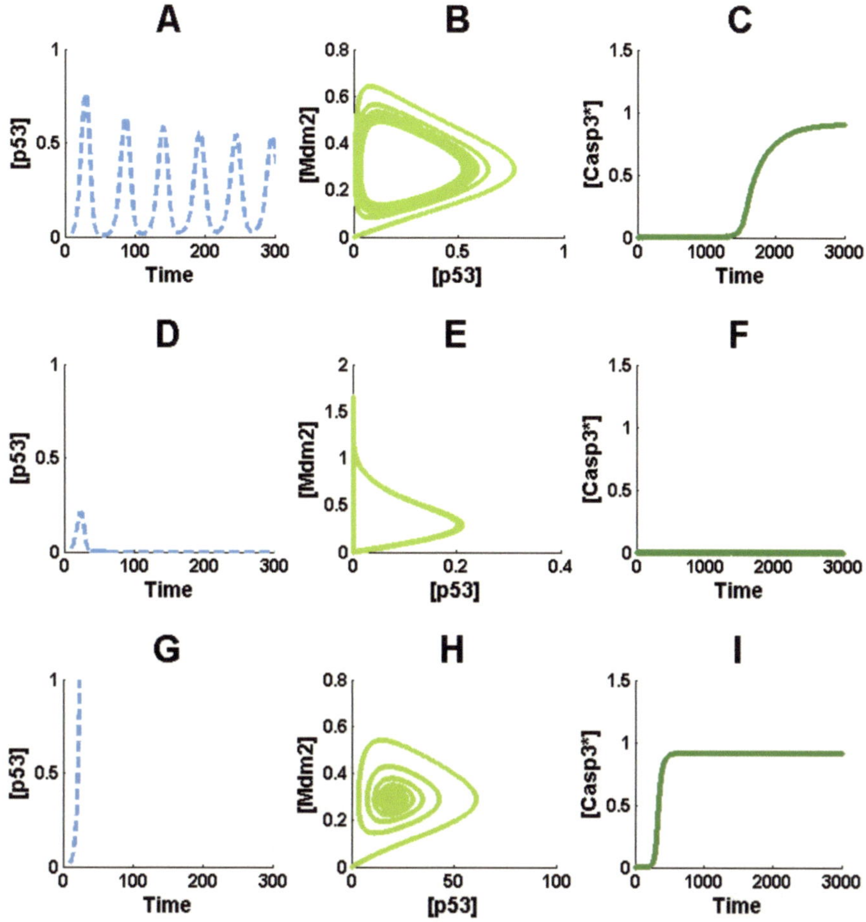


**Figure 11.** Variation of the basal degradation of Mdm2 ($\delta_3$). Compared with the normal cell response to severe DNA damage (A, B, and C), the effect of increasing in two orders the basal degradation of Mdm2 $\delta_3$ abrogates p53 oscillations (D), as well as its limit-cycle (E), avoiding apoptosis (F). However, the effect of decreasing $\delta_3$ in two orders augments the magnitude of p53 concentration (G), making quicker the activation of apoptosis (I) but destroys the limit cycle dynamics (H). For all simulations, the time scale is in minutes and concentrations are in micromolar units.

The model shows that decreasing $k_1$ allows a faster induction of apoptosis in damaged cells (Figures 10D to 10F). In biological terms, this effect can be produced when damaged cells are treated with nutlin-3 [86] and proteins of the influenza A virus [87]. Moreover, augments on $k_1$ avoid apoptosis triggering (Figures 10G to 10I). In the nature, blocking such an interaction is a common strategy used by pathogens like Epstein–Barr virus [88] and Chlamydia [89] in order to ensure their replication inside the host.

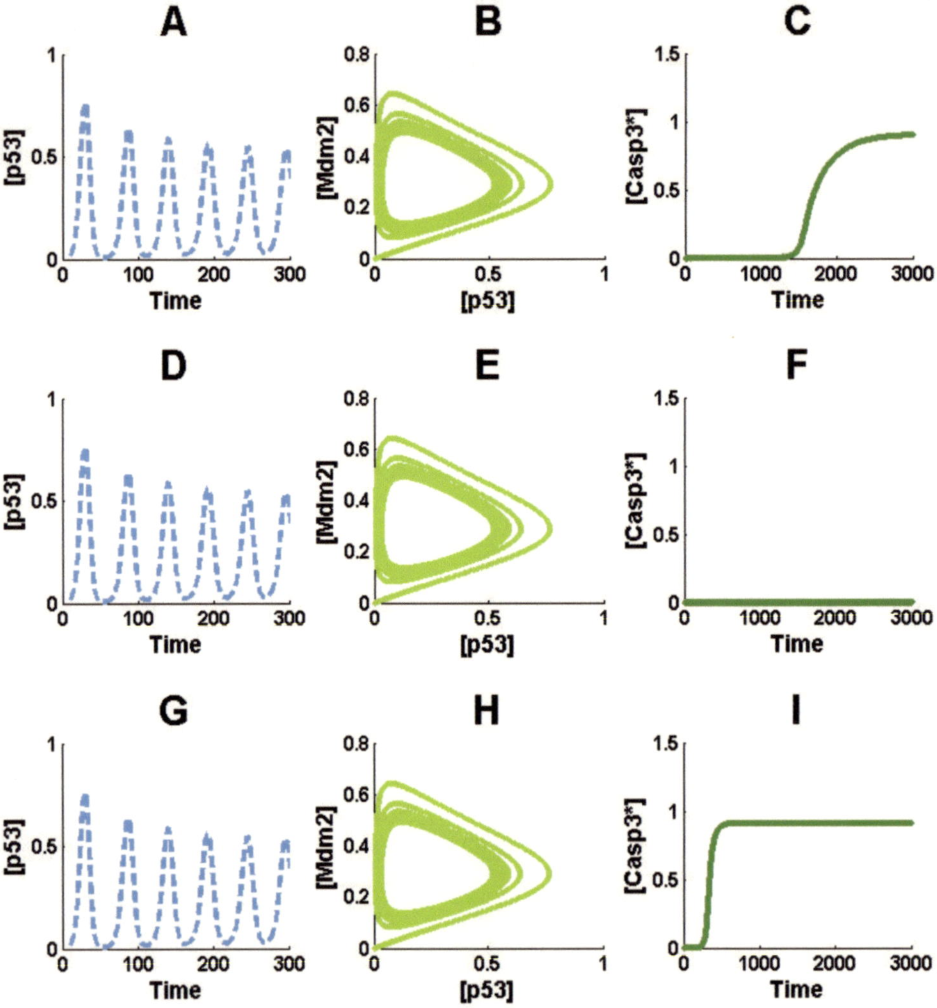


**Figure 12.** Variation of the rate of Bax releasing due to p53 activity ($k_2$). Compared with the normal cell response to severe DNA damage (A, B, and C), the effect of decreasing in two orders the rate of Bax releasing due to p53 activity $k_2$ does not interfere with the normal behavior of p53 (D), as well as its limit-cycle (E), but avoids apoptosis activation (F). Furthermore, the effect of increasing $k_2$ in two orders only accelerates the activation of apoptosis (I). For all simulations, the time scale is in minutes and concentrations are in micromolar units.

According to previous results, the increase of $\delta_3$ makes the cell more sensitive to DNA damage, allowing a quicker activation of apoptosis (Figures 11D to 11F). At the molecular level, this effect can be obtained by augmenting p14ARF activity [90], using substances like capsaicin [91], or viral proteins like hepatitis C [92]. Another way to achieve this effect is destabilizing Mdm2 protein by affecting NEDD-4-1 functioning [93]. On the other hand, reducing $\delta_3$ destroys Hopf bifurcation on p53–Mdm2 interaction, blocking definitively the initiation of apoptosis (Figures 11G to 11I).

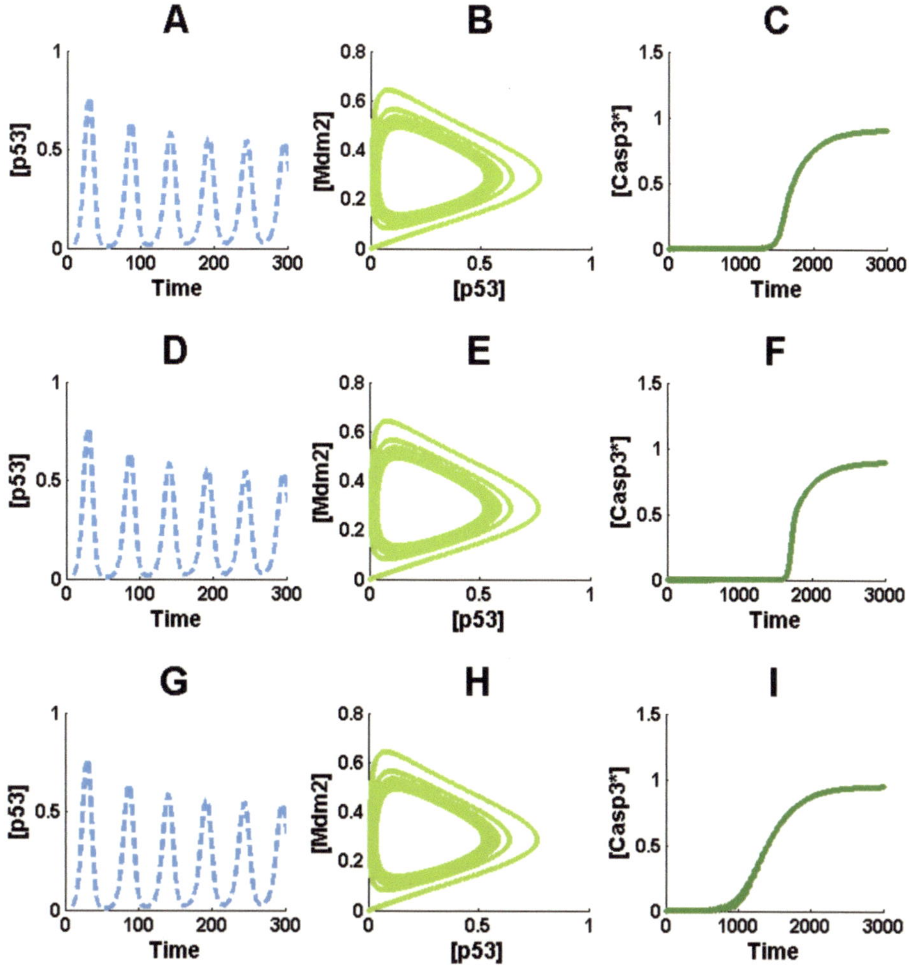


**Figure 13.** Variation of the rate of procaspase-9 activation ($k_3$). Compared with the normal cell response to severe DNA damage (A, B, and C), the effect of decreasing in two orders the rate of procaspase-9 activation $k_3$ does not interfere with the normal behavior of p53 (D), its limit-cycle (E), and apoptosis activation (F). However, the effect of increasing $k_3$ in two orders only increases slightly apoptosis activation (I). For all simulations, the time scale is in minutes, and concentrations are in micromolar units.

According to the model, when the rate of Bax releasing ($k_2$) increases, the cell is more sensitive to genotoxic damage as is observed in cells that overexpress Bax [94] (Figures 12G to 12I). However, when $k_2$ is reduced, the cell lacks its damage responsiveness (Figures 12D to 12F). Paradoxically, decreasing in $k_3$ has no significant effect on caspase-3 activation (Figures 13D to 13F). This phenomenon can be explained by the fact that there are many bistable switches on caspase cascade, sustained by positive feedback loops [59]. Thus, when those alterations are maintained for long time, the cell tends to accumulate several mutations that destabilize its genome, which eventually allows its transformation into a tumor cell.

## 5. Therapeutics approaches against cancer

Treatments against all types of cancer aim to interfere the functioning of transformed cells. Traditionally, this interference was induced by using drugs that block the cell growth together with radiotherapy [95, 96]. This approach has been the most successful; however, produces serious collateral effects on patients.

For this motive, in recent years, the use of more selective drugs for inducing apoptosis in cancerous cells has been proposed. These drugs comprise inhibitors of IAPs, BCL2, Mdm2, p53, and constitutive activators of death receptors [97–100]. Similarly, another new type of drugs derived from extract of *Emilia sonchifolia* has been used to activate caspases through the induction of reactive oxygen species (ROS) [101]. Polyphenolic curcumin is another promising natural compound that is capable to trigger apoptosis by increasing the expression and clustering of FasR in tumor cells. Another approach is based on the construction of nanoparticles that transport anticancerous compounds to tumor cells [102]. Moreover, more revolutionary approaches to treat cancer use phototherapy, which consists in administrate of nontoxic drugs known as "photosensitizers" to cancerous cells. After that, cells receive posterior illumination of visible light at specific wavelength, triggering the activation of photosensitizers that produce high amounts of ROS to eliminate the tumor cells by apoptosis [103].

Another novel approach to treat cancer is gene therapy, which uses many genes to alter the functioning of transformed cells blocking survival mechanisms and triggering apoptosis [1, 96, 104]. Other original strategies related with gene therapy use micro-RNA specifically designed to avoid the overexpression of oncogenes like BCL2 family or p53 in cancerous cells [105, 106].

In addition to gene therapy, the use of cytokines to treat cancer as an effective approach was proposed because there are reports indicating that the cancer cell lines treated with TNF$\alpha$ have an increase in DR5 as well of interferon-$\gamma$ and interferon-$\alpha$ [95]. It has been suggested that the utilization of some hormones like estrogen in controlled doses can be used as a treatment because it was reported that cancer breast cells exposed to high doses of estrogen died by apoptosis [107].

The multiple origins of cancer are the main obstacle for all of these approaches. In order to show the importance of this point on cancer treatment, the model of apoptosis for simulating two groups of cancerous cells is used. Both groups overexpress IAPs, but the second group also has an important deletion on caspase-3 gene. Mathematically, for both groups, Equation (24) of model 5.1 are modified by adding an oncogenic perturbation parameter ($\varepsilon_{\text{onco}}$) to overexpress IAPs [108] as follows:

$$\frac{\mathrm{d}[I]}{\mathrm{d}t} = \varepsilon_{\text{onco}} - k_3[B^*][I] \tag{27}$$

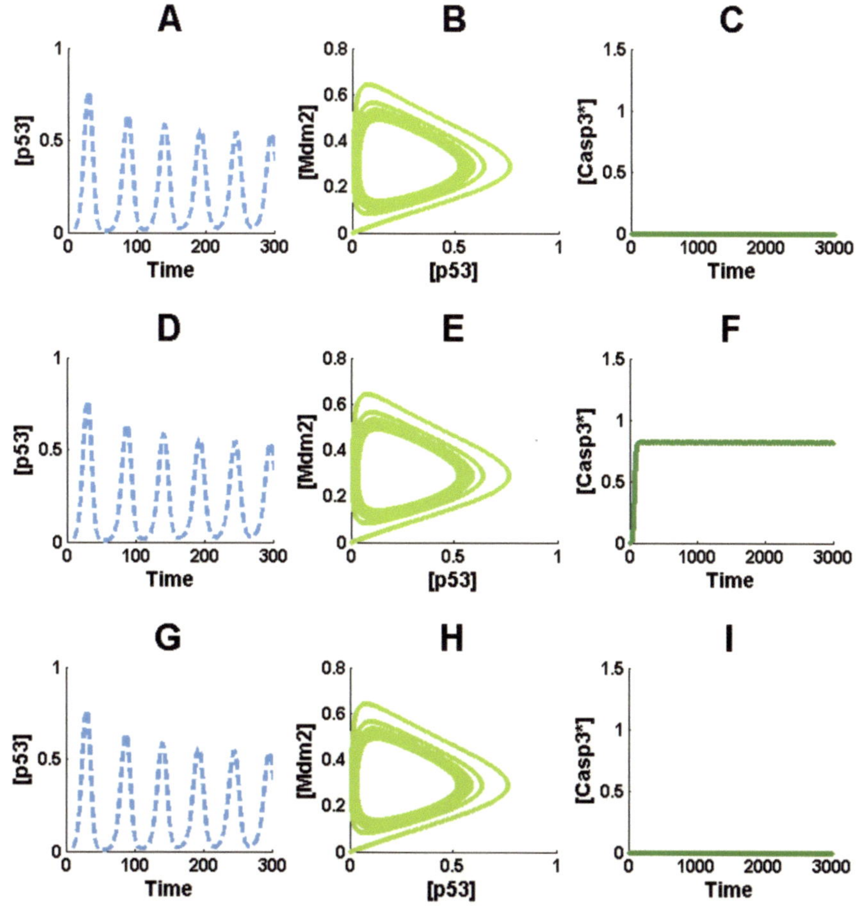


**Figure 14.** The subclonal heterogeneity as a therapeutics challenge. Cancerous cells that have deletions on the caspase-3 gene and/or overexpress IAPs present the same phenotype, and dynamically such mutations do not affect the p53 behavior when the cell receives severe DNA damage (A, B) but avoid apoptosis initiation (C). When cancerous cells that overexpress IAPs are treated with an inhibitor of IAPs activity and radiotherapy, the apoptosis starts normally (F). However, this approach does not work with cells that lack caspase-3 gene (I). For all simulations, the time scale is in minutes and concentrations are in micromolar units.

For simulating caspase-3 deletion in the second group of cancerous cells, Equation (24) must be modified as follows:

$$\frac{d[C3*]}{dt} = 0 \quad (28)$$

With respect to their phenotype, both groups of cells look very similar, and neither the oncogenic perturbation nor the deletion of caspase-3 interferes with the Hopf bifurcation on p53–Mdm2 feedback loop (Figures 14A to 14C).

The effect of treatments on both cells groups with drugs that inhibit IAPs together with radiotherapy is represented as follows:

$$\frac{d[I]}{dt} = \varepsilon_{onco} - k_3[B^*][I] - k_d[D][I] \tag{29}$$

where $k_d$ is the rate of IAPs repression due to inhibitor drugs, represented by [*D*]. After the theoretical treatment, the model shows that the first type of cancerous cells is successfully eliminated by apoptosis (Figures 14D to 14F) but not the second one (Figures 14G to 14I). Clinically, this problem is known as subclonal heterogeneity [109] and is very common to several types of tumors, in which transformed cells have many different damages, and thus, they do not response equally to treatments.

One way to overcome this problem is testing hypothesis and planning strategies to manipulate the cell functioning with mathematical models. For example, nowadays, it is possible designing genetic programs with specific functions like self-regulatory systems to attack cancerous cells [110].

## 6. Conclusions

The p53–Mdm2 module has a central role to prevent the development of tumor cells. This module has different modes of operation, depending upon the degree o DNA damage by pathogens, chemicals, or genotoxic signals. When DNA damage is severe, p53–Mdm2 takes the control of the apoptotic pathway by activating Bax, which in turn switches on the caspase cascade. From a dynamical point of view, this process implies the coupling of a supercritical Hopf bifurcation produced in the nucleus, with a saddle-node bifurcation produced in the cytoplasm that induces bistability in Bax [111], and caspases activation [59]. Thus, even if the DNA damage signal is intense, the p53–Mdm2 oscillatory dynamics must sustain enough time to increase Bax concentration to a threshold value that initiates apoptosis in an irreversible form; otherwise, the damaged cell can survive.

Mdm2 overexpression disrupts the p53–Mdm2 limit cycle in the nucleus blocking the transmission of the death signal to the caspase cascade, allowing the survival of the mutated cell. The suppression of caspase-9 gene causes the uncoupling of the p53–Mdm2 limit cycle and the caspase cascade, leading to the survival of the mutated cell despite that the DNA damage signal activates the p53 mechanism of damage suppression. Accordingly, mutated cells can escape from apoptosis when mutations in cells either disrupt the p53–Mdm2 limit cycle or uncouple it from the caspase cascade.

As a consequence, tumor cells can arise when they can overcome the apoptotic pathway by different mechanisms, leading to different strategies for their survival [110]. This variety in the causes that can originate and allow the survival of transformed cells is one of the obstacles to find a definitive cure for cancer. The combination of treatments seems to be the best strategy against cancer at the moment. Mathematical models of different aspects of cancer can be a guidance to find betters strategies to optimize and design novel treatments against cancer.

## Acknowledgements

We thank PRODEP and Secretaría Académica UAEM for financial support for the publication of this paper. We also thank to Erika Juárez Luna by logistic support.

## Author details

Antonio Bensussen and José Díaz*

*Address all correspondence to: biofisica@yahoo.com

Grupo de Biología Teórica y Computacional, Centro de Investigación en Dinámica Celular, Instituto de Investigación en Ciencias Básicas y Aplicadas, Universidad Autónoma del Estado de Morelos, Cuernavaca, Morelos, México

## References

[1] Gee AP. Manufacturing genetically modified T cells for clinical trials. Cancer Gene Ther. 2015. doi: 10.1038/cgt.2014.71.

[2] Cao Y, DePinho RA, Ernst M, Vousden K. Cancer research: past, present and future. Nat Rev Cancer. 2011. 11:749–54. doi: 10.1038/nrc3138.

[3] Bensussen A, Díaz J. Dynamical aspects of apoptosis. In: Nenoi M, editor. Current Topics in Ionizing Radiation Research. InTech. ISBN: 978-953-51-0196-3.

[4] Polager S, Ginsberg D. p53 and E2f: partners in life and death. Nat Rev Cancer. 2009. 9:738–48. doi: 10.1038/nrc2718

[5] Sun Y, Jiang X, Chen S, Fernandes N, Price BD. A role for the Tip60 histone acetyltransferase in the acetylation and activation of ATM. Proc Natl Acad Sci U S A. 2005. 102:13182-7.

[6] Tanaka T, Huang X, Halicka HD, Zhao H, Traganos F, Albino AP, Dai W, Darzynkiewicz Z. Cytometry of ATM activation and histone H2AX phosphorylation to estimate extent of DNA damage induced by exogenous agents. Cytometry A. 2007. 71:648–61.

[7] Vissers JHA, Nicassio F, van Lohuizen M, Di Fiore P, Citterio E. The many faces of ubiquitinated histone H2A: insights from the DUBs. Cell Div. 2008. 3:8. doi: 10.1186/1747-1028-3-8.

[8] Pommier Y, Weinstein JN. Chk2 molecular interaction map and rationale for Chk2 inhibitors. Clin Cancer Res. 2006. 12:2657–61.

[9] Yoda A, Toyoshima K, Watanabe YY, Onishi N, Hazaka Y, Tsukuda Y, Tsukada J, Kondo T, Tanaka Y, Minami Y. Arsenic trioxide augments Chk2/p53-mediated apoptosis by inhibiting oncogenic Wip1 phosphatase. J Biol Chem. 2008. 283:18969–79. doi: 10.1074/jbc.M800560200.

[10] Bergamaschi D, Samuels Y, Jin B, Duraisingham S, Crook T and Lu X. ASPP1 and ASPP2: common activators of p53 family members. Mol Cell Biol. 2004. 24: 1341–50.

[11] Schmid J, Birbach A. IκB kinase β (IKKβ/IKK2/IKBKB)—a key molecule in signaling to the transcription factor NF-κB. Cytokine Growth Factor Rev. 2008. 19:157–65. doi: 10.1016/j.cytogfr.2008.01.006.

[12] Zhu B, Xing C, Lin F, Fan X, Zhao K, Qin Z. Blocking NF-κB nuclear translocation leads to p53-related autophagy activation and cell apoptosis. World J Gastroenterol. 2011. 17:478–87. doi: 10.3748/wjg.v17.i4.478.

[13] Thapa RJ, Basagoudanavar SH, Nogusa S, Irrinki K, Mallilankaraman K, Slifker MJ, Beg AA, Madesh M, Balachandran S. NF-κB protects cells from gamma interferon-induced RIP1-dependent necroptosis. Mol Cell Biol. 2011. 31:2934–46. doi: 10.1128/MCB.05445-11.

[14] Méndez O, Martín B, Sanz R, Aragüés R, Moreno V, Oliva B, Stresing V, Sierra A. (2006), Underexpression of transcriptional regulators is common in metastatic breast cancer cells overexpressing Bcl-xL. Carcinogenesis. 2006. 27:1169–1179

[15] Chen C, Edelstein LC, Gélinas C. The Rel/NF-κB family directly activates expression of the apoptosis inhibitor BCL-XL. Mol Cell Biol. 2000. 20:2687–95.

[16] Dogu Y, Díaz J. Mathematical model of a network of interaction between p53 and Bcl-2 during genotoxic-induced apoptosis. Biophys Chem. 2009. 143:44–54. doi: 10.1016/j.bpc.2009.03.012.

[17] Robinson RA, Lu X, Jones EY. Biochemical and structural studies of ASPP proteins reveal differential binding to p53, p63, and p73. Structure. 2008. 16:259–68. doi: 10.1016/j.str.2007.11.012.

[18] Pietsch EC, Sykes SM, McMahon SB, Murphy ME. The p53 family and programmed cell death. Oncogene. 2008. 27:6507–21. doi: 10.1038/onc.2008.315.

[19] Agata N, Ahmad R, Kawano T, Raina D, Kharbanda S, Kufe D. MUC1 oncoprotein blocks death receptor–mediated apoptosis by inhibiting recruitment of caspase-8. Cancer Res. 2008. 68:6136–44. doi: 10.1158/0008-5472.

[20] Ahmad R, Raina D, Trivedi V, Ren J, Rajabi H, Kharbanda S, Kufe D. MUC1 oncoprotein activates the IkappaB kinase beta complex and constitutive NF-kappaB signaling. Nat Cell Biol. 2007. 9:1419–27.

[21] Bafna S, Kaur S, Batra S. Membrane-bound mucins: the mechanistic basis for alterations in the growth and survival of cancer cells. Oncogene. 2010. 29:2893–904. doi: 10.1038/onc.2010.87.

[22] Huang L, Chen D, Liu D, Yin L, Kharbanda S, Kufe D. MUC1 oncoprotein blocks glycogen synthase kinase 3 beta-mediated phosphorylation and degradation of beta-catenin. Cancer Res. 2005. 65:10413–22.

[23] Pochampalli M, el Bejjani R, Schroeder J. MUC1 is a novel regulator of ErbB1 receptor trafficking. Oncogene. 2007. 26:1693–701.

[24] Raina D, Ahmad R, Kumar S, Ren J, Yoshida K, Kharbanda S, Kufe D. MUC1 oncoprotein blocks nuclear targeting of c-Abl in the apoptotic response to DNA damage. EMBO J. 2006. 25:3774–83.

[25] Braithwaite AW. Some p53-binding proteins that can function as arbiters of life and death. Cell Death Differ. 2006. 13: 984–93.

[26] Teufel DP, Freund SM, Bycroft M, Fersht AR. Four domains of p300 each bind tightly to a sequence spanning both transactivation subdomains of p53. Proc Natl Acad Sci U S A. 2007. 10:7009–14.

[27] Ahmad R, Alam M, Rajabi H, Kufe. The MUC1-C oncoprotein binds to the BH3 domain of the pro-apoptotic BAX protein and blocks BAX function. J. Biol. Chem. 2012. 287:20866–75. doi: 10.1074/jbc.M112.357293.

[28] Raina D, Ahmad R, Chen D, Kumar S, Kharbanda S, Kufe D. MUC1 oncoprotein suppresses activation of the ARF-MDM2-p53 pathway. Cancer Biol Ther. 2008. 7:1959–67.

[29] Yamamori T, DeRicco J, Naqvi A, Hoffman TA, Mattagajasingh I, Kasuno K, Jung S, Kim C, Irani K. SIRT1 deacetylates APE1 and regulates cellular base excision repair. Nucleic Acids Res. 2010. 38:832–45. doi: 10.1093/nar/gkp1039.

[30] Pediconi N, Guerrieri F, Vossio S, Bruno T, Belloni L, Valeria Schinzari V, Scisciani C, Fanciulli M, Levrero M. hSirT1-dependent regulation of the PCAF-E2F1-p73 apoptotic pathway in response to DNA damage. Mol Cell Biol. 2009. 29:1989–98. doi: 10.1128/MCB.00552-08.

[31] Ikenoue T, Inoki K, Zhao B, Guan K. PTEN acetylation modulates its interaction with PDZ domain. Cancer Res. 2008. 68:6908–12. doi: 10.1158/0008-5472.CAN-08-1107.

[32] Huang WC, Ju TK, Hung MC, Chen CC. Phosphorylation of CBP by IKK$\alpha$ promotes cell growth by switching the binding preference of CBP from p53 to NF-$\kappa$B. Mol Cell. 2007. 26:75–87.

[33] Jänicke RU, Sohn D, Schulze-Osthoff K. The dark side of a tumor suppressor: anti-apoptotic p53. Cell Death Differ. 2008. 15:959–76. doi: 10.1038/cdd.2008.33.

[34] Youle RJ, Strasser A. The BCL-2 protein family: opposing activities that mediate cell death. Nat Rev Mol Cell Biol. 2008. 9:47–59.

[35] Candeias MM, Malbert-Colas L, Powell DJ, Daskalogianni C, Maslon MM, Naski N, Bourougaa K, Calvo F, Fähraeus R. p53 mRNA controls p53 activity by managing Mdm2 functions. Nat Cell Biol. 2008. 10: 1098–105. doi: 10.1038/ncb1770.

[36] Kobet E, Zeng X, Zhu Y, Keller D, Lu H. Mdm2 inhibits p300-mediated p53 acetylation and activation by forming a ternary complex with the two proteins. Proc Natl Acad Sci U S A. 2000. 97:12547–52.

[37] Lahav G, Rosenfeld N, Sigal A, Geva-Zatorsky N, Levine AJ, Elowitz MB, Alon. Dynamics of the p53–Mdm2 feedback loop in individual cells. Nat Genet. 2004. 36:147–150

[38] Bernards R. Wip-ing out cancer. Nat Genet. 2004. 36: 319–20.

[39] Hu W, Feng Z, Ma L, Wagner J, Rice JJ, Stolovitzky G, Levine AJ. A single nucleotide polymorphism in the MDM2 gene disrupts the oscillation of p53 and Mdm2 levels in cells. Cancer Res. 2007. 67:2757–65.

[40] Shreeram S, Demidov ON, Hee WK, Yamaguchi H, Onishi N, Kek C, Timofeev ON, Dudgeon C, Fornace AJ, Anderson CW, Minami Y, Appella E, Bulavin DV. Wip1 phosphatase modulates ATM-dependent signaling pathways. Mol Cell. 2006. 23:757–64.

[41] Lu X, Nguyen T, Moon S, Darlington Y, Sommer M, Donehower LA. The type 2C phosphatase Wip1: an oncogenic regulator of tumor suppressor and DNA damage response pathways. Cancer Metastasis Rev. 2008. 27:123–35. doi: 10.1007/s10555-008-9127-x.

[42] Patel S, George R, Autore F, Fraternali F, Ladbury JE, Nikolova PV. Molecular interactions of ASPP1 and ASPP2 with the p53 protein family and the apoptotic promoters PUMA and Bax. Nucleic Acids Res. 2008. 36:5139–51. doi: 10.1093/nar/gkn490.

[43] Dai C, Tang Y, Jung S, Qin J, Aaronson S, Gu E. Differential effects on p53-mediated cell cycle arrest vs. apoptosis by p90. Proc Natl Acad Sci U S A. 2011. 108:18937–42. doi: 10.1073/pnas.1110988108.

[44] Sakamaki J, Daitoku H, Ueno K, Hagiwara A, Yamagata K, Fukamizu A. Arginine methylation of BCL-2 antagonist of cell death (BAD) counteracts its phosphorylation and inactivation by Akt. Proc Natl Acad Sci U S A. 2011. 108:6085–90. doi: 10.1073/pnas.1015328108.

[45] Tanno M, Sakamoto J, Miura T, Shimamoto K, Horio Y. Nucleo-cytoplasmic shuttling of NAD+ -dependent histone deacetylase SIRT1. J Biol Chem. 2007.282:6823–32

[46] Green D, Kroemer G. Cytoplasmic functions of the tumour suppressor p53. Nature. 2009. 458:1127–30. doi: 10.1038/nature07986.

[47] Elmore S. Apoptosis: a review of programmed cell death. Toxicol Pathol. 35: 495–516.

[48] Wang P, Yu J, Zhang L. The nuclear function of p53 is required for PUMA-mediated apoptosis induced by DNA damage. Proc Natl Acad Sci U S A. 2007. 104:4054–9.

[49] Mondal S, Bhattacharya K, Mallick A, Sangwan R and Mandal C. Bak Compensated for Bax in p53-null cells to release cytochrome *c* for the initiation of mitochondrial signaling during with anolide D-induced apoptosis. PLoS One. 2012. 7: e34277

[50] Tsuda H, Ning Z, Yamaguchi Y, Suzuki N. Programmed cell death and its possible relationship with periodontal disease. J Oral Sci. 2012. 54:137–49.

[51] Field N, Low W, Daniels M, Howell S, Daviet L, Boshoff C, Collins M. KSHV vFLIP binds to IKK-γ to activate IKK. J Cell Sci. 2003. 116:3721–28.

[52] Srinivasula S, Ashwell J. IAPs: what's in a name? Mol Cell. 2008. 30:123–35. doi: 10.1016/j.molcel.2008.03.008.

[53] Altieri D. Survivin and IAP proteins in cell death mechanisms. Biochem J. 2010. 430:199–205. doi: 10.1042/BJ20100814.

[54] Clem R, Miller L. Control of programmed cell death by the baculovirus genes p35 and IAP. Mol Cell Biol. 1994. 14:5212–22.

[55] Kataoka T, Budd R, Holler N, Thome M, Martinon F, Irmler M, Burns K, Hahne M, Kennedy N, Kovacsovics M, Tschopp J. The caspase-8 inhibitor FLIP promotes activation of NF-κB and Erk signaling pathways. Curr. Biol. 2000. 10:640–648

[56] Lu M, Lin S, Huang Y, Kang Y, Rich R, Lo Y, Myszka D, Han J, Wu H. XIAP induces NF-κB activation via the BIR1/TAB1 interaction and BIR1 dimerization. Mol Cell. 26:689–702

[57] Safa A, Pollok K. Targeting the anti-apoptotic protein c-FLIP for cancer therapy. Cancers (Basel). 2011. 3:1639–71. doi: 10.3390/cancers3021639.

[58] Ghavami S, Hashemi M, Ande SR, Yeganeh B, Xiao W, Eshraghi M, Bus CJ, Kadkhoda K, Wiechec E, Halayko AJ, Los M. Apoptosis and cancer: mutations within caspase genes. J Med Genet. 2009. 46:497–510. doi: 10.1136/jmg.2009.066944.

[59] Legewie S, Blüthgen N, Herzel H. Mathematical modeling identifies inhibitors of apoptosis as mediators of positive feedback and bistability. PLoS Comput Biol. 2006. 2:e120.

[60] Bagci EZ, Vodovotz Y, Billiar TR, Ermentrout GB, Bahar I. Bistability in apoptosis: roles of BAX, BCL-2, and mitochondrial permeability transition pores. Biophys J. 2006. 90:1546–59.

[61] Feng Z, Hu W, de Stanchina E, Teresky AK, Jin S, Lowe S, Levine AJ. The regulation of AMPK beta1, TSC2, and PTEN expression by p53: stress, cell and tissue specificity,

and the role of these gene products in modulating the IGF-1-AKT-mTOR pathways. Cancer Res. 2007. 67: 3043–53.

[62] Kitano H. Biological robustness. Nat Rev Genet. 2004. 5:826–37.

[63] Cosentino C, Bates D. Feedback control in systems biology. Boca Raton: CRC Press; 2012. ISBN-13: 9781439816912.

[64] Kremling A, Saez-Rodriguez J. Systems Biology—An Engineering Perspective. J Biotechnol. 2007. 129:329–51.

[65] Skogestad S, Postlethwaite I. Multivariable Feedback Control: Analysis and Design. 2nd edition. Hoboken, NJ: John Wiley; 2005.

[66] Ogata K. Modern Control Engineering. Boston, MA: Prentice-Hall; 2010. ISBN-10: 0136156738.

[67] Iglesias P, Ingalls B, editors. Control Theory and Systems Biology. Cambridge, MA: MIT Press. 2010. ISBN-13: 9780262013345.

[68] Malumbres M, Barbacid M. Cell cycle, CDKs and cancer: a changing paradigm. Nat Rev Cancer. 2009. 9:153–66. doi: 10.1038/nrc2602.

[69] Abukhdeir AM, Park BH. p21 and p27: roles in carcinogenesis and drug resistance. Expert Rev Mol Med. 2008. 10:e19. doi: 10.1017/S1462399408000744.

[70] Lacroix M, Toillon R, Leclercq G. p53 and breast cancer, an update. Endocr Relat Cancer. 2006. 13:293–325.

[71] Maddocks O, Vousden K. Metabolic regulation by p53. J Mol Med (Berl). 2011. 89:237–45. doi: 10.1007/s00109-011-0735-5.

[72] Baillat D, Laitem C, Leprivier G, Margerin C and Aumercier C. Ets-1 binds cooperatively to the palindromic Ets-binding sites in the p53 promoter. Biochem Biophys Res Commun. 2009. 378:213–7. doi: 10.1016/j.bbrc.2008.11.035.

[73] Wang S, El-Deiry WS. p73 or p53 directly regulates human p53 transcription to maintain cell cycle checkpoints. Cancer Res. 2006. 66:6982–9.

[74] Boogs K, Reisman D. C/EPBβ participates in regulating transcription of the p53 gene in response to mitogen stimulation. J Biol Chem. 2007. 282:7982–90.

[75] Wu H, Lozano G. NF-κB activation of p53 a potential mechanism for suppressing cell growth in response to stress. J Biol Chem. 1994. 269:20067–74.

[76] Noble C, Dong J, Manser E, Song H. BCL-XL and UVRAG cause a monomer–dimer switch in beclin1. J Biol Chem. 2008. 283:26274–82. doi: 10.1074/jbc.M804723200.

[77] Perry ME. Mdm2 in the response to radiation. Mol Cancer Res. 2004. 2:9–19.

[78] Phelps M, Darley M, Primrose JN, Blaydes JP. p53-independent activation of the Hdm2-P2 promoter through multiple transcription factor response elements results

in elevated Hdm2 expression in estrogen receptor a-positive breast cancer cells. Cancer Res. 2003. 63:2616–23.

[79] Ma L, Wagner J, Rice JJ, Hu W, Levine AJ, Stolovitzky GA. A plausible model for the digital response of p53 to DNA damage. Proc Natl Acad Sci U S A. 2005. 102:14266–71.

[80] Geva-Zatorsky N, Rosenfeld N, Itzkovitz S, Milo R, Sigal A, Dekel E, Yarnitzky T, Liron Y, Polak P, Lahav G, Alon U. Oscillations and variability in the p53 system. Mol Syst Biol. 2006. 2:2006.0033.

[81] Proctor CJ, Gray DA. Explaining oscillations and variability in the p53–Mdm2 system. BMC Syst Biol. 2008. 2:75. doi: 10.1186/1752-0509-2-75.

[82] Liu B, Bhatt D, Oltvai ZN, Greenberger JS, Bahar I. Significance of p53 dynamics in regulating apoptosis in response to ionizing radiation, and polypharmacological strategies. Sci Rep. 2014. 4:6245. doi: 10.1038/srep06245.

[83] Strogatz SH. Nonlinear Dynamics and Chaos: With Applications to Physics, Biology, Chemistry, and Engineering. 1st ed. Westview Press; 2001. ISBN-13: 978-0738204536

[84] Dongiovanni P, Fracanzani AL, Cairo G, Megazzini CP, Gatti S, Rametta R, Fargion S, Valenti L. Iron-dependent regulation of MDM2 influences p53 activity and hepatic carcinogenesis. Am J Pathol. 2010. 176:1006–17. doi: 10.2353/ajpath.2010.090249.

[85] Rayburn E, Zhang R, He J, Wang H. MDM2 and human malignancies: expression, clinical pathology, prognostic markers, and implications for chemotherapy. Curr Cancer Drug Targets. 2005. 5:27–41.

[86] Verma R, Rigatti MJ, Belinsky GS, Godman CA, Giardina C. DNA damage response to the Mdm2 inhibitor nutlin-3. Biochem Pharmacol. 2010. 79:565–74. doi: 10.1016/j.bcp.2009.09.020.

[87] Wang X, Deng X, Yan W, Zhu Z, Shen Y, Qiu Y, Shi Z, Shao D, Wei J, Xia X, Ma Z. Stabilization of p53 in influenza A virus-infected cells is associated with compromised MDM2-mediated ubiquitination of p53. J Biol Chem. 2012. 287:18366–75. doi: 10.1074/jbc.M111.335422.

[88] Saha A, Murakami M, Kumar P, Bajaj B, Sims K, Robertson ES. Epstein–Barr virus nuclear antigen 3C augments Mdm2-mediated p53 ubiquitination and degradation by deubiquitinating Mdm2. J Virol. 2009. 83:4652–69. doi: 10.1128/JVI.02408-08.

[89] González E, Rother M, Kerr MC, Al-Zeer MA, Abu-Lubad M, Kessler M, Brinkmann V, Loewer A, Meyer TF. Chlamydia infection depends on a functional MDM2-p53 axis. Nat Commun. 2014. 5:5201. doi: 10.1038/ncomms6201.

[90] Horiguchi M, Koyanagi S, Hamdan AM, Kakimoto K, Matsunaga N, Yamashita C, Ohdo S. Rhythmic control of the ARF-MDM2 pathway by ATF4 underlies circadian

accumulation of p53 in malignant cells. Cancer Res. 2013. 73:2639–49. doi: 10.1158/0008-5472.CAN-12-2492.

[91] Jin J, Lin G, Huang H, Xu D, Yu H, Ma X, Zhu L, Ma D, Jiang H. Capsaicin mediates cell cycle arrest and apoptosis in human colon cancer cells via stabilizing and activating p53. Int J Biol Sci. 2014. 10:285–95. doi: 10.7150/ijbs.7730.

[92] Seo YL, Heo S, Jang KL. Hepatitis C virus Core protein overcomes $H_2O_2$-induced apoptosis by down-regulating p14 expression via DNA methylation. J Gen Virol. 2014. doi: 10.1099/vir.0.000032.

[93] Xu C, Fan CD, Wang X. Regulation of Mdm2 protein stability and the p53 response by NEDD4-1 E3 ligase. Oncogene. 2015. 34:281–9. doi: 10.1038/onc.2013.557.

[94] Guo B, Cao S, Tóth K, Azrak RG, Rustum YM. Overexpression of Bax enhances antitumor activity of chemotherapeutic agents in human head and neck squamous cell carcinoma. Clin Cancer Res. 2000. 6:718–24.

[95] Elrod H, Sun SY. Modulation of death receptors by cancer therapeutic agents. Cancer Biol Ther. 2008. 7:163–73.

[96] Norian L, James B, Griffith T. Advances in viral vector-based TRAIL gene therapy for cancer. Cancers (Basel). 2011. 3:603–20. doi: 10.3390/cancers3010603.

[97] Fesik S. Promoting apoptosis as a strategy for cancer drug discovery. Nat Rev Cancer. 2005. 5: 876–885

[98] Fulda S. Targeting apoptosis signaling pathways for anticancer therapy. Front Oncol. 1:23. doi: 10.3389/fonc.2011.00023.

[99] Vitali R, Mancini C, Cesi V, Tanno B, Mancuso MT, Bossi G, Zhang Y, Martinez R, Calabretta B, Dominici C, Raschella G. Slug (SNAI2) down-regulation by RNA interference facilitates apoptosis and inhibits invasive growth in neuroblastoma preclinical models. Clin Cancer Res. 2008. 14:4622–30. doi: 10.1158/1078-0432.CCR-07-5210.

[100] Plati J, Bucur O, Khosravi-Far R. Apoptotic cell signaling in cancer progression and therapy. Integr Biol (Camb). 2011. 3:279–96. doi: 10.1039/c0ib00144a

[101] Lan YH, Chiang JH, Huang WW, Lu CC, Chung JG, Wu TS, Jhan JH, Lin KL, Pai SJ, Chiu YJ, Tsuzuki M, Yang JS. Activations of both extrinsic and intrinsic pathways in HCT 116 human colorectal cancer cells contribute to apoptosis through p53-Mediated ATM/Fas signaling by *Emilia sonchifolia* extract, a folklore medicinal plant. Evid Based Complement Alternat Med. 2012. 2012:178178. doi: 10.1155/2012/178178

[102] Bisht S, Feldmann G, Soni S, Ravi R, Karikar C, Maitra A, Maitra A. Polymeric nanoparticle-encapsulated curcumin ("nanocurcumin"): a novel strategy for human cancer therapy. J Nanobiotechnology. 2007. 5:3

[103] Pavarina A, Dias Ribeiro A, Nordi Dovigo L, De Andrade C, De Souza Costa C, Vergani C. Photodynamic therapy to eradicate tumor cells, cell metabolism. In: Paula

Bubulya, editor. Cell Homeostasis and Stress Response. InTech. ISBN: 978-953-307-978-3.

[104] Wu TL, Zhou D. Viral delivery for gene therapy against cell movement in cancer. Adv Drug Deliv Rev. 2011. 63:671–7. doi: 10.1016/j.addr.2011.05.005.

[105] Bhardwaj A, Singh S, Singh A. MicroRNA-based cancer therapeutics: big hope from small RNAs. Mol Cell Pharmacol. 2004. 2:213–219

[106] Boominathan L. The tumor suppressors p53, p63, and p73 are regulators of micro-RNA processing complex. PLoS One. 2010. 5: e10615

[107] Lewis-Wambi JS, Jordan VC. Estrogen regulation of apoptosis: how can one hormone stimulate and inhibit? Breast Cancer Res. 2009. 11:206. doi: 10.1186/bcr2255.

[108] Safa A, Pollok K. Targeting the anti-apoptotic protein c-FLIP for cancer therapy. Cancers (Basel). 2011. 3:1639–71. doi: 10.3390/cancers3021639.

[109] Marusyk A, Tabassum DP, Altrock PM, Almendro V, Michor F, Polyak K. Non-cell-autonomous driving of tumour growth supports sub-clonal heterogeneity. Nature. 2014. 514:54–8. doi: 10.1038/nature13556.

[110] Kalos M, June CH. Adoptive T cell transfer for cancer immunotherapy in the era of synthetic biology. Immunity. 2013. 39:49–60. doi: 10.1016/j.immuni.2013.07.002.

[111] Sun T, Lin X, Wei Y, Xu Y, Shen P. Evaluating bistability of Bax activation switch. FEBS Lett. 2010. 584:954–60. doi: 10.1016/j.febslet.2010.01.034.

Chapter 2

# On the Dynamical Approach of Quantitative Radiation Biology

Noriyuki B. Ouchi

Additional information is available at the end of the chapter

http://dx.doi.org/10.5772/60859

**Abstract**

A quantitative approach in radiation biology based on the clonogenic method and cell survival curves in various conditions are introduced. The cell survival curves seem to have universality with regard to its functional form; in other words, functional form of survival curve seems to be unchanged under various conditions including different species. Various factors affecting the radiosensitivity have been introduced to find macroscopic nature of living organisms. Mathematical models that describe cell survival curves have been presented for discussing the derivation of the mathematical form based on biological mechanism. Finally, the possibility that the structural change of chromosome affects the repair process is discussed.

**Keywords:** Cell survival curves, Mathematical model, Target theory, Quantitative radiation biology, DNA repair

## 1. Introduction

Now, over ten years have passed since initial sequencing and analysis of the human genome [1]. The human genome is thought to contain approximately 20,000 protein-coding genes, which are supposed to drive a human being as a living system. The successful sequencing and analysis of the human genome made a major step forward in the understanding of life. The progress in decoding the human genome reveals aspects of the components of life; however, it is difficult to study dynamical principle of living organisms, which are supposed to emerge via many interactions between many proteins (30,000 or more) in a collective manner.

In fact, a major method in modern molecular biology is each event involved in the corresponding process will be examined by breaking the whole system into small pieces, and biological activity in each small part of the system is explained by the activation of corresponding genes. This is the method so-called reductionism. This kind of reductionism has made great accomplishments in the physical sciences; however, only the knowledge of the breaking elements would not help to understand life, e.g. its origin, evolution, and persistence (Fig. 1). In other words, so-called emergent properties of the system would not be impossible to predict from knowledge of the breaking elements of the system [2]. Therefore, the understanding of life itself requires research technique which handles the whole system.

**Figure 1.** Karakuri ningyou (Japanese mechanized doll). This is a tea serving mechanized robot. The karakuri ningyou serves a cup of tea, and its sequence is roughly as follows: 1) When a cup of tea is placed on the hands, they start to move toward a guest by moving its feet like walking. 2) It moves, setting distance, and bows its head. 3) When a cup of tea is taken from the hands, it stops and waits for a next action. 4) It turns around and returns to the starting position (host's place) when an empty cup is placed on the hands. It works by springs. Only the examination of each mechanical gear would not help to understand the entire behavior of the doll. Picture is taken from Wikimedia Commons [3] and converted to gray scale

There is a well-known logic of research methods that have been used to derive the theory or some formula without breaking the corresponding system. For example, linear response theory is based on the idea that the global behavior of the system is obtained by investigating the various responses of the system against the given perturbation/stimulation. Especially, this kind of research design, i.e., investigating the global response of the system to a spatiotemporally varying perturbation, e.g., electromagnetic field, temperature, etc., for analyzing dynamical properties of the system has been well adopted in nonequilibrium statistical physics [4].

http://dx.doi.org/10.5772/60859

Turning now to the case of biological science, ionizing radiation seems to be a good example of such kind of "perturbation" to examine system's global behavior from the action to the living organisms. Generally, different types of scientific index of the biological responses to ionizing radiation are used to study the action of ionizing radiation, depending on the corresponding systems. For most cases, cancer incidence is used as the biological response if the corresponding system is a human being. This type of study, known as epidemiology, is basically the statistics.

On the other hand, major advances have been made in the mechanism-based study of molecular radiation biology, and these advances shed light on the relationship between carcinogenesis and radiation-induced DNA damage. Biological studies on the various radiation responses are basically the cellular-scale investigation. By considering ionizing radiation as an example of external stimulus, many types of cellular responses, e.g., induction of chromosomal aberrations, gene mutations, cell apoptosis, cell transformation, and tumorigenesis, are known to occur. In many cases, quantification of these responses can help discuss these issues mathematically. In the next section, a very effective quantification experimental technique in radiation biology is discussed.

## 2. Quantitative radiation biology

Early in the twentieth century, it is known that the ionizing radiation may cause harmful effects on the biological organisms. Most of the experiments in the early radiation biology research are aimed to study the effects of X-ray irradiation on various types of living organisms: bacteria, virus, and unicellular organism. In short, unicellular organisms were mostly used to study the action of ionizing radiation on living organisms.

These early experiments have revealed the amount of radiation doses that needed to kill or inactivate the various species. For example, the mean lethal dose, $D_0$ (dose required to reduce the population to the 37% level, i.e., fraction $1/e$) is over 150 Gy for *Chilomonas paramecium* and 400 Gy for virus (Table. 1). Here, Gy is a unit absorbed radiation dose, and 1 Gy=1 J $kg^{-1}$, in SI unit. Therefore, the fact that there is a big difference in mean lethal dose between species, or, in other words, there is a big difference in radiosensitivity between species, has been recognized.

However, irradiation experiment of human cells was needed to study the radiosensitivity of humans. Therefore, experimental technique that possibly cultures the separated cells *in vitro* was needed. In 1907, a new technique, known as cell culture, was successfully introduced by Harrison [6] to study the argument of development process of the nervous system, whether the nervous system was composed of many cells (syncytial theory) or made up of a single seamless, continuous cell (reticular theory) [7]. Nowadays, this simple experiment, cell culture, is becoming a more and more important tool not only for radiation biology but also for other wide varieties of life sciences (Fig. 2). In this way, establishment of the cell culture technique has made possible to study the radiation effects using cultured human cells.

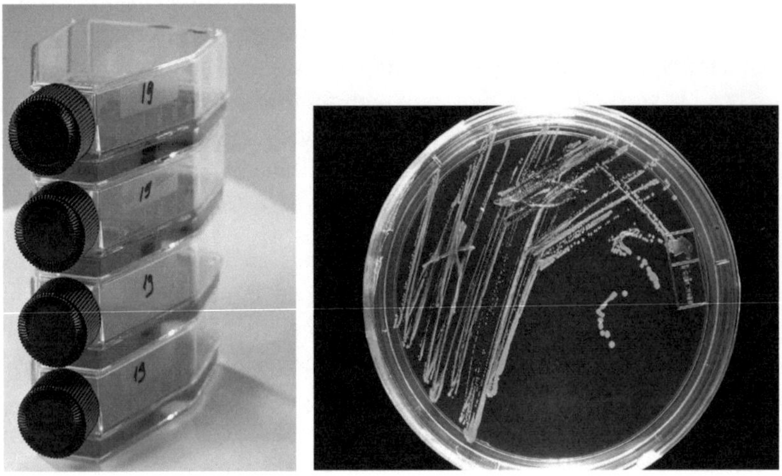

**Figure 2.** Two examples of the cell culture system. Left: Flasks containing some growth medium for cell growth. Right: Petri dish containing growth medium, called agar plate. It contains cultured microorganisms. Taken from Wikimedia Commons [8, 9] and converted to gray scale

| **Species** | **$D_0$ (Gy)** |
|---|---|
| *E. coli* | 40 |
| T2 bacteriophage | 400 |
| Newcastle's disease virus | 400 |
| Yeast | 50–180 |
| *Chilomonas paramecium* | 150 |
| HeLa cell | 1 |

**Table 1.** Roughly estimated value of $D_0$ for various biological species. After [5]

According to the early human cell irradiation experiments, the $D_0$ value for isolated human cells was found to have about 100 Gy; it is almost the same dose as that of *Chilomonas paramecium*. Afterward, it was found not to be a true. Then, statistical scientific data taken from the investigation of A-bomb survivors of Hiroshima and Nagasaki and nuclear accident in the United States have revealed a scientific fact that the $D_0$ of unicellular organisms (over 100 Gy) and human beings (1–4 Gy) is quite different. Furthermore, a big difference on lethal dose of human individuals (1–4 Gy) and its constituent cells (over 100 Gy!) was recognized. At that time, this big difference on lethal dose was one of the unresolved questions in radiation biology.

Here, another important aspect of the radiation is discussed. It is important to emphasize that the total absorbed energy by a human body estimated from the lethal radiation dose for human (4 Gy) seems to be extremely small compared to the other harmful sources.

http://dx.doi.org/10.5772/60859

For a person who has a mass of 60 kg, the total absorbed energy from 4 Gy X-ray is calculated as follows:

$$4\text{Gy} \times 60\text{kg} = 240\text{J}, \tag{1}$$

where 240 J=57 cal. Energy of 1 J equals to the work done by a constant force of 1 N and moves a 1 m displacement. Therefore, 240J equals to the work that lifts a 60 kg mass to a height of about 40 cm (Fig. 3). Thus, the energy level of lethal radiation dose seems sufficiently small compared to the other harmful sources. Naturally, this scientific fact suggests that there is a small target which controls cell life and death inside the living organisms and viruses, i.e., DNA.

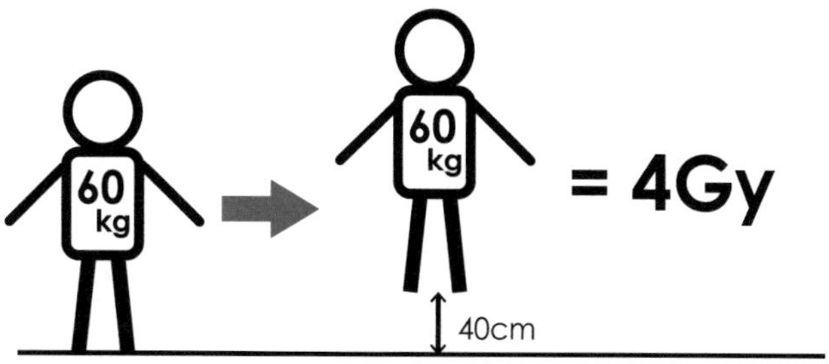


**Figure 3.** Lethal radiation dose explanation. Total energy imparted to the 60 kg mass by a 4 Gy radiation, which is the mean lethal dose $D_0$ for human, is roughly equivalent to lifting 60 kg mass to a 40 cm height

Now, return the discussion to the problem of the difference in radiosensitivity between human individuals (1–4 Gy) and its constituent cells (over 100 Gy). This problem has been clarified due to the successful development of new experimental technique. More precisely, for the sake of clonogenic assay method developed by Puck and Marcus [5], quantitative study of cellular response of mammalian cells to an external stimulus is established. Thus, the first survival curve, which will be described later, for X-ray irradiated mammalian (HeLa) cells *in vitro* was obtained by the method. The cell culture technique made the observation of proliferation of cells isolated from tissues easier; therefore, cell surviving fraction based on the ability of a single cell to grow into a large colony was successfully conceived (Fig. 4).

The most important concept in the development of clonogenic assay is the definition of cell death. In other words, the cell survival is defined as whether the cell has a proliferation ability or not. In this situation, "dead cell" means the cell which loses its reproductive integrity, and this kind of cell death is called "reproductive cell death." The cells which cause reproductive cell death may still be present physically and morphologically intact, may even be able to make proteins or synthesize DNA, and may even be able to progress a few cell cycles and a few divisions may still occur. Generally, a dose of over 100 Gy is required to destroy basic cell function, in contrast to the 1–4 Gy for causing reproductive cell death. Quantifying the pure cell survival number by employing single-cell-based culturing technique has enabled to

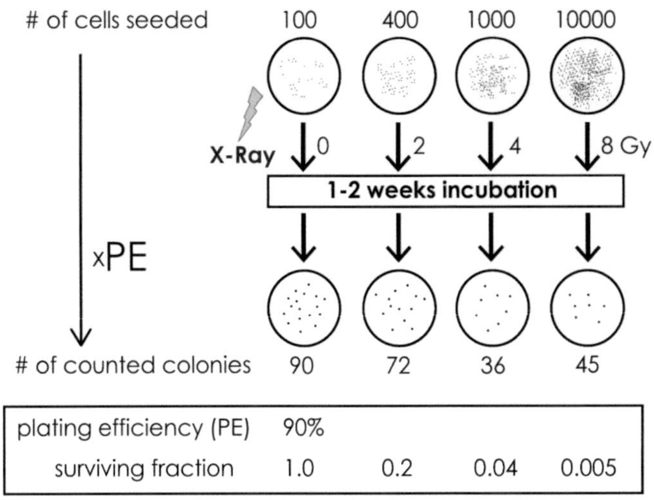


**Figure 4.** Illustration of the clonogenic assay method. Cell surviving fraction is finally calculated by dividing the counted number of colonies by the number of seeded cells which multiplied by plating efficiency (PE). PE is defined as the growing probability under the control condition (dose=0, in this case)

distinguish the reproductive cell death from the cell death in the narrow sense. Thus, clonogenic assay method may clarify the problem of the difference in radiosensitivity between human individuals and its constituent cells. Human individuals and its constituent cells may have almost the same value of mean lethal dose. Establishment of their work is often considered to mark the beginning of quantitative cellular radiation biology [10].

Generally, cell survival curves are used for the quantitative representation of biological cellular responses. Cell survival curves, or in short, survival curves, are defined by the proportion of surviving cells ($S$) as a function of radiation dose ($D$), as in Fig. 5, and known to be different, depending on corresponding biological systems (mammalian cells, virus, yeast, bacteria, etc.), both in terms of shape and absolute value of $S$ at a given dose. In this sense, "slope" of each survival curve implies degree of radiation sensitivity of corresponding biological systems. Interestingly, its slope is known to vary with DNA content [11].

Until now, many experiments to measure radiation survival curves have been performed for various species, including human cells. As a consequence of these experiments, many new findings have been revealed. For example, radiosensitivities for the corresponding species A, mammalian cells; B, *E. coli*; C, *E.coli* B/r; D, yeast; E, phage staph E; F, *Bacillus megaterium*; G, potato virus; and H, *M. radiodurans*, are comparable in the slope of the cell survival curves, and they have the relation in radiosensitivity:

$$A > B > C > D > E > F > G > H. \tag{2}$$

Thus, the mammalian cells are most radiosensitive than other species. The differences of the radiosensitivity of various species are said to be correlated with its DNA content and efficiency of the DNA repair system [12].

http://dx.doi.org/10.5772/60859

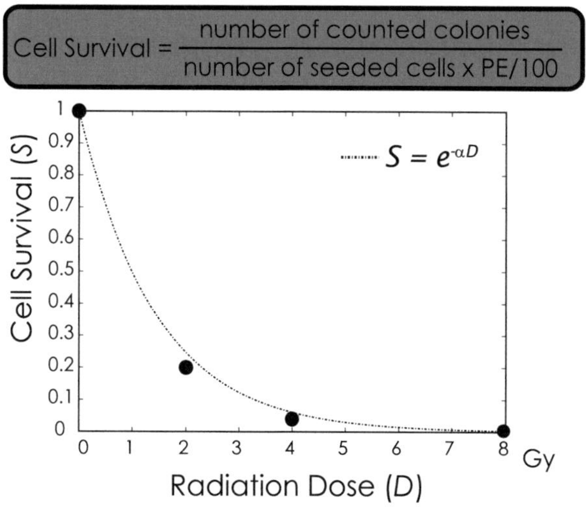


**Figure 5.** Plot of cell survival curve for the surviving fraction data in Fig. 4. The survival curve is well fitted with $S = e^{-\alpha D}$, where $S$ is cell survival and $D$ is the radiation dose. $\alpha = 0.7$

Generally, differences on the biological or experimental conditions make the radiosensitivity change, even among the same type of cells. A lot of experiments have been performed to study various factors that affect the radiosensitivity until now; some of the major factors are described in the following. Here, various factors affecting the shape of survival curves (radiosensitivity) are summarized.

The radiosensitivity is known to vary with various conditions, for example:

1. species [11]
2. type of radiation [13]
3. dose rate [14, 15]
4. oxygen concentration [16]
5. extent of apoptosis [12]
6. AT cell lines [17]
7. type of mutation [18]
8. cell cycle (cell time) [19, 20]

An affecting factor regarding the difference of species can be interpreted as the dependence of DNA content, and that of the radiation type and oxygen concentration can be interpreted as the dependence of the linear energy transfer (LET) values, i.e., average energy deposition per unit length. Here, the definition of the LET is described later. Clearly, these factors (2–4) are the external factors not relating to the biological activities inside the cell, and the following factors (5–8) seem to be the intracellular factors depending on the biological activities.

The extent of apoptosis seems to have been associated with the abrogation of *p53* function. It is well known that the gene *p53* mediates cellular responses to DNA damage, and results show that radiosensitivity is inversely correlated with the amount of DNA double-strand breaks (DSBs). In this case, the amount of DSBs should be the important feature to elucidate the affecting factor on cellular radiosensitivity. Moreover, cell lines derived from subjects with *ataxia telangiectasia* (AT) are three or four times sensitive to radiation than the normal cell lines, and no or very small shoulders appeared in survival curves for AT cell lines [17]. The genetic disorder responsible for the AT subject is the mutation of *ATM* gene, which is involved in the cellular response to DNA damage, especially in the DSB damage. Results of the *E. coli* mutant experiment clearly show that the *rec-* mutants have a higher radiosensitivity than that of *rec+*. The *rec* gene is involved in genetic recombination and DNA repair. These experimental facts also imply that the ability or capacity of DNA repair in specific cellular system plays an important role in radiosensitivity of cell.

The most complicated factor is the cell cycle. Generally, radiosensitivity is cyclically changed during cell cycle (Fig. 6), depending on the position of cell-cycle stages. To put it more precisely, radiation sensitivity is minimal when cells are irradiated in the early postmitotic (G1) and the premitotic (G2) phases of the cell-cycle and maximal in the mitotic (M) phase and late G1 or early synthesis (S) phases [19]. However, the pattern of the response cycle may vary depending on cell lines, especially on the length of G1 period [20]; typically, it is known to indicate a bimodal pattern. Generally, the origin of cell-cycle-dependent radiation sensitivity is supposed to be a consequence of some intracellular dynamics, e.g., the regulation mechanism of cell-cycle checkpoint, repair ability of DNA damage, and higher-order structure of chromosomes; no explicit theoretical explanation exists presently for this cyclic response.

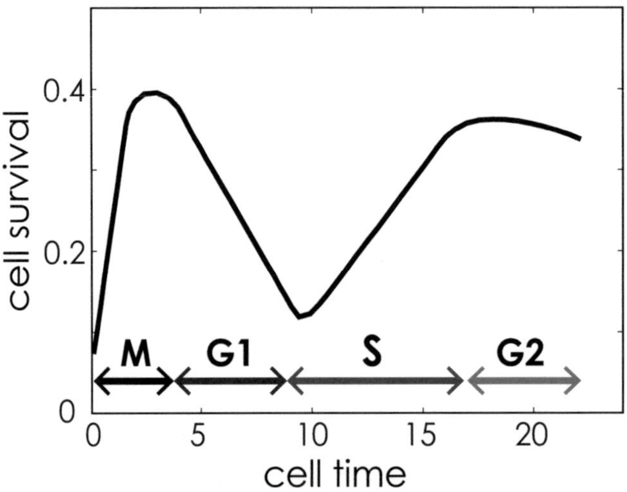


**Figure 6.** Plot of the surviving fraction of synchronized HeLa cells irradiated by X-ray. The horizontal axis "cell time" shows irradiated time of cell cycle. Time zero represents mitosis (M in figure), and G1, S, and G2 represent corresponding cell-cycle stages. Adapted from [19]

Finally, the affecting factors will be summarized using quantitative indices (Table 2). It should be noted that the term "repair ability" in Table 2 seems not to be a quantitative index, but rather a qualitative index, it is thought that it can be measured, although not directly, but through other quantities, e.g., the DSB yield. Mathematical methods to analyze these experimental data and many attempts to understand the biological mechanisms involved in the variation of radiosensitivity under a variety of conditions are introduced in the next section.

| Factor | Radiosensitivity |
|---|---|
| DNA content | ↗ |
| LET | ↗ |
| Dose rate | ↗ |
| Repair ability | ↘ |
| Cell cycle (cell time) | ↘↗↘↗ |

**Table 2.** Summary of the factors that affect the radiosensitivity. Arrows inside the table show variation of the radiosensitivity. Radiosensitivity dependence has a linear relation with respect to the corresponding affecting quantities, except for the cell-cycle dependence, which has a cyclic relation

# 3. Mathematical representations

Many efforts have been done to understand a priori the actions of radiations on the living organisms from the early 1920s, and most of these types of studies have been done by physicists. The logical structure of this theory is basically based on the simple idea, that is, studying the behavior of the living organisms as the "response" to the radiation as the "action"; actually, it is a so-called dose-response relationship in today's radiation biology. Most of experiments in such studies have mainly used mortality or survival as the corresponding biological "effects." The surviving fraction of the various living organisms, including bacteria, virus, and drosophila egg, have been measured by irradiation experiments. It should be noted that the inactivation is used for the survival endpoint of virus or drosophila egg.

In a radiation biology, a plot of the surviving fraction of cells vs radiation dose is called survival curve. Generally, radiation dose is plotted along the horizontal axis (logarithm in some cases), and surviving fraction is plotted along the vertical axis. This kind of dose-response relationship obtained by measuring cell surviving fraction is a momentous indicator of the radiosensitivity of corresponding living organisms, up to now.

As functional forms of survival curves, two general types of functions are well known to represent various survival curves, exponential and sigmoid (Fig. 7). However, as described later, there is no theory that can clearly explain the difference of the functional form of these survival curves from biological mechanisms.

In such a situation, the most famous mathematical model to derive the function that has a good fit with the experimentally obtained survival data is the target theory that has been summar-

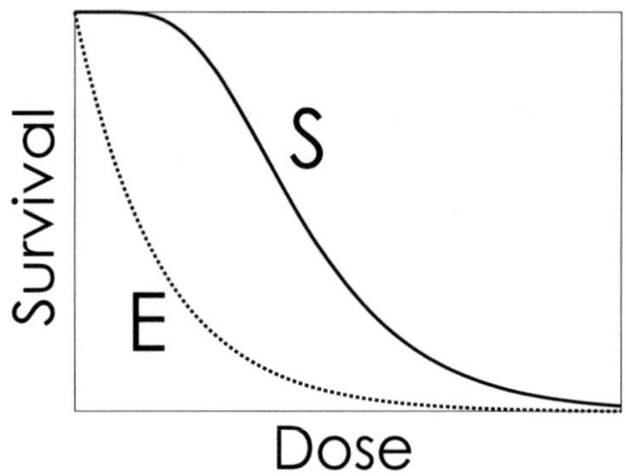


**Figure 7.** Two examples of survival curves. "E" means exponential and "S" means sigmoid

ized by Lea [21]. Target theory assumes that the "hit" of the discrete radiation to the radiation-sensitive site called "target" may induce cell death in a broad sense; moreover, the radiation hit is assumed a random process which has a Poisson probability distribution. (Fig. 8).

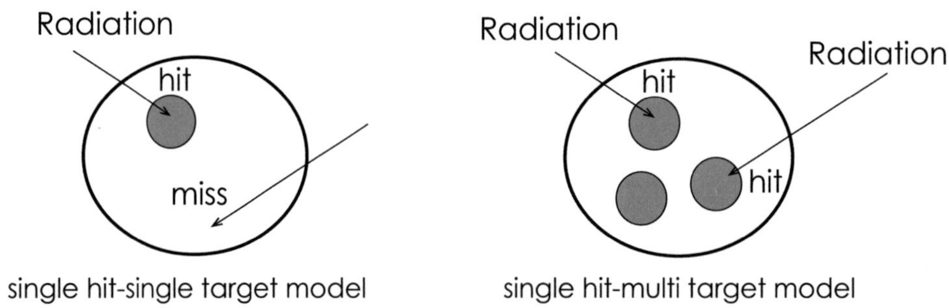


**Figure 8.** Schematic explanation of the target theory. Target theory assumes that each cell has some targets and radiation hits of all target may induce cell death. Left: Single radiation hit to the single target kills cell. Right: Single radiation hit to all of three targets kills cell

Let $P(n)$ be the probability of $n$ hits with expected value of hits $m$; then

$$P(n) = e^{-m} m^{n} n! \tag{3}$$

Let $V$ be a target volume, and suppose that $p$ denotes hit number per unit volume; then the expected value of hit number is $m=Vp$. Then, the hit number $p$ may be proportional to the radiation dose $D$; thus, $p=\alpha D$, where $\alpha$ is a proportional coefficient. Therefore, the expected hit number is written as

$$m = V\alpha D \tag{4}$$

There are several model types in target theory corresponding to the combination of required target and hit number for cell death; for simplicity, single-hit-single-target model and single-hit-multi-target model are described.

At first, suppose each cell contains only one single target and single radiation hit kills (inactivates) the cell, then the cell survival probability $S$, i.e., non-hit ($n=0$) probability, is

$$S = P(0) = e^{-m} = e^{-V\alpha D} \tag{5}$$

Let $D_0$ be a radiation dose that has an expected hit number $m = V\alpha D_0 = 1$ ; thus,

$$V\alpha = 1/D_0 \tag{6}$$

therefore,

$$S = e^{-D/D_0} \tag{7}$$

This survival curve equation (7) has a good fit with exponential-shaped curve (denoted as "E") in Fig. 7. Consequently, the probability of cell death for single-target cell is

$$1 - S = 1 - e^{-D/D_0}. \tag{8}$$

Next, suppose each cell contains two targets, and single radiation hit of both targets kills (inactivates) the cell. Moreover, suppose these two hit events are stochastically independent, then the cell death probability of two targets case $D_2$ equals the product of cell death probability of single target case, equation (8), therefore

$$D_2 = (1 - S)(1 - S) = \left(1 - e^{-D/D_0}\right)^2 \tag{9}$$

Thus, the cell survival probability of two targets case, $S_2$, is

$$S_2 = 1 - \left(1 - e^{-D/D_0}\right)^2; \tag{10}$$

consequently, cell survival probability of $k$ targets case, $S_k$, is

$$S_k = 1 - \left(1 - e^{-D/D_0}\right)^k \tag{11}$$

Alternatively, this survival curve, equation (11), has a good fit with sigmoid-shaped survival curve (denoted as "S") in Fig. 7.

Studying the validity of target theory, various inactivation experiments of viruses by radiation have been done and examined [21]. Various measurement results have shown that there is a strong correlation between virus size (diameter) and inactivation dose (of X- or $\gamma$-rays); moreover, in some cases, target volume $V$ equals the size of the virus itself. It should be noted that the word "target" used in the theory does not give a specific object; it is just a concept. Today, many experimental evidences suggest the DNA as the primary target for radiation.

Although the target theory can derive the functions to fit well with the survival curves of various experimental data, it is known that target theory gives higher surviving probability in very low-dose region than the real experimental data [22]. That is to say, actually, many cell deaths have occurred than expected from target theory in the very low-dose region. Moreover, it has been found that survival function obtained by target theory do not fit well with the experimental data not only in the low-dose region but in low-dose-rate irradiation.

In target theory, $D_0$ is generally called "mean lethal dose"; in a sense, one hit will kill the living organisms; however, it is just a parameter that can be estimated from fitting of experimental data to the target theory. On the contrary, by tuning this $D_0$ parameter, target theory can derive any functions to fit well with almost all experimental data. It is clear that this kind of high possibility of application to fit experimental data represents the arbitrariness of the model; thus, the model is not supposed to capture the essential dynamics of phenomenon.

Here, another mathematical model which will explain the shape of cell survival curves is introduced. The so-called linear-quadratic model (L-Q model) has been proposed for the candidate expression of the cell survival curves,

$$S = e^{-\alpha D - \beta D^2} \tag{12}$$

where $S$ is the surviving fraction for radiation dose $D$ and $\alpha$ and $\beta$ are proportionality constants. This kind of mathematical expression of the dose-response relationship can be found in many articles, e.g., [23, 24, 25, 26, 27].

L-Q model is often used to explain dose-response relationship of unstable chromosomal aberrations, e.g., dicentrics and rings ((b) and (d) in Fig. 9), which is thought to be the major cause of cell death. Rings and dicentrics can be seen and counted under the light microscope, and about 3 years of the half-life of lymphocytes carrying these aberrations have made it possible to estimate the accumulated radiation dose during a long period. Generally, this

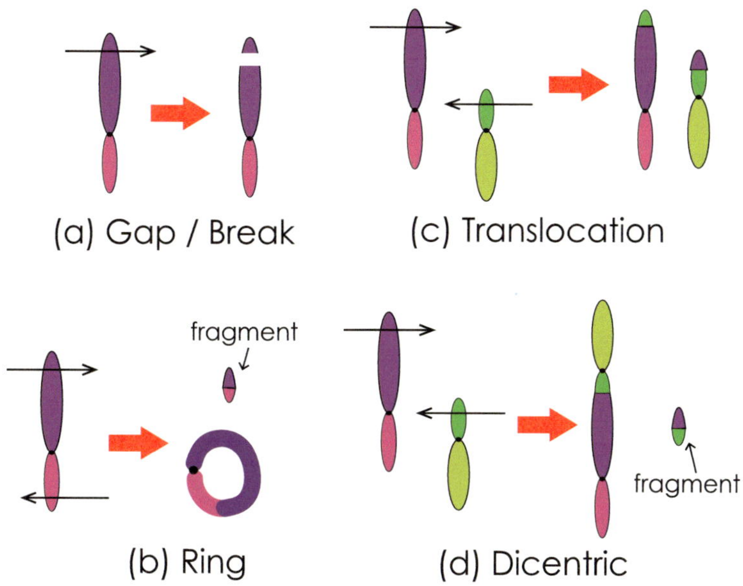


**Figure 9.** Schematic illustrations of the induction mechanism of chromosome aberrations. The four typical types of radiation-induced chromosome aberrations are presented. (a) One brake or cut is produced in one arm. "Gap" is defined as an achromatic lesion which has a smaller width than that of one chromatid and not separating. (b) Two breaks are produced in each arm of one chromatid. Then the broken ends may rejoin and make one ring and one fragment. (c)–(d) Two breaks are produced, one break in each arm of two chromosomes. Translocation (c) is made by rejoining of the proximal part, which has one centromere, and acentric fragment. Dicentric (d) is made by rejoining of the two proximal parts. Chromosome aberrations (c) and (d) are classified as the "exchange"

method is so-called biological dosimetry. Further examples of the chromosome aberration structure can be seen in the NIH Web site [28].

What is the phenomenological meaning of this function? It has been shown that the frequency of the chromosomal aberrations caused only by one break or damage has a linear dose dependence, $\alpha D$ ; however, that of induced by two chromosome breaks has a quadratic dose dependence, $\beta D^2$ (Fig. 10). In other words, frequency of the chromosome aberrations made from one break (one hit event) has a linear dose-response relationship, and then, frequency of the two-break (requires two hit)-induced aberrations has a quadratic dose-response relationship. Consequently, frequency of the chromosome aberrations per cell $X$ may have the dose-response relationship

$$X = \alpha' D + \beta' D^2 \tag{13}$$

where $D$ is the radiation dose and $\alpha'$ and $\beta'$ being parameters. Moreover, it has been found that the relationship between chromosome aberrations per cell $X$ and surviving fraction $S$ has a relation [30]:

$$S = e^{-\gamma X} \tag{14}$$

therefore,

$$S = e^{-\gamma\alpha' D - \gamma\beta' D^2} = e^{-\alpha D - \beta D^2} \tag{15}$$

where $\alpha = \gamma\alpha'$, $\beta = \gamma\beta'$.

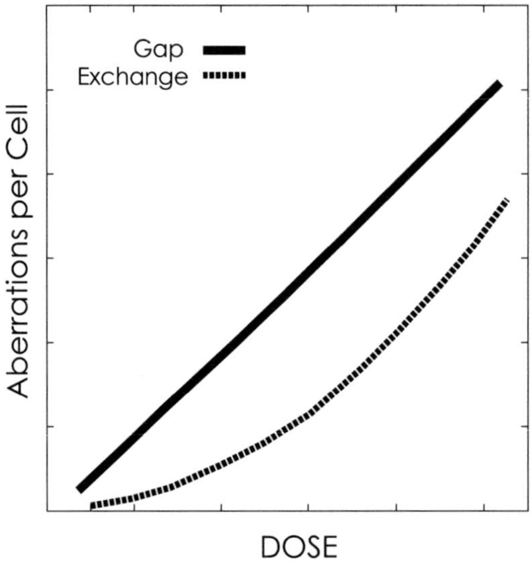


**Figure 10.** Dose effect curves for the different types of chromosome aberrations (gap and exchange) induced by X-ray in the *Vicia faba*. Clearly, gap shows a linear dose dependence while the exchange shows quadratic. Adapted from[29]

As stated above, the dose-response relationship between chromosome aberrations and radiation dose can be well explained from the relationship between the number of chromosome breaks and radiation dose using L-Q model. However, it may be said that L-Q model is rather descriptive than the dynamical.

Finally, one more interesting experimental data corresponding to the relation between radiation quality and dose is shown in Fig. 11. Here, each type of radiation has a different radiation quantity, linear energy transfer (LET) which is defined as the linear density of imparted energy. Therefore, LET is written mathematically as

$$\mathrm{LET} = dEdl \tag{16}$$

where $dE$ is the averaged energy deposition to the medium by charged particles in traversing a total length $dl$. In short, it is a quantity that represents averaged energy deposition around the tracks of charged particles (Fig. 12). Some examples of the LET values for different types of radiation are shown in Table 3. Moreover, Fig. 11 clearly shows that quality of the radiation determines the number of aberrations. It should be noted that the LET is not an actual, experimentally obtained value but just a practical, stochastically estimated value.

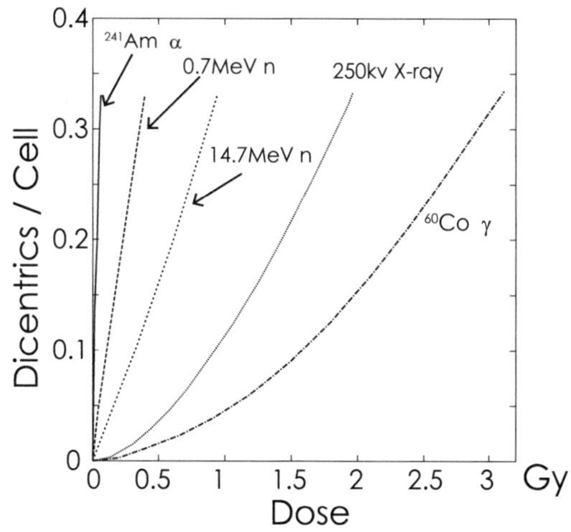


**Figure 11.** Dose-response curves for dicentric chromosome aberration induction in human lymphocytes with various LET radiations. Adapted from [31]

The effects of ionizing radiation on the living organisms have been studied in many years to understand three principal physical aspects: 1) **quantity of radiation,** 2) **temporal distribution of radiation,** and 3) **radiation quality** [32]. Studies on the dose-response relationship including cell survival curves are considered to belong to the research field of quantity of radiation, and studies on the dose rate and fractionation [33] are considered to belong to the research field of temporal distribution.

Historically, the concept of radiation quality has been established to study the biological effectiveness of ionizing radiation depending not only on the amount of absorbed radiation dose but also on the spatial distribution of energy deposition [34]. It is thought that the beginning of such study may have been closely related to the discovery of scientific evidence; i.e., DNA is the principal target for the biological radiation responses. For the sake of the presence of a small target in the cell, scientific knowledge of ionizing radiation, not only the macroscopic property (amount of absorbed radiation dose) but also detailed track structure (spatial distribution of energy deposition), has been required to study biological radiation responses.

| Radiation | LET keV/μm |
|---|---|
| Cobalt-60 $\gamma$-rays | 0.2 |
| -kV X-rays | 2.0 |
| -MeV protons | 4.7 |
| -MeV protons | 0.5 |
| .5 MeV $\alpha$-particles | 166 |
| -GeV Fe ions | 1000 |

**Table 3.** Typical values of LET. Values taken from [12]

In other words, A detailed information on how much damage on the DNA are produced by the different types of radiations with the same amount of dose has been necessary to estimate the radiation effects. Estimation or calculation of LET is one of such studies.

As mentioned above, requirement of the detailed information concerning radiation quality leads to plenty of studies that correspond to radiation track structure [35]; moreover, many studies regarding radiation action that has a more broad timescale including very early physical ($10^{-18}$-$10^{-12}$ sec) and chemical ($10^{-12}$-$10^{-0}$ sec) process have been performed [36].

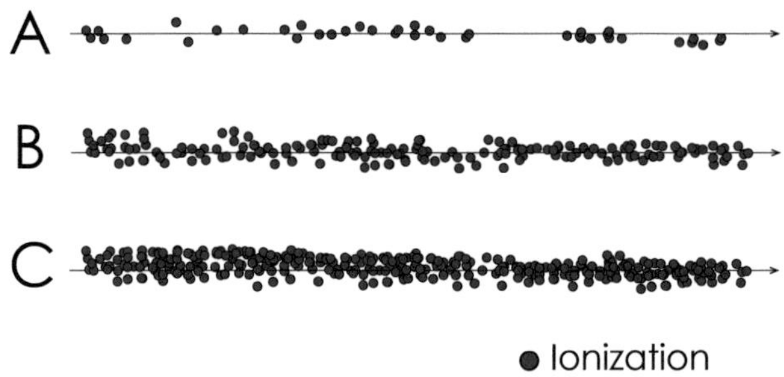


**Figure 12.** Schematic explanation of the variation of the ionization density. LET is increased in the order from A to C (top to bottom). Black circles show positions of ionization

## 4. Analyzing the various survival curves based on the knowledge of modern molecular biology

Survival curves of cells obtained under various conditions and mathematical expressions describing the survival curves have been introduced until now. It seems that the radiosensitivity and most of affecting factors have a simple proportional relationship; however, the interpretation of the cell-cycle-dependent nature of the radiosensitivity seems to be the most difficult. Can such a radiosensitivity really be explained by the mathematical model mentioned above? Here, the cell-cycle-dependent cellular response has been briefly discussed from the viewpoint of mathematical model and expected intracellular affecting factors.

At first, factors that can affect the shape of cell survival curve are discussed from the target theory, equations (7) and (11). Clearly, these equations show that the radiation dose $D$, required target number to cell death $k$, and target volume $V$ are all the affecting factors; moreover, radiosensitivity is inversely proportional to a required target number $k$ and in proportion to the target volume $V$ and radiation dose $D$. In order to determine a target volume and number, the information concerning the substance of the target should be needed. What is the actual target of radiation? Generally, it is believed that DNA is the principal target of radiation.

In case of cell-cycle-dependent radiosensitivity, the radiation dose $D$ is not the control parameter, and the required target number $k$ for cell death (reproductive cell death) seems to be unchanged, because one chromosome aberration may induce reproductive cell death. Thus, the required target number to cell death $k$ and target volume $V$ are the factors that need to be considered now. The volume of the target seems to have almost the same value from the M-phase of the cell cycle to the early S-phase; occupying the volume of all chromosomes for the most condensed stage (M-phase) is $\sim 91.6\mu m^3$ [37], whereas that of the interphase is estimated as the volume of cell nucleus; in case of $5.4\mu m$ diameter, interphase volume is almost the same as that of metaphase. On the contrary, target volume (DNA content) is doubled at the late S phase. By the interpretation from a target theory, increase in target volume causes increase in cell death. From Fig. 6, cell death seems to decrease during S-phase; this is not consistent with the analysis from the target theory.

As seen in the above discussion, time variation in the target number and volume is not recognized to have a bimodal change in the cell cycle. In other words, it is difficult to discuss a cell-cycle-dependent cyclic change of radiosensitivity by using a target theory. Then, does any kind of intracellular biochemical molecule which is involved in the DNA repair vary periodically during cell cycle?

There are not so many quantitative time-series data about the intracellular density change of proteins involved in DNA repair; however, the experimental data for cyclin B1 [38] shows that the variations in radiosensitivity (several tens fold in range) during the cell cycle cannot be satisfactorily explained by the temporal variation in cellular molecules; the functional form is completely different from a radiosensitivity curve during cell cycle (Fig. 6). Then, what kind of dynamics will be related with this characteristic bimodal periodic variation? One of the possible reasons for the radiosensitivity variation during cell cycle is that some kind of biological changes as a consequence of the structural change of the chromosome may affect the radiosensitivity variation.

Chromosomes are known to change their structure and volume corresponding to cell cycle stages. Generally, higher-order chromosome structure is important for many intracellular dynamics including DNA repair [39, 40]. Regarding spatial structure, a part of the aspects are not yet understood, but it is thought to have roughly four hierarchical levels, i.e., nucleosome, 30 nm chromatin fiber, lampbrush structure, and chromosomal structure. These structures are also known to exhibit cell-cycle-dependent structural changes. The shortest human autosome, chromosome 21, contains $48\times10^6$ base pairs (bps) of DNA, and its extended length is about $16\mu m$; however, its length is about $1\mu m$ during the metaphase, the most condensed state. Comparing the length of metaphase chromosomes to linear DNA, the packing ratio of DNA in metaphase chromosome is about 10,000.

As the first level of higher order compaction, 147 bps of DNA wraps 1.67 times around the histone octamer and makes unique unit structure called nucleosome core particle, and these units are connected by small amount of DNA, linker DNA. This basic repeating structure consists of nucleosome core particle and linker DNA is called "nucleosome," and it looks like a "beads-on-a-string" structure that corresponds to the state of interphase chromosome and the most basic unit structure of DNA packaging. Moreover, nucleosome is coiled into a 30 nm diameter helical structure known as the 30 nm chromatin fiber; however, their detailed packing manner is still unknown, and this is one of the expected basic structures in interphase. The most condensed metaphase chromosome is expected to have further two-step compaction from this 30 nm chromatin fiber, i.e., 300 nm "loop-like" (lampbrush) structure and 700 nm coiled structure (chromatid). Interphase chromosomes are known to occupy cell nucleus in spatially organized manner, i.e., chromosome territories (CTs), and on the other hand, their internal organized structures are poorly understood [41]. CTs have irregular shapes and occupy discrete compartments with little overlap; so interphase chromosomes are clearly separated inside a cell nucleus, respectively. Through the process of condensation, chromatin fiber increases its number of linkages and makes a condensed state, i.e., M-phase chromosome [42]. This process is very similar to the kind of phase transition from liquid to solid, freezing.

In addition to its unique structural properties, chromosome is known to have extraordinary physical properties. Observations of stretching in the region of DSB using partially broken X-irradiated chromosomes led to the proposal of a governing equation for chromosome movement by Nicklas [43] as follows:

$$F = Kl + \Gamma \frac{dl}{dt} \tag{17}$$

where $F$ is the force; $K$, the elastic proportionality constant between the force and stretch $l$; and $\Gamma$, the proportionality constant between force and velocity. Figure 13 is a schematic description of the structural changes of chromosome during cell cycle. From this figure, structural parameters of chromosome, $L$ and $R$ are found to change periodically. Similarly, elastic coefficient of chromosome $K$ in eq. (17) is also expected to change during cell cycle.

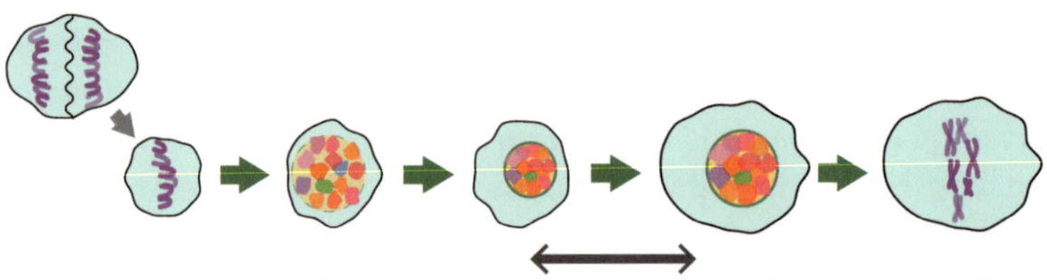

| | M | G1 | S | G2 | NEB |
|---|---|---|---|---|---|
| L | $L_M$ | $L_I$ | $L_{30nm}$ | $2L_I$ | $L_M$ |
| R | $R_M$ | $R_I$ | $R_{15nm}$ | $R_I$ | $2R_M$ |

**Figure 13.** Schematic explanation of the structural change of chromosome during cell cycle. M, G1, S, G2, and NEB show phases of cell cycle. "NEB" is an abbreviation for "nuclear envelope breakdown." L and R are the "length" and "radius (thickness)" of chromosome, respectively. The subscripts M and I for the L and R show length and radius for mitosis and interphase, respectively, and $L_{30nm}$ shows a length of 30 nm chromatin fiber and $R_{15nm}$ shows that of radius

In the recent years, many simulations using molecular dynamics (MD) on the DNA damage sites induced by ionizing radiation have been performed to study the detailed mechanics of the process of DNA damage and repair. These MD simulations have found some detailed states of the damage sites, though it is a limited condition, and one of the impressive findings is the very fast separating movement of the atoms that constitute damaged sites [44, 45]. Generally, the structure of the DNA damage sites is expected to be important for the repair probability which determines its ability. Thus, the dynamical aspects of the damaged sites accompanied by the structural transition of chromosome during cell cycle may affect the cell survival through a process of repair.

## 5. Conclusion

Some quantitative approaches in radiation biology based on the clonogenic method and obtained cell survival curves as a dose-response relationship are introduced. Cell survival curves seem to have a **universality** of its function, i.e., functional form of survival curves seems to be unchanged under various conditions including different species. Generally, the functional form of cell survival curve, especially by its slope, represents a quantity of radiosensitivity. Various factors affecting the radiosensitivity have been investigated to find macroscopic nature of living organisms, not to divide the system but to investigate the whole of it. Many mathematical models that describe cell survival curves have been presented; however, functional form of cell survival curves derived from, based on biological mechanism, does not yet exist. Mathematical or theoretical derivation of the functional form of cell survival may lead to understand the general theory of cellular responses to ionizing radiation, especially for the low-dose region.

Finally, the possibility that the structural change of chromosome affects the repair process is discussed. The survival curves that have a universality with respect to the dose-response relationship of cell should be studied further, to derive more analytically with the biological mechanism, also in order to progress the research on the secret of life.

## Author details

Noriyuki B. Ouchi

Address all correspondence to: ouchi.noriyuki@jaea.go.jp

Research Group for Radiation and Biomolecular Science, Radiation Biology Research Division, Quantum Beam Science Center, JAEA, Japan

## References

[1] International Human Genome Sequencing Consortium. Initial Sequencing and Analysis of the Human Genome. Nature. 2001 February;409(6822):860–921.

[2] Anderson PW. More Is Different. Science. 1972;177(4047):393–396.

[3] Wikimedia Commons [Image]. Wikimedia;. TeaAutomatAndMechanism.jpg. Available from: http://commons.wikimedia.org/.

[4] Kubo R. Statistical-Mechanical Theory of Irreversible Processes. I. General Theory and Simple Applications to Magnetic and Conduction Problems. Journal of the Physical Society of Japan. 1957;12(6):570–586.

[5] Puck TT, Marcus PI. Action of X-Rays on Mammalian Cells. The Journal of Experimental Medicine. 1956 May 1;103(5):653–666.

[6] Harrison RG, Greenman MJ, Mall FP, Jackson CM. Observations of the Living Developing Nerve Fiber. The Anatomical Record. 1907;1(5):116–128. Available from: http://dx.doi.org/10.1002/ar.1090010503.

[7] Keshishian H. Ross Harrison's "The Outgrowth of the Nerve Fiber as a Mode of Protoplasmic Movement". Journal of Experimental Zoology. 2004 March;301A(3):201–203.

[8] Wikimedia Commons [Image]. Wikimedia;. Tissue_culture_vials_nci-vol-2142-300.jpg. Available from: http://commons.wikimedia.org/.

[9] Wikimedia Commons [Image]. Wikimedia;. Agar_plate_with_colonies.jpg. Available from: http://commons.wikimedia.org/.

[10] Tubiana M, Dutreix J, Wambersie A. Introduction to Radiobiology. London: Taylor & Francis; 1990.

[11] Kaplan HS, Moses LE. Biological Complexity and Radiosensitivity: Radiation Lethality in Cells and Viruses is Correlated with Nucleic Acid Content, Structure, and Ploidy. Science. 1964 July;164(3627):21–25.

[12] Hall EJ, Giaccia AJ. Radiobiology for the Radiologist. 7th ed. Philadelphia: Lippincott Williams & Wilkins; 2012.

[13] Lloyd DC, Purrott RJ. Chromosome Aberration Analysis in Radiological Protection Dosimetry. Radiation Protection Dosimetry. 1981;1(1):19–28.

[14] Bedford JS, Mitchell JB. Dose-Rate Effects in Synchronous Mammalian Cells in Culture. Radiation Research. 1973;54(2):316–327.

[15] Fu KK, Phillips TL, Kane LJ, Smith V. Tumor and Normal Tissue Response to Irradiation In Vivo: Variation with Decreasing Dose Rates. Radiology. 1975;114(3):709–716.

[16] Town CD, Smith KC, Kaplan HS. Repair of DNA Single-Strand Breaks in E coli K-12 x-Irradiated in the Presence or Absence of Oxygen: The Influence of Repair on Cell Survival. Radiation Research. 1973;55(2):334–345.

[17] Tobias CA, Blakely EA, Chang PY, Lommel L, Roots R. Response of Sensitive Human Ataxia and Resistant T-1 Cell Lines to Accelerated Heavy Ions. The British Journal of Cancer. 1984;6:175–185.

[18] Kapp DS, Smith KC. Repair of Radiation-Induced Damage in *Escherichia coli* II. Effect of *rec* and *uvr* Mutations on Radiosensitivity, and Repair of X-Ray-Induced Single-Strand Breaks in Deoxyribonucleic Acid. Journal of Bacteriology. 1970 Jul;103(1):49–54.

[19] Terashima T, Tolmach LJ. Variations in Several Responses of HeLa Cells to X-Irradiation During the Division Cycle. Biophysical Journal. 1963 January;3(1):11–33.

[20] Sinclair WK. Cyclic X-Ray Responses in Mammalian Cells in Vitro. Radiation Research. 1968;33(3):620–643.

[21] Lea DE. Actions of Radiations on Living Cells. 2nd ed. Cambridge: Cambridge University Press; 1956.

[22] Kiefer J. Biological Radiation Effects. Berlin Heidelberg: Springer-Verlag; 1990.

[23] Douglas BG, Fowler JF. The Effect of Multiple Small Doses of X Rays on Skin Reactions in the Mouse and a Basic Interpretation. Radiation Research. 1976;66:401–426.

[24] Neary GJ. Chromosome Aberrations and the Theory of RBE. International Journal of Radiation Biology. 1965;9(5):477–502.

[25] Kellerer AM, Rossi HH. A Generalized Formulation of Dual Radiation Action. Radiation Research. 1978;75(3):471–488.

[26] Chadwick KH, Leenhouts HP. A Molecular Theory of Cell Survival. Physics in Medicine and Biology. 1973;18(1):78–87.

[27] Sinclair WK. Biophysical Aspects of Radiation Quality. Vienna: IAEA; 1966. Series 58.

[28] Genetics Home Reference. U.S. National Library of Medicine; [cited 2015 March 3]. Mutations and Health; Can Changes in the Structure of Chromosomes Affect Health and Development?. Available from: http://ghr.nlm.nih.gov/handbook.

[29] Revell SH. Evidence for a Dose-Squared Term in the Dose-Response Curve for Real Chromatid Discontinuities Induced by X-Rays, and Some Theoretical Consequences Thereof. Mutation Research/Fundamental and Molecular Mechanisms of Mutagenesis. 1966;3(1):34–53.

[30] Dewey WC, Stone LE, Miller HH, Giblak RE. Radiosensitization with 5-Bromodeoxyuridine of Chinese Hamster Cells X-Irradiated during Different Phases of the Cell Cycle. Radiation Research. 1971 September;47(3):672–688.

[31] Dufrain RJ, Littlefield GL, Joiner EE, Frome EL. Human Cytogenetic Dosimetry: A Dose-Response Relationship for Alpha Particle Radiation from 241Am. Health Physics. 1979;37(3):279–289.

[32] Rossi HH. Biophysical Aspects of Radiation Quality. Vienna: IAEA; 1966. Series 58.

[33] Elkind MM, Sutton H. Radiation Response of Mammalian Cells Grown in Culture: I. Repair of X-Ray Damage in Surviving Chinese Hamster Cells. Radiation Research. 1960;13(4):556–593.

[34] Rossi HH. Specification of Radiation Quality. Radiation Research. 1959;10(5):522–531.

[35] Nikjoo H, Uehara S, Emfietzoglou D, Cucinotta FA. Track-Structure Codes in Radiation Research. Radiation Measurements. 2006;41(9-10):1052–1074.

[36] Adams GE, Jameson DG. Time Effects in Molecular Radiation Biology. Radiation and Environmental Biophysics. 1980;17(2):95–113.

[37] Heslop-Harrison JS, Leitch AR, Schwarzacher T, Smith JB, Atkinson MD, Bennett MD. The Volumes and Morphology of Human Chromosomes in Mitotic Reconstructions. Human Genetics. 1989;84(1):27–34.

[38] Frisa PS, Jacobberger JW. Cell Cycle-Related Cyclin B1 Quantification. PLoS ONE. 2009 09;4(9):e7064.

[39] Alberts B, Johnson A, Lewis J, Raff M, Roberts K, Walter P. Molecular Biology of The Cell. 5th ed. New York: Garland Science; 2007.

[40] Mosesso P, Palitti F, Pepe G, Pinero J, Bellacima R, Ahnstrom G, et al. Relationship Between Chromatin Structure, DNA Damage and Repair Following X-Irradiation of Human Lymphocytes. Mutation Research/Genetic Toxicology and Environmental Mutagenesis. 2010;701(1):86 – 91.

[41] Cremer T, Cremer C. Chromosome Territories, Nuclear Architecture and Gene Regulation in Mammalian Cells. Nature Reviews Genetics. 2001;2:292–301.

[42] Thadani R, Uhlmann F, Heeger S. Condensin, Chromatin Crossbarring and Chromosome Condensation. Current Biology. 2012;22(23):R1012–R1021.

[43] Nicklas RB. A Quantitative Study OF Chromosomal Elasticity and Its Influence on Chromosome Movement. Chromosoma. 1963;14:276–295.

[44] Fujimoto H, Pinak M, Nemoto T, O'Neill P, Kume E, Saito K, et al. Molecular Dynamics Simulation of Clustered DNA Damage Sites Containing 8-Oxoguanine and Abasic Site. Journal of Computational Chemistry. 2005;26(8):788–798.

[45] Higuchi M, Pinak M, Saito K. Effects of Abasic Site and 8oxoG Lesions on DNA Molecule. Japanese Journal of Health Physics. 2007;42(2):130–130.

Chapter 3

# Dietary Modification of Mouse Response to Total-Body-Irradiation

Bing Wang, Kaoru Tanaka, Takanori Katsube, Kouichi Maruyama, Yasuharu Ninomiya and Mitsuru Nenoi

Additional information is available at the end of the chapter

http://dx.doi.org/10.5772/60653

### Abstract

Exposure to ionizing radiation (IR) could induce deleterious effects including cancer. Diet, as one of the major factors to influence susceptibility to many diseases, plays a critical role in maintaining human heath. It is known that unbalanced diet could result in health consequences, for example, high-calorie diet could lead to obesity, which could increase the risk of diabetes, heart disease, fatty liver, and some forms of cancer. Although the impact of diet on susceptibility to IR is thought to be big, the evidence is not clear due to lack of study. In this work, effects from dietary fat on modulation of mouse responses to total-body-irradiation (TBI) were studied. The mice were fed after weaning at postnatal age of 4 weeks with a standard diet (MB-1), a very high-fat diet (HFD32), and a very low-fat diet (CE-2 Low Fat), containing of 4.4%, 32.0%, and 0.4% of crude fat, respectively. A mouse model for radiation-induced adaptive response (AR) was applied to this work. The priming low-dose TBI at a dose of 0.5 Gy from X-rays was given at postnatal age of 6 weeks, and the challenge high dose of TBI was given at postnatal age of 8 weeks. The mouse response to low dose of TBI was evaluated by the efficacy of the priming low dose to rescue the animals from bone marrow death induced by the challenge high dose in the 30-day survival test. The mouse response to high dose of TBI was evaluated by comparing the LD50 in the 30-day survival test. In addition, dietary modulation of the residual (late) genotoxic effect from TBI was also evaluated by comparing the incidence of micronucleated erythrocytes in bone marrow using micronucleus test. Results showed that for the mice fed with the MB-1, a successful AR was demonstrated. While for the mice fed with either HFD32 or CE-2 Low Fat, no AR was observed, and all the animals died within 15 days after TBI with the challenge high dose at 7.5 Gy regardless the priming low dose at

0.5 Gy. When comparing the LD50 in the 30-day survival test, the LD50 values for the animals fed with the MB-1, HFD32 diet, and CE-2 Low Fat were 7.1 Gy, 6.0 Gy, and 6.2 Gy, respectively. As to the micronucleus test, for the mice fed with MB-1, the priming low dose at 0.5 Gy could significantly reduce the incidence of micronucleated erythrocytes in bone marrow that were caused by a challenge high dose at 4.0 Gy, while for the mice fed with either HFD32 or CE-2 Low Fat no such effect was observed. These findings indicated that under an unbalanced diet, namely, either of very high fat or of very low fat, alterations in mouse responses to TBI were induced. These findings confirmed that diet played a pivotal role in the response of the animals to radiation exposure, and suggested the possibility to modulate radiosensitivity through diet intervention in humans.

**Keywords:** Total body irradiation, diet, adaptive response, bone marrow death, micronucleated erythrocytes, mice

---

## 1. Introduction

Studies on radiation risk have been increasingly highlighted for special attention from both the academic and the public. Ionizing radiation (IR) could induce both genetic changes and epigenetic alterations. IR-induced genomic instability is a well-documented phenomenon, which can be observed even in the progeny of irradiated animals, suggesting the involvement of epigenetic mechanisms, such as the modulation of genome methylation or the regulation of micro-RNA expression [1]. Recent data suggest that exposures to IR even at low dose could also result in epigenetic modifications [2]. On the other hand, though being still small in the advances in understanding, to more it would have been known that responses to IR are epigenetically regulated. Epigenetic mechanisms may play a key role in the response of our body to IR. For examples, in the experimental animal models, diet-induced obesity modulated epigenetic responses (DNA methylation and microRNA regulation) to IR in mice [3]; antioxidant diets containing blueberry or strawberry extract were capable of mitigating the effects from exposure to heavy particles on behavioral alterations in rats [4, 5]. Multiple dietary, lifestyle and environmental factors, in addition to genetic factors, have a big impact on the health of our body [6]. Nutrients come from the diet, which include carbohydrates, fats, dietary fiber, minerals, proteins, vitamins, and water. Dietary factors have a profound effect on many aspects of health including aging and do so, at least partly, through interactions with the genome, which result in altered gene expression with consequences for cell function and health throughout the life course. In fact, dietary factors were associated with varied diseases and disease risk factors. Studies show that nutrition has a strong impact upon epigenetic processes, nutritional epigenetics emerges as a novel mechanism underlying gene-diet interactions, holding promise in having important roles in regulating health state, including age-related disease development, aging and longevity [7, 8].

http://dx.doi.org/10.5772/60653

Dietary, lifestyle and the environmental factors are critical to the state of health. Epigenetic mechanisms play an important role in shaping our phenotype via mediating between the nutrient inputs and the ensuing phenotypic changes throughout our entire life. Some epigenetic alterations may lead to accumulation of gene expression dysregulation and thus they are important in age-related diseases. In fact, interaction between nutrients and genes is responsible for regulating metabolic processes; and dietary and other environmental factors induce epigenetic alterations that may have important consequences for the initiation and development of pathological conditions such as obesity, metabolic syndrome, cancer, and alterations of the biological responses [6–8]. A recent study on twins even shows that our environment, more than our heredity, plays the starring role, especially as we age, in determining the state of our immune system which is the primary defense of the body against disease [9]. The pioneering work by Doll and Peto provided strong evidence that dietary factors may be as important as smoking behavior in explaining variation in cancer risk [10]. A healthy well-balanced diet contributes to a good quality of life, including both the health inside the body (i.e., the body growth, mental development, prevention of many diseases and infections, maintenance of the health state, longevity, etc.) and the way we look externally. For examples, individuals consuming a diet containing high amount of fruits and vegetables exhibit fewer age-related diseases: a greater intake of high-antioxidant foods such as berries may increase health span and enhance cognitive and motor function in aging [11], and the Mediterranean diet, which is rich in fruits, vegetables, nuts, legumes, unrefined grains, olive oil, and fish, with a moderate amount of alcohol intake and low intake of dairy products, meat, and poultry, could benefit health, namely, reduction of overall mortality, increased longevity and reduced incidence of chronic diseases [12]; diets containing a high proportion of plant foods are associated with lower risk of several common cancers [6]. A poor unbalanced diet could cause malnutrition, a condition due to eating a diet in which nutrients are not enough (i.e., starvation and a deficiency of one or more particular nutrients) or too much (i.e., intake of too many calories) for proper function of the cells, leading to health problems ranging from mild to severe and life-threatening. For example, a diet containing high amount of fat, resulting in overweight and obesity, leading consequently to high risks especially for cardiovascular disease, cancer, diabetes, osteoarthritis, and chronic kidney disease (13–16); higher intakes of red processed meats appear to be causal for colorectal cancer in contrast to an inverse correlation between higher intakes of fish and colorectal cancer risk [17]; high consumption of instant noodles is associated with a higher prevalence of metabolic syndrome in women [18]; and even regular consumption of sugar-sweetened sodas might influence metabolic disease development through accelerated cell aging [19].

Malnutrition includes both undernutrition and overnutrition. According to the International Federation of the Red Cross, in 2011, there were 1.5 billion people who suffered obesity worldwide while 925 million were undernourished [20]. Because of the established health risks and substantial increases in prevalence (increased risk of diabetes, heart disease, fatty liver, and some forms of cancer), obesity is now a global major health problem in developed countries and a growing one in the developing world [21]. In the United States, around half of the population are overweight, and one-third are obese. Epidemiological studies also show that the poor unbalanced diet is a major contributor to the leading causes of chronic disease and death [22]. The aggregate economic cost of obesity in this one country is estimated to be in excess of US$ 60 billion per year [21]. Worldwide, in 2010, overweight and obesity were estimated to cause 3.4 million deaths [23]. Metabolic disorders are among the fastest growing health problems worldwide, with a tendency for manifestation at earlier ages in recent years and with a higher rate in women than men [24]. The proportion of overweight adults (including obese) worldwide increased between 1980 and 2013 from 28.8% to 36.9% in men, and from 29.8% to 38.0% in women. In 2013, the prevalence of overweight and obesity increased in children and adolescents: in developed countries, 23.8% in boys and 22.6% in girls; in developing countries, 12.9% in boys and 13.4% in girls [23]. In the meantime, undernourishment (undernutrition) remains an important concern in the developing countries. Because severe acute undernutrition is associated with loss of a person's body fat and wasting of their skeletal muscle, causing many of those affected susceptible to disease, the infants and young children are the most vulnerable as they require extra nutrition for growth and development but have comparatively limited energy reserves. Studies present overwhelming evidence that early childhood nutritional status affects both the short- and long-term health status and development [25]. In children, undernutrition could have drastic and wide-ranging health consequences, such as increased gastrointestinal and respiratory infections and mortality risk; undernutrition is also closely associated with immunological alterations, development of noncommunicable diseases in adulthood, and cognitive and behavioral impairment in childhood and adolescence [26–29]. Epidemiological studies suggest that excessive adiposity, decreased physical activity, and unhealthy poor diets are key players in the pathogenesis and prognosis of many common cancers. As a fact, poor diet (including its resultant obesity and protein-energy malnutrition) is also a major risk factor for premature death, anemia, cardiovascular diseases, cancers, respiratory diseases, and injuries that are the most prominent causes of mortality [30, 31].

The potential role of epigenetic elements in the regulation of radiation effects deserves to be further investigated as such studies would give new insights into the mechanistic study on radiation effects and its risk [32]. Although the impact of diet on susceptibility to IR is thought to be big, the evidence is not clear due to lack of study. Elucidating the diet-related epigenetic mechanisms would facilitate a better understanding of radiation risk and prompt the development of more efficient strategies against radiation. There is a crucial need in better understanding the interactions between IR effects and dietary factors. Fat is one of the three main macronutrients in addition to carbohydrate and protein. It is an important foodstuff for many forms of life and serves both structural and metabolic functions: insulating body organs against shock; providing sources of essential fatty acids; providing energy sources and stores; being essential for digestion, absorption, and transportation of fat-soluble vitamins; playing a critical role in maintaining body temperature, healthy skin and hair [33–36]. Fat also serves as a useful buffer toward a host of diseases, and visceral fat is a significant producer of several hormones involved in inflammatory tissue responses and obesity, insulin resistance, and diabetes [37]. It is known that high-fat diet is responsible for most of the obese cardiovascular diseases, and cancer [38], and on the other hand, low-fat diet also shows potential health risk [34, 39, 40]. In this study, to understand the mechanisms that link nutritional factors to alterations in responses to IR, as the first approach, effects from dietary fat on possible alteration in response to total-body-irradiation (TBI) were comparatively studied in young female mice fed with a standard diet, a very high-fat diet, and a very low-fat diet.

## 2. Materials and methods

### 2.1. Animals and diets

Three-weeks-old C57BL/6J Jms strain female mice, wean just from breastfeeding, were purchased from SLC, Inc. (Hamamatsu, Japan). To avoid possible effects from the developmental condition of the animals, any mouse with a significantly different body weight (more or less than the mean ± 2 SD) upon arrival was omitted from this study. The selected mice were maintained in a conventional animal facility under a 12-h light/12-h dark photoperiod (lights on from 7:00 a.m. to 7:00 p.m.). Animals housed in autoclaved cages (1 mouse per cage) with sterilized wood chips were randomly assigned to three experimental groups and allowed free access to acidified water (pH = 3.0 ± 0.2) and a standard laboratory chow MB-1 (Funabashi Farm Co., Funabashi, Japan), a high-fat diet HFD32 (CLEA Japan, Inc. Tokyo, Japan), or a low-fat diet CE-2 Low Fat (CLEA Japan, Inc. Tokyo, Japan), respectively. The percentages of crude fat in the ingredient of MB-1, HFD32, and CE-2 Low Fat were 4.4%, 32.0%, and 0.4%, respectively. The metabolizable energy in kcal/100g for MB-1, HFD32, and CE-2 Low Fat was 354.0, 507.6, and 309.2, respectively. Animals were acclimatized to the laboratory conditions for 3 weeks before use. Based on our previous studies and preliminary trials, in the present study at least 20 mice were used in each experimental group and the experiment was repeated at least once. All experimental protocols involving mice were reviewed and approved by The Institutional Animal Care and Use Committee of the National Institute of Radiological Sciences (NIRS). The experiments were performed in strict accordance with the NIRS *Guidelines for the Care and Use of Laboratory Animals.*

### 2.2. Irradiation

X-rays were generated with an X-ray machine (Pantak-320S, Shimadzu, Japan) operated at 200 kVp and 20 mA, using a 0.50 mm Al + 0.50 mm Cu filter. An exposure rate meter (AE-1321M, Applied Engineering Inc, Japan) was used for the dosimetry. The dose rate for delivering irradiations at a low dose at 0.50 Gy and high doses ranging from 5.0 Gy to 8.0 Gy was at about 0.30 and 0.85 Gy/min, respectively. The mice held in acryl containers without anesthesia were exposed to TBI at room temperature.

### 2.3. Biological endpoints

Diet intake, body weight gain, organ weight, and intra-abdominal fat weight: The diet intake and body weight gain of the animals in each experiment group were recorded daily. The animals at age of postnatal 13 weeks were anesthetized by inhalation of gaseous isoflurane (2-chloro-2-(difluoromethoxy)-1, 1, 1-trifluoro-ethane) (CDS019936, Sigma-Aldrich, Japan) and then euthanized by cervical dislocation. Weight of some main organs and intra-abdominal fat

(IA fat) was weighed. These parameters obtained from mice fed with different diets were comparatively studied.

Bone marrow nucleated cells, the peripheral blood hemogram, and serum biochemical examination: The mice at postnatal age of 6 weeks and/or 8 weeks, or 13 weeks were anesthetized for collection of peripheral blood from the right femoral artery under anesthesia and then killed by cervical dislocation. Bone marrow cells were collected from both humeri and femora and the bone marrow nucleated cells were counted. As peripheral blood is the only tissue routinely available from human subjects, the peripheral blood hemogram and serum biochemical parameters were assessed in the present work to provide information for possible comparative clinical study in the future. Values of blood hemogram and blood biochemistry were comparatively studied in mice fed with different diets. The blood collected with a heparinized syringe in vacutainer blood collection tubes containing EDTA (Venoject II, Terumo Co., Japan) were immediately subjected to differential blood cell count (erythrocytes, leucocytes, and thrombocytes) and measurement of blood hemoglobin concentration using a blood cell differential automatic analyzer (SYSMEX K-4500, Sysmex Corporation, Japan). For biochemical examination, the serum of the blood collected without anticoagulant treatment was assessed using a biochemical automatic analyzer (DRI-CHEM 7000V, Fujifilm Corporation, Japan). The data for each experimental group were from at least five mice.

The 30-day survival test: The number of deaths that occurred within the 30-day period after TBI at high doses (from 5.0 Gy to 8.0 Gy) was recorded. The median lethal dose (lethal dose 50%, LD50) was used to comparatively study the radiosensitivity in mice fed with different diets. As a good in vivo model to study the response of mice to low-dose-induced adaptive response (AR), the mouse AR model for rescue of IR-induced bone marrow death [41] was adopted, verified, and confirmed under the experimental conditions in our research facilities, and finally applied to this study. In brief, the efficient priming dose of X-rays was 0.50 Gy. The timing for delivery of the priming dose and challenge dose was on postnatal ages of 6 and 8 weeks of the mice, respectively. When the priming dose induced a significant suppression of the mortality caused by the challenge dose, AR was considered as being successfully induced. Different challenge doses were chosen depended on the endpoint. A challenge dose at 7.50 Gy was used for verification and confirmation of the experimental conditions ensuring the successful induction of AR in mice fed with the standard diet MB-1. This dose was also tested as the challenge dose in mice fed with the high-fat diet HFD32 and low-fat diet CE-2 Low Fat. A challenge dose at 4.0 Gy was used to obtain more survivors in the 30-day survival test for investigations on the residual damage in the hematopoietic system.

Micronucleus test: The bone marrow micronucleus test was carried out accordingly [42] with minor modifications [43, 44]. Mice were sacrificed the following day after the 30-day survival test. Bone marrow smears prepared from both femora were processed for the enumeration of micronucleated polychromatic erythrocytes (MNPCEs) and micronucleated normochromatic erythrocytes (MNNCEs). The slides were coded to avoid any observer bias. The micronuclei were scored using a light microscope at a magnification of 1000×. At least 5000 cells per mouse were counted and the data for each experimental point were from at least 5 mice.

### 2.4. Statistical analysis

For LD50 determination, curvilinear regression of second degree was applied to the survival data using the programs embedded in KaleidaGraph Software (Version 4.1.2, Synergy Software, Hulinks Inc., Tokyo, Japan). Statistical evaluation of all the data was done using Student's *t*-test except for the micronucleus data where the $\chi^2$ test was performed. Statistical significance was assigned to $P < 0.05$.

## 3. Results

### 3.1. Intake of diet and metabolizable energy

The amount (weight in gram, g) of diet consumption by each mouse was recorded daily, based on which the mean values of both diet intake (Figure 1A) and metabolizable energy intake (Figure 1B) daily per mouse in each experimental group was calculated. When compared to the weight of the diet intake by the control animals fed with the standard diet MB-1, the weight of the diet intake by the mice fed with the low-fat diet CE-2 Low Fat in the first 2 weeks was significantly lower while it was markedly increased in the last week in the period of the experiment. On the other hand, mice fed with the high-fat diet HFD32 consumed significantly less amount of diet from the second week to the tenth week after diet onset (Figure 1A). As for the amount of metabolizable energy daily intake, it was significantly bigger in the mice fed with the high-fat diet throughout the experimental period while it was markedly smaller in the mice fed with the low-fat diet for most of the time in the period of the experiment (Figure 1B). Generally, there was a statistically significant difference on the mean metabolizable energy daily intake between the control group fed with MB-1 diet and that fed with an unbalanced diet (either HFD32 or CE-2 Low Fat). The animals fed with HFD32 diet took a significantly bigger amount of energy while the animals fed with the CE-2 Low Fat diet took a markedly smaller amount of energy.

### 3.2. Body weight gain, main organ weight, and IA fat weight

Body weight gain, main organ weight, and IA fat weight of the mice were studied to evaluate the effects from eating a diet containing different amount of dietary fat on physiological development. Body weight of each mouse was recorded daily until the end of the study at postnatal age 13 weeks (10 weeks after the diet onset). The mean body weight of mice in each experimental group continuously increased throughout the experiment (Figure 1C). When compared to the mice fed with the standard diet MB-1, significantly higher body weight was observed from 1.5 weeks after onset of the high-fat diet HFD32 to the end of the study while markedly lower body weight was recorded from as early as half week after onset of the low-fat diet CE-2 Low Fat to the end of the study. Ten weeks after the animals fed with different diets, the weight of some main organs and the IA fat were measured. As shown in Figure 1D, the weight of brain, liver, and kidney of the animals fed with the high-fat diet HFD32 were significantly higher compared to that of the animals fed with the standard diet MB-1. Notably,

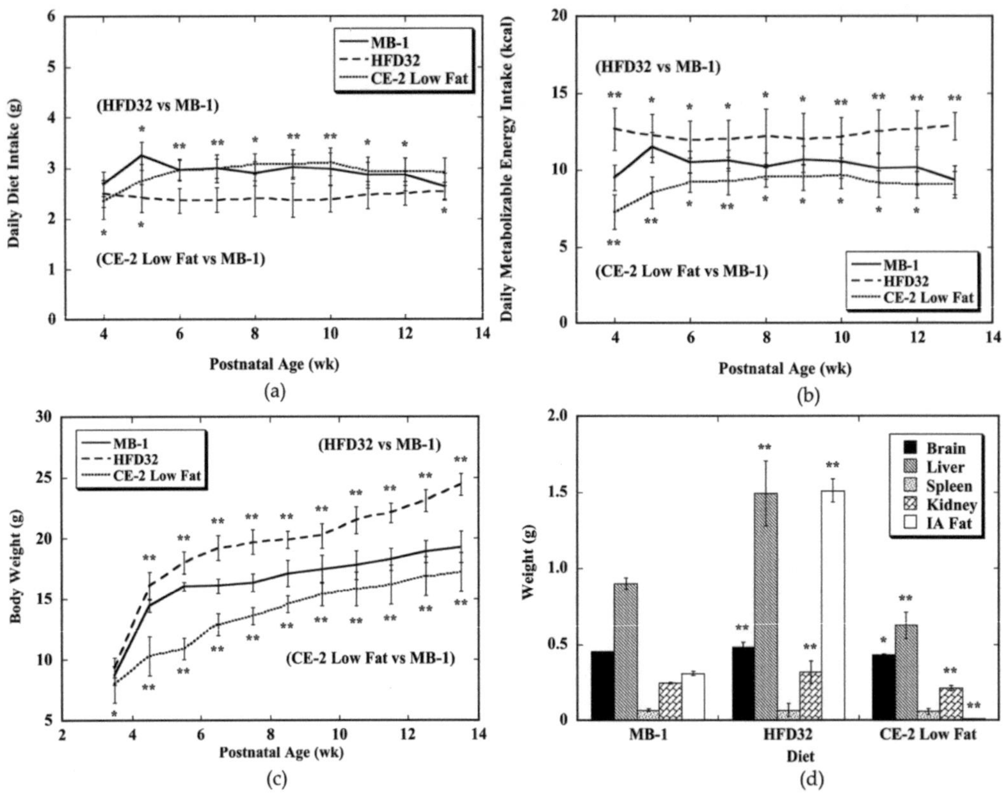


**Figure 1.** (a) Daily Intake of Different Types of Diets *ad libitum* after Weaning in Mice (b). Daily Metabolizable Energy Intake from Different Types of Diets ad libitum after Weaning in Mice (c). Effect from Consuming Different Types of Diets on Body Weight Gain of Mice (d). Effect from Consuming Different Types of Diets on Organ Weight and Intra-abdominal Fat Weight of Mice at Postnatal Age 13 Weeks

the weight of IA fat from the mice fed with HFD32 and from mice fed with CE-2 Low Fat were dramatically higher and lower than from the mice fed with MB-1. In addition, pathological study on liver showed characteristics of fatty liver and fatty tissue was noticeable in these organs. On the contrary, the weight of these organs and the IA fat were markedly lower in the animals fed with the low-fat diet CE-2 Low Fat.

### 3.3. Bone marrow nucleated cell count, peripheral blood hemogram, and biochemical examination

The peripheral blood was collected from animals in each experimental group at age of postnatal 13 weeks (10 weeks after onset of different diets). The hemogram was measured and biochemical examination of the serum was performed. For peripheral blood hemogram (Figure 2), mice subjected to the high-fat diet HFD32 displayed a significant decrease in both red blood cell count (RBC) and white blood cell count (WBC) when compared to the control that fed with the standard diet MB-1. Mice fed with the low-fat diet CE-2 Low Fat showed a significant

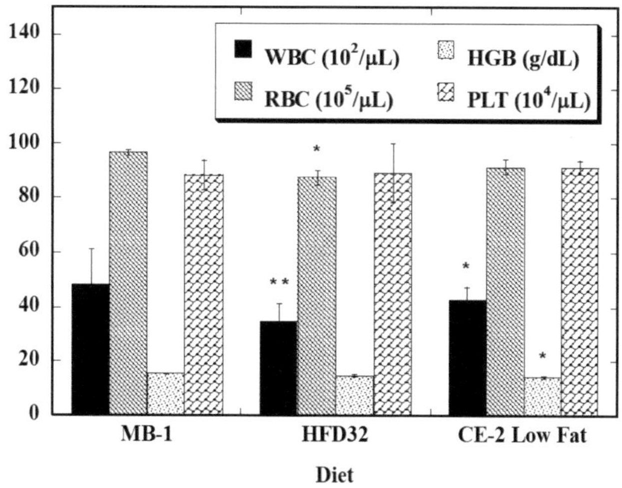


**Figure 2.** Effect from Consuming Different Types of Diets on Peripheral Blood Hemogram of Mice at Postnatal Age 13 Weeks.

reduction of both WBC count and hemoglobin concentration (HGB). No difference on blood platelet count (PLT) among the experimental groups was observed. For the bone marrow nucleated cells in both humeri and femora, the number ($\times10^7$) per mouse was 5.1 ± 0.8, 4.4 ± 0.5, 4.8 ± 1.0 in MB-1, HFD32, and CE-2 Low Fat group, respectively. The decrease in the number of bone marrow nucleated cells from the animals of HFD32 and CE-2 Low Fat groups was not of statistic significance when compared to that from MB-1 group. For biochemical examination of the serum collected from mice at postnatal age of 6 weeks (Table 1A) and postnatal age of 8 weeks (Table 1B), animals subjected to the unbalanced diet of either the high-fat diet HFD32 or the low-fat diet CE-2 Low Fat showed a significant alteration in may parameters. For example, at postnatal age of 6 weeks (3 weeks after diet onset), increase in glucose concentration (GLU-PIII) and total cholesterol concentration (TCHO-PIII) and decrease in the creatinine concentration (CRE-PIII) and alkaline phosphatase activity (ALP-PIII) were observed in mice fed with HFD32, while increase in CRE-PIII, uric acid concentration (UA-PIII), TCHO-PIII, ammonia concentration ($NH_3$), glutamic oxaloacetic transaminase (aspartate aminotransferase) activity (GOT/AST), glutamic pyruvic transaminase (alanine aminotransferase) activity (GPT/ALT-PIII), and $K^+$ concentration and decrease in GLU-PIII, triglyceride concentration (TG-PIII), total bilirubin concentration (TBIL-PIII), total protein concentration (TP-PIII), albumin concentration (ALB-P), gamma-glutamyl transferase activity (GGTP), and leucine aminopeptidase activity (LAP-P) were detected in mice fed with CE-2 Low Fat. At postnatal age of 8 weeks (5 weeks after diet onset), significant alteration in concentration of most parameters in general chemistry test in mice fed with either HFD32 or CE-2 Low Fat was detected when compared to that in the mice consuming the standard diet MB-1. For example, markedly increased GLU-PIII was observed in the animals fed with either of the unbalanced diet HFD32 or CE-2 Low Fat; increased TCHO-PIII, TG-PIII, TBIL-PIII, and TP-PII were detected in the mice fed with high-fat diet HFD32; and decreased TG-PIII, TBIL-PIII, inorganic phosphorus concentration (IP-P), TP-PIII, ALB-P, magnesium concentration (Mg), and $NH_3$ were observed in mice fed with low-fat diet CE-2 Low Fat. Marked alteration in enzyme activity was also detected in most of the parameters, such as GGTP, GOT/AST, GTP/ALT-PIII, etc., in enzymology test in the mice fed with the unbalanced diet HFD32 and/or CE-2 Low Fat.

| Biochemical examination | Experimental group MB-1 | Experimental group HFD32 | Experimental group CE-2 Low Fat |
|---|---|---|---|
| **General chemistry** | | | |
| Glucose concentration (GLU-PIII, mg/dL) | 143.0 ± 12.7 | 176.7 ± 5.8* | 54.7 ± 5.8** |
| Urea nitrogen concentration (BUN-PIII, mg/dL) | 31.6 ± 3.8 | 26.6 ± 0.8 | 27.2 ± 3.9 |
| Creatinine concentration (CRE-PIII, mg/dL) | 0.2 ± 0.0 | 0.1 ± 0.0** | 0.4 ± 0.1** |
| Uric acid concentration (UA-PIII, mg/dL) | 5.4 ± 1.9 | 4.6 ± 1.0 | 8.4 ± 2.1* |
| Total cholesterol concentration (TCHO-PIII, mg/dL) | 67.5 ± 0.7 | 103.0 ± 9.7* | 10.2 ± 1.1** |
| Triglyceride concentration (TG-PIII, mg/dL) | 272.0 ± 51.3 | 309.3 ± 43.5 | 112.0 ± 18.1** |
| Total bilirubin concentration (TBIL-PIII, mg/dL) | 5.9 ± 0.7 | 5.1 ± 2.4 | 1.9 ± 0.6** |
| Calcium concentration (Ca-PIII, mmol/L) | 9.0 ± 1.0 | 5.1 ± 2.4 | 7.7 ± 0.1 |
| Inorganic phosphorus concentration (IP-P, mg/dL) | 23.1 ± 2.9 | 21.3 ± 1.8 | 18.0 ± 1.3 |
| Total protein concentration (TP-PIII, g/dL) | 7.7 ± 0.7 | 7.2 ± 0.7 | 6.1 ± 0.3** |
| Albumin concentration (ALB-P, g/L) | 4.2 ± 0.5 | 4.5 ± 1.2 | 3.0 ± 0.2** |
| Magnesium concentration (Mg, mg/dL) | 3.2 ± 0.2 | 2.7 ± 0.2 | 3.3 ± 0.2 |
| Ammonia concentration ($NH_3$, μg/dL) | 737.0 ± 98.1 | 781.0 ± 71.6 | 1434.0 ± 42.3** |
| **Enzymology** | | | |
| Gamma-glutamyl transferase activity (GGTP, U/L) | 169.5 ± 58.5 | 128.0 ± 54.8 | 21.4 ± 15.0** |
| Glutamic oxalacetic transaminase (aspartate aminotransferase) activity (GOT/AST, U/L) | 66.3 ± 16.3 | 54.3 ± 14.0 | 185.0 ± 26.9** |
| Glutamic pyruvic transaminase(alanine aminotransferase) activity (GPT/ALT-PIII, U/L) | 33.3 ± 6.5 | 27.0 ± 4.4 | 28.7 ± 4.7 |
| Creatine phosphokinase activity (CPK-PIII, U/L) | 6183.3 ± 2664.4 | 6122.7 ± 3660.2 | 10263.8 ± 3750.2* |
| Lactate dehydrogenase activity (LDH-PIII, U/L) | 4048.0 ± 544.9 | 4011.7 ± 401.9 | 3763.7 ± 420.4 |
| Alkaline phosphatase activity (ALP-PIII, U/L) | 868.0 ± 32.5 | 778.7 ± 27.0* | 801.7 ± 38.4 |
| Leucine aminopeptidase activity (LAP-P, U/L) | 277.0 ± 12.7 | 216.3 ± 67.0 | 108.8 ± 32.3** |
| Creatine phosphokinase isozyme MB activity (CKMB-P, U/L) | 242.5 ± 81.3 | 181.3 ± 86.8 | 423.8 ± 85.7 |
| **Electrolyte** | | | |
| $Na^+$ (mEq/L) | 148.0 ± 2.8 | 142.7 ± 6.7 | 142.3 ± 8.4 |
| $K^+$ (mEq/L) | 11.2 ± 1.1 | 9.2 ± 0.3 | 22.9 ± 0.9** |
| $Cl^-$ (mEq/L) | 126.0 ± 8.5 | 118.3 ± 2.1 | 115.7 ± 1.5 |

*: P < 0.05, **: P < 0.01

(a)

| Biochemical examination | Experimental group | | |
|---|---|---|---|
| | MB-1 | HFD32 | CE-2 Low Fat |
| **General chemistry** | | | |
| Glucose concentration (GLU-PIII, mg/dL) | 169.7 ± 11.1 | 295.7 ± 45.6* | 225.0 ± 14.7** |
| Urea nitrogen concentration (BUN-PIII, mg/dL) | 32.8 ± 2.5 | 23.7 ± 3.3* | 31.2 ± 3.0 |
| Creatinine concentration (CRE-PIII, mg/dL) | 0.2 ± 0.0 | 0.1 ± 0.0** | 0.3 ± 0.1* |
| Uric acid concentration (UA-PIII, mg/dL) | 8.4 ± 0.7 | 7.3 ± 1.2 | 3.8 ± 0.2 |
| Total cholesterol concentration (TCHO-PIII, mg/dL) | 64.0 ± 3.0 | 104.3 ± 11.2** | 61.3 ± 2.5 |
| Triglyceride concentration (TG-PIII, mg/dL) | 256.3 ± 12.2 | 299.7 ± 17.9* | 66.0 ± 20.1** |
| Total bilirubin concentration (TBIL-PIII, mg/dL) | 6.7 ± 1.2 | 15.3 ± 4.4* | 1.8 ± 0.1** |
| Calcium concentration (Ca-PIII, mmol/L) | 8.0 ± 0.3 | 6.4 ± 0.4** | 8.7 ± 0.4** |
| Inorganic phosphorus concentration (IP-P, mg/dL) | 15.7 ± 2.5 | 22.5 ± 5.2 | 10.8 ± 0.6** |
| Total protein concentration (TP-PIII, g/dL) | 6.8 ± 1.0 | 10.9 ± 2.1* | 5.8 ± 0.3* |
| Albumin concentration (ALB-P, g/L) | 4.0 ± 0.7 | 6.9 ± 2.3 | 2.6 ± 0.2** |
| Magnesium concentration (Mg, mg/dL) | 2.9 ± 0.1 | 2.5 ± 0.1** | 2.5 ± 0.1** |
| Ammonia concentration ($NH_3$, μg/dL) | 796.3 ± 145.5 | 877.3 ± 115.2 | 464.0 ± 89.1** |
| **Enzymology** | | | |
| Gamma-glutamyl transferase activity (GGTP, U/L) | 139.0 ± 80.0 | 409.7 ± 102.4 | 5.3 ± 4.5** |
| Glutamic oxalacetic transaminase (aspartate aminotransferase) activity (GOT/AST, U/L) | 42.3 ± 12.9 | 84.3 ± 27.9* | 161.5 ± 60.1** |
| Glutamic pyruvic transaminase(alanine aminotransferase) activity (GPT/ALT-PIII, U/L) | 25.3 ± 7.8 | 84.3 ± 27.9** | 29.3 ± 17.9 |
| Creatine phosphokinase activity (CPK-PIII, U/L) | 7722.8 ± 3123.5 | 12597.3 ± 7131.1 | 6554.0 ± 3169.2* |
| Lactate dehydrogenase activity (LDH-PIII, U/L) | 3476.0 ± 500.4 | 7082.5 ± 2107.8* | 1380.8 ± 122.2** |
| Alkaline phosphatase activity (ALP-PIII, U/L) | 637.7 ± 41.5 | 15837.3 ± 3645.5** | 763.5 ± 31.5** |
| Leucine aminopeptidase activity (LAP-P, U/L) | 200.8 ± 58.1 | 417.3 ± 178.1* | 78.5 ± 7.9** |
| Creatine phosphokinase isozyme MB activity (CKMB-P, U/L) | 226.3 ± 107.0 | 326.7 ± 67.5 | 222.6 ± 64.0 |
| **Electrolyte** | | | |
| $Na^+$ (mEq/L) | 146.0 ± 1.4 | 140.0 ± 2.8 | 148.3 ± 1.2 |
| $K^+$ (mEq/L) | 8.9 ± 0.2 | 9.1 ± 1.5 | 5.6 ± 0.3** |
| $Cl^-$ (mEq/L) | 116.5 ± 0.7 | 113.0 ± 1.4 | 116.3 ± 2.3 |

*: P < 0.05, **: P < 0.01

(b)

**Table 1.** A. Biochemical Examination on the Serum of Mice at Postnatal Age of 6 Weeks Table 1B. Biochemical Examination on the Serum of Mice at Postnatal Age of 8 Weeks

These results showed that dietary fat had a big impact on the development of mice. These results indicated that consuming an unbalanced diet containing different amount of dietary fat could cause a series of detrimental health consequence, manifesting as alterations in body weight, organ weight, IA fat weight, peripheral blood hemogram, and serum biochemistry.

### 3.4. LD50 in the 30-day survival test

Alteration in sensitivity to TBI-induced bone marrow death due to mice consuming different diets was comparatively studied using LD50 in the 30-day survival tests. The curvilinear regression of second degree was applied to the data analysis; survival curve for each group fitted well a quadratic polynomial expression (Figure 3). The regression analysis yielded LD50 as 7.1 Gy, 6.0 Gy, and 6.2 Gy, respectively, for the animals fed with the standard diet MB-1, the high-fat diet HFD32 and the low-fat diet CE-2 Low Fat. These results indicated that mice fed with an unbalanced diet containing different amount of dietary fat became highly sensitive to the killing effect from TBI at high doses.

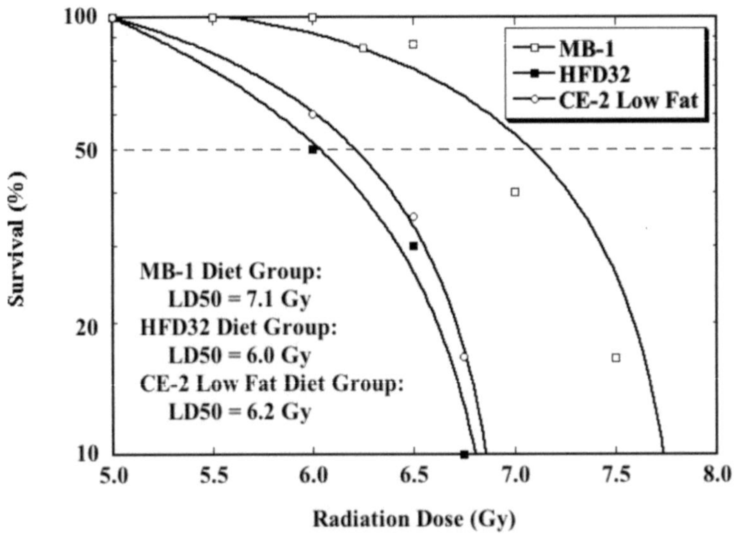


**Figure 3.** 30-Day Survival after Total Body Irradiation of Mice Fed with Different Kinds of Diets at Postnatal Age 8 Weeks

### 3.5. Induction of AR

Reproducibility of the mouse model for AR induction was verified and confirmed by delivery of the priming low dose at 0.5 Gy and challenge high dose at 7.50 Gy using mice fed with the standard diet MB-1. The priming dose markedly increased the survival rate from 16.7% to 80.0% in the 30-day survival test (Figure 4A). These results clearly indicated that AR was successfully induced with efficient reliability and reproducibility in our experimental setup.

Serving as the positive control, the effect from consuming an unbalanced diet containing different amount of dietary fat on AR induction was comparatively studied. As for the mice fed with the high-fat diet HFD32, all animals died within 15 days after the challenge dose, regardless of receiving the priming dose (Figure 4B). As for the mice fed with the low-fat diet CE-2 Low Fat, all animals died within 13 days after the challenge dose, regardless of receiving the priming dose (Figure 4C). Considering the sensitivity to TBI-induced bone marrow death was higher in mice fed with an unbalanced diet and 7.5 Gy would be too high to be used as the challenge dose in the AR induction study; a dose at 6.5 Gy which resulted in a survival rate at 30.0% and 35.0%, respectively, in mice fed with the high-fat diet HFD32 and low-fat diet CE-2 Low Fat in the 30-day survival test (Figure 3) was also tested as a challenge dose. The survival rate was respectively 29.4% and 33.3% for these mice, showing that no AR was induced in the mice fed with an unbalanced diet containing different amount of dietary fat. These results indicated that the response to low dose of TBI altered in mice fed with either a high-fat diet or a low-fat diet when compared to that fed with the standard diet.

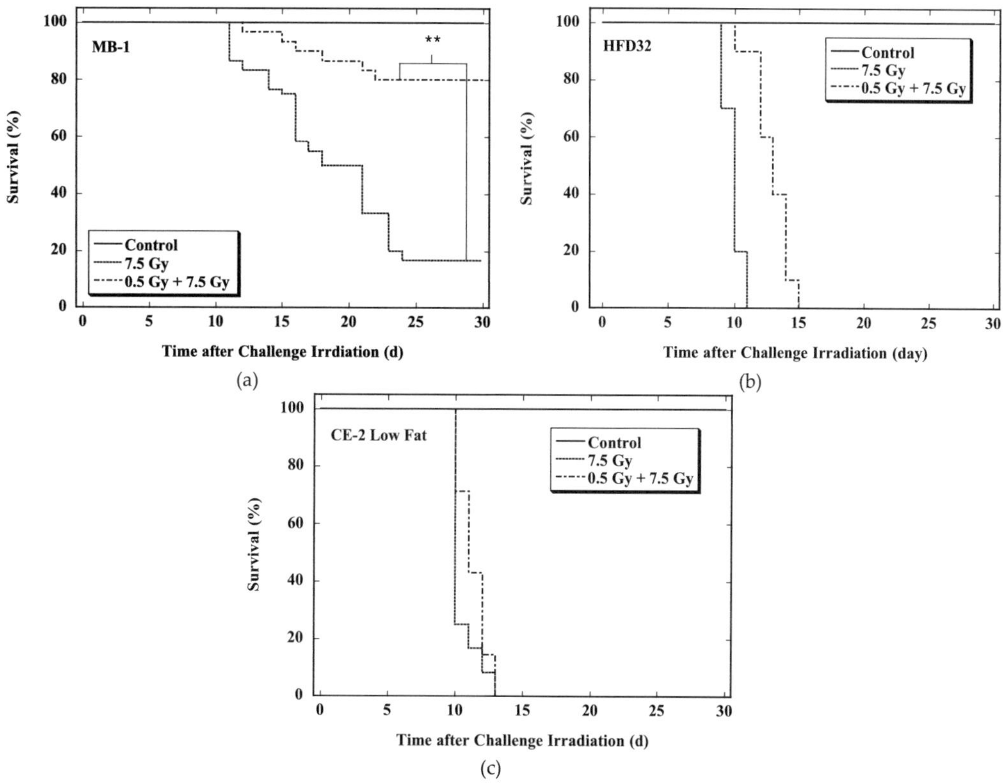


**Figure 4.** (a). Induction of Adaptive Response in Mice Fed with the Standard Diet MB-1. (b).Induction of Adaptive Response in Mice Fed with the High-Fat Diet HFD32. (c). Induction of Adaptive Response in Mice Fed with the Low-Fat Diet CE-2 Low Fat.

### 3.6. Residual damage in bone marrow erythrocytes

As bone marrow failure was the main cause for the animal death in the AR mouse model, to evaluate alterations in TBI-induced cytotoxicity and genotoxicity due to consuming a diet containing different amount of dietary fat, the residual damage in the bone marrow cells of the animals was measured 1 day after the 30-day survival test. The percentage of polychromatic erythrocytes (PCEs) to the sum of PCEs and normochromatic erythrocytes (NCEs) is an indicator for mutagen-induced cytotoxicity to bone marrow proliferation in the erythroid lineage [45], and micronucleus test is a tool for genotoxic assessment. Though the percentage of PCEs to the sum of PCEs and NCEs was lower for mice fed with the low-fat diet CE-2 Low Fat, it was not of statistical difference when compared to that for the mice fed with the standard diet MB-1 (data not shown). Results obtained in mice fed with the MB-1 diet (the positive control group for AR induction) showed that the priming low dose at 0.5 Gy significantly reduced the occurrences of both MNPCEs in PCEs (Figure 5A) and MNNCEs in NCEs (Figure 5B) in the femur bone marrow when respectively compared to that receiving the challenge dose alone. On the other hand, in the animals fed with either a high-fat diet or a low-fat diet, the priming dose failed to induce a marked reduction of the occurrences of MNPCEs and MNNCEs (Figure 5C, 5D, 5E, and 5F) when respectively compared to that receiving the challenge dose alone. These results were consistent with the results obtained on AR induction in the 30-day survival test. In addition, though the increase was not of a statistical significance, when compared to the incidences of MNPCEs and MNNCEs induced by the challenge dose alone in the animals fed with the standard diet MB-1, increased incidences was always recorded in mice fed with an unbalanced diet containing different amount of dietary fat. These results indicated that consuming a diet containing different amount of dietary fat had a significant impact on the cytotoxic and genotoxic effect on the bone marrow erythrocytes.

## 4. Discussion

The mouse model for IR-induced AR was applied to present work and alterations in response of mice to TBI were investigated under diets containing different amount of dietary fat. The mouse response to low dose of TBI was evaluated by the efficacy of the priming low dose to rescue the animals from bone marrow death induced by the challenge IR at higher doses in the 30-day survival test. The mouse response to high doses of TBI was studied by comparing the LD50 values in the 30-day survival test. In addition, dietary modulation of the residual (late) genotoxic effect from TBI was also evaluated by comparing the incidence of micro-nucleated erythrocytes in bone marrow using micronucleus test. Results demonstrated that under an unbalanced diet, namely, either of very high fat or of very low fat, alterations in the response of mice to TBI were induced at both low dose and high doses: abolishment of AR induction by the low dose which was efficient in mice fed by the standard diet, increase in the radiosensitivity to bone marrow failure induced by high doses, and increase in genomic instability after high dose. These findings confirm that dietary fat plays a pivotal role in the response of the animals to IR exposure and provide new insight into the study on the epigenetic

http://dx.doi.org/10.5772/60653

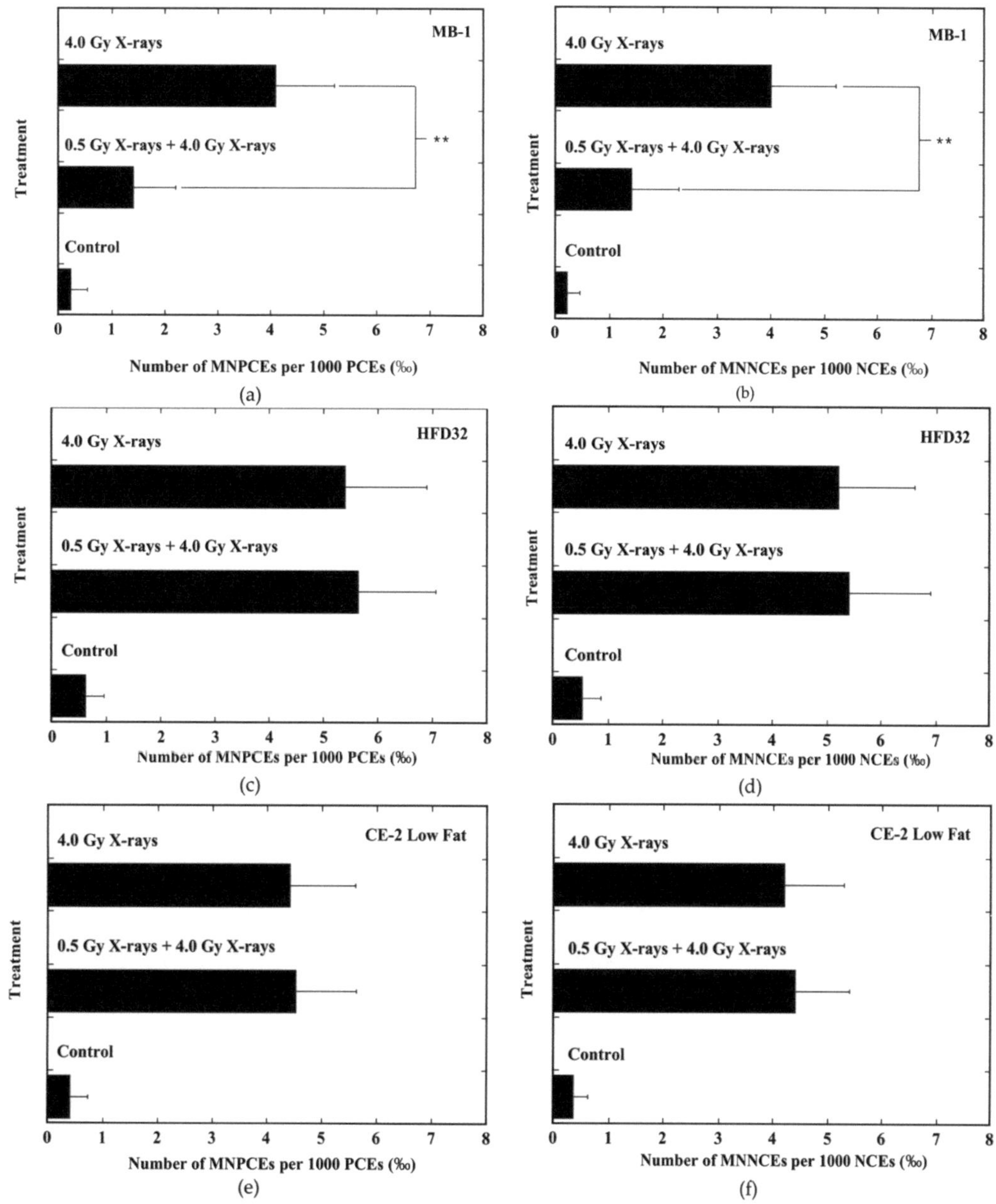


**Figure 5.** (a) Incidence of MNPCEs induced by TBI in Femur Bone Marrow Erythrocytes of Mice Fed with the Standard Diet MB-1. (b). Incidence of MNNCEs induced by TBI in Femur Bone Marrow Erythrocytes of Mice Fed with the Standard Diet MB-1. (c). Incidence of MNPCEs induced by TBI in Femur Bone Marrow Erythrocytes of Mice Fed with the High-Fat Diet HFD32. (d). Incidence of MNNCEs induced by TBI in Femur Bone Marrow Erythrocytes of Mice Fed with the High-Fat Diet HFD32. (e). Incidence of MNPCEs induced by TBI in Femur Bone Marrow Erythrocytes of Mice Fed with the Low-Fat Diet CE-2 Low Fat. (f).Incidence of MNNCEs induced by TBI in Femur Bone Marrow Erythrocytes of Mice Fed with the Low-Fat Diet CE-2 Low Fat.

contribution to radiation risk. These findings aid knowledge for radiation risk reduction and suggest the possibility to modulate radiosensitivity through diet intervention in humans.

Although the results obtained in this phenomenal study clearly show that dietary fat has a big impact on mouse response to TIB, the underlying mechanisms remain unclear. It is well known that disturbance of hormonal balance, production and secretion of cytokines and growth factors could result in alterations in radiation susceptibility as these intercellular and intracellular messengers play important roles in preserving and restoring functions of tissues compromised by IR [44–48]. As dietary fat plays a critical role in such as physiological development and maintaining metabolic and immune function of the body [33–37], disturbed hormonal level, metabolic environment, and immune functions due to eating an unbalanced diet containing either very high fat or very low fat would be responsible for the altered response of the mice to TBI. In fact, results obtained in this study on biochemical examination of serum showed clearly that an unbalanced diet could result in alterations in many parameters and these findings could suggest the resultant disturbed metabolism and physiological development of the animals. It is known that consumption of a high-fat diet correlates with increased oxidative stress and chronic inflammation in many organs [49, 50]. Oxidative stress could result in increased sensitivity to IR [51–54]. Cumulative evidences also show that high-fat diet causes disturbance of nutrient balance, leading to dysregulation of the adipoinsular axis, a dual hormonal feedback loop involving insulin and leptin, and alterations in hormone production and secretion, i.e., increased insulin and decreased leptin [55]. On the other hand, as dietary fat clearly serves a number of essential functions, diet with adequate energy from fat is needed to promote normal growth and normal sexual maturation; maternal energy deficiency due to very low-fat intakes is one key determinant in the etiology of low birth weight [34]. C57BL/6J strain mice fed with a very high-fat diet provide a good experimental model of diet-induced obese in a way closely matching the development of human metabolic syndrome, and a good experimental model for studying the molecular mechanisms as well [56]. The present work indicates that C57BL/6J strain mice could be also used as a good model to study the effect from an unbalanced diet containing very low fat. There is growing evidence showing that diet can strongly influence epigenetic processes [57]. Given the importance of dietary fat on the alterations in responses to IR at whole body level, in the future, in particular for this reason, a significant amount of interest should be given to the molecular analysis of mechanisms involving epigenetic regulation such as metabolic and immune disruptions under conditions with very high-fat diet and very low-fat diet.

The present study reinforces the importance of understanding the dietary factors for that there is a striking modification effect on the response to IR. Developing active prevention strategies would be a practical approach to meeting the critical need to reduce the radiation risk, in addition to the improvement of overall health and the quality of life. In fact, understanding of the ability and mechanisms of dietary modification will fuel the development of effective countermeasures to reduce radiation risk. Further work is required to understand the mechanisms through which specific dietary factors produce epigenetic

http://dx.doi.org/10.5772/60653

changes and to identify those changes that are likely to be causal in the alterations in response to radiation. As to the diet fat, it is known that a healthy low-fat diet is not a diet containing fat approximately down to zero. More attention needs to be devoted to the effect of dietary fat reduction on the nutrient density, especially for children [34]. Notably, recent study shows that it is more than just the amount of fat, the types of fat really matter. Bad fats may increase the risk of certain diseases, while good fats support overall health. Future study may focus on the types of fat in the diet and their possible modification effect on response to IR. In addition, there are many factors that impact eating behaviors in humans, i.e., the profile of food choice depends on the gender and age. Studies described remarkable differences between genders in food choice: women had higher intakes of fruit and vegetables, higher intakes of dietary fiber and lower intakes of fat. The motivation of weight control was more prominent in women who were more likely to diet or restrain their eating behavior. Age-related changes in the chemosensory perceptual systems motivated the choice of other foods, and therefore a varied diet [58]. Eating takes place in a context of environmental stimuli known as ambience, which has influence on nutritional health. Various external moderators such as social and physical surroundings and internal factors such as food variables also affect food intake and food choice [59, 60]. In humans, as the relationship among stress, dietary restraint, and food preference is complicated, for an example, it was reported that high-stressed women preferred sweet, high-fat food more than did low-stressed women, whereas low-stressed women ate more low-fat than high-fat food [61, 62]. As many factors that impact the eating behaviors, given the complexity of the epigenetic machinery, it is also important to unravel the differential role of the various epigenetic participants in a given physiopathological condition [24]. In further study, it is important to further decipher for various nutritional factors the role and the mechanisms involved in driving epigenetic-related alterations in responses to radiation, as well as to assess the role of the presence of other factors at different ages and in both sexes.

Dietary, lifestyle, and environmental factors could affect many biological and pathological processes. It is known that making positive diet and lifestyle changes, namely, eating a healthy diet, getting enough exercise, and refraining from tobacco and excessive alcohol use, could confer numerous health benefits including possibly preventing the onset of chronic diseases. This suggests that these processes are interventable through dietary intervention. In fact, there is a move to improve nutritional status in malnourished patients through the use of multimodal interventions including nutritional supplements in the prevention and management of disease-related malnutrition [63]. It is shown that dietary intervention by distribution of nutritious supplementary foods to young children in conjunction with household support is an appropriate strategy for the prevention of moderate acute malnutrition and severe acute malnutrition in young children [64], and adequate treatment by dietary intervention is important for reversing these effects [65]. Nutrition intervention (nutritional counseling and oral nutritional supplements) is also used to prevent therapy-associated weight loss and interruption of chemotherapy and/or radiotherapy in cancer patients [66, 67]. Adequate nutrition during cancer plays a decisive role in treatment response and quality of life in humans [68]. On the other hand, calorie restriction (CR) has been shown to be effective as one of the dietary interventions. Although it still requires intense efforts for the interventions to unravel the complexity of the epigenetic, genetic, and environment interactions and to evaluate their potential reversibility with minimal side effects, encouraging trials suggest the prevention and therapy of age- and lifestyle-related diseases such as by individualized tailoring to optimal epigenetic diets [24]. In addition to the extension of the maximal lifespan of a diverse group of species, CR without malnutrition was demonstrated to reduce the morbidity of a host of diseases including broadly effective in cancer prevention in laboratory rodents [69, 70]. In animals, in addition to cardiovascular-specific effects, CR caused a variety of improvements related to overall health; in humans, studies noted favorable changes in multifarious biomarkers, particularly those related to cardiovascular and glucoregulatory function [71]. CR could decelerate the rate of aging and inhibit tumor formation in mammals [72–74]. It could also decrease the tumors that were induced by chemicals and IR in rodents and nonhuman primates [70, 75–83]. Even after exposure to IR, CR could still be effective for reducing cancer incidence in mice [82, 83]. These studies showed that dietary intervention could alter the phenotype and epigenotype in animal models [84], and demonstrated the feasibility of an active means to prevent or reverse the adverse effects from unbalanced diet or malnutrition. All these experimental and clinical works indicate that dietary intervention would hold special potential to treat certain diseases as both useful support to conventional therapy and countermeasure as well against IR-induced detrimental effects. Notably, dietary intervention is not limited to administration of nutritious supplementary foods only, some biologically active chemicals, such as calorie restriction mimetics, should be taken into account based on the mechanism studies [85]. For an example, extensive studies indicated that resveratrol, a natural biologically active polyphenol present in different plant species, has anti-diabetic action in animal models and in diabetic humans [49, 50, 86, 87]. A healthy diet is the key to maintaining well-being and preventing health problems. Health consequences due to poor unbalanced diet-induced malnutrition are getting worse, as an example for obesity, not only is it increasing but no national prevention success has made so far [23]. Making healthy food choices is more important than ever. A global action with simple and effective countermeasures against poor unbalanced diet-induced health consequences is urgently needed. This is of great significance and importance for prevention of diet-related health problems and reversibility of the altered biological responses including responses to IR.

## Acknowledgements

The authors would like to thank Ms. Yasuko Morimoto, Mr. Sadao Hirobe, Ms. Mikiko Nakajima, and Ms. Hiromi Arai for their expert technical assistance and administrative support. The critical and constructive comments on manuscript preparation from Dr. Yi Shang are gratefully acknowledged. Great appreciation is especially given to the anonymous peer reviewers for providing the constructive comments that strengthened the presentation of this work.

This work was supported by the National Institute of Radiological Sciences, Japan.

## Author details

Bing Wang*, Kaoru Tanaka, Takanori Katsube, Kouichi Maruyama, Yasuharu Ninomiya and Mitsuru Nenoi

*Address all correspondence to: jp2813km@nirs.go.jp

Radiation Risk Reduction Research Program, Research Center for Radiation Protection, National Institute of Radiological Sciences, Inage-ku, Chiba, Japan

The authors have declared no conflicts of interest.

## References

[1] Ilnytskyy Y, Kovalchuk O. Non-targeted radiation effects-an epigenetic connection. Mutat Res 2011;714(1–2):113–25.

[2] Bernal AJ, Dolinoy DC, Huang D, et al. Adaptive radiation-induced epigenetic alterations mitigated by antioxidants. FASEB J 2013;27(2):665–71.

[3] Vares G, Wang B, Ishii-Ohba H, et al. Diet-induced obesity modulates epigenetic responses to ionizing radiation in mice. PLoS One 2014;9(8):e106277.

[4] Rabin BM, Joseph JA, Shukitt-Hale B. Effects of age and diet on the heavy particle-induced disruption of operant responding produced by a ground-based model for exposure to cosmic rays. Brain Res 2005;1036(1–2):122–9.

[5] Rabin BM, Shukitt-Hale B, Joseph J. et al. Diet as a factor in behavioral radiation protection following exposure to heavy particles. Gravit Space Biol Bull 2005;18(2):71–7.

[6] Mathers JC, Strathdee G, Relton CL. Induction of epigenetic alterations by dietary and other environmental factors. Adv Genet 2010;71:3–39.

[7] Niculescu MD, Lupu DS. Nutritional influence on epigenetics and effects on longevity. Curr Opin Clin Nutr Metab Care 2011;14(1):35–40.

[8] Park LK, Friso S, Choi SW. Nutritional influences on epigenetics and age-related disease. Proc Nutr Soc 2012;71(1):75–83.

[9] Brodin P, Jojic V, Gao T, et al. Variation in the human immune system is largely driven by non-heritable influences. Cell 2015;160:37–47.

[10] Doll R, Peto R. The causes of cancer: Quantitative estimates of avoidable risks of cancer in the United States today. J Natl Cancer Inst 1981;66:1191–308.

[11] Joseph JA, Shukitt-Hale B, Willis LM. Grape juice, berries, and walnuts affect brain aging and behavior. J Nutr 2009;139:1813S–7.

[12] Crous-Bou M, Fung TT, Prescott J, et al. Mediterranean diet and telomere length in Nurses' Health Study: population based cohort study. BMJ 2014;349:g6674.

[13] Ni Mhurchu C, Rodgers A, Pan WH, et al. Body mass index and cardiovascular disease in the Asia-Pacific region: an overview of 33 cohorts involving 310000 participants. Int J Epidemiol 2004;33:751–8.

[14] Renehan AG, Tyson M, Egger M, et al. Body-mass index and incidence of cancer: a systematic review and meta-analysis of prospective observational studies. Lancet 2008;371:569–78.

[15] Whitlock G, Lewington S, Sherliker P, et al. Body-mass index and cause-specific mortality in 900000 adults: collaborative analyses of 57 prospective studies. Lancet 2009;373:1083–96.

[16] Wormser D, Kaptoge S, Di Angelantonio E, et al. Separate and combined associations of body-mass index and abdominal adiposity with cardiovascular disease: collaborative analysis of 58 prospective studies. Lancet 2011;377:1085–95.

[17] Norat T, Bingham S, Ferrari P, et al. Meat, fish, and colorectal cancer risk: the European prospective investigation into cancer and nutrition. J Natl Cancer Inst 2005;97:906–16.

[18] Shin HJ, Cho E, Lee HJ, et al. Instant noodle intake and dietary patterns are associated with distinct cardiometabolic risk factors in Korea. J Nutr 2014;144(8):1247–55.

[19] Leung CW, Laraia BA, Needham BL, et al. Soda and cell aging: associations between sugar-sweetened beverage consumption and leukocyte telomere length in healthy adults from the national health and nutrition examination surveys. Am J Public Health 2014;104(12):2425–31.

[20] International Federation of the Red Cross. A world of hunger amid plenty. Opinions and Positions. 2011. Available from: http://www.ifrc.org/en/news-and-media/opinions-and-positions/opinion-pieces/2011/a-world-of-hunger-amid-plenty/?print=true [Accessed: 2015-02-26].

[21] Friedman JM. Causes and control of excess body fat. Nature 2009;459:340–2.

[22] Lin JS, O'Connor E, Whitlock EP, et al. Behavioral counseling to promote physical activity and a healthful diet to prevent cardiovascular disease in adults: a systematic

review for the U.S. Preventive Services Task Force. Ann Intern Med 2010;153(11):736–50.

[23] Ng M, Fleming T, Robinson M, et al. Global, regional, and national prevalence of overweight and obesity in children and adults during 1980–2013: a systematic analysis for the global burden of disease study 2013. Lancet 2014;384(9945):766–81.

[24] Attig L, Gabory A, Junien C. Nutritional developmental epigenomics: immediate and long-lasting effects. Proc Nutr Soc 2010;69(2):221–31.

[25] Klimek P, Leitner M, Kautzky-Willer A, et al. Effect of fetal and infant malnutrition on metabolism in older age. Gerontol 2014;60(6):502–7.

[26] Rodríguez L, Cervantes E, Ortiz R. Malnutrition and gastrointestinal and respiratory infections in children: a public health problem. Int J Environ Res Public Health 2011 April; 8(4):1174–205.

[27] Galler JR, Bryce CP, Zichlin ML, et al. Infant malnutrition is associated with persisting attention deficits in middle adulthood. J Nutr 2012;142(4):788–94.

[28] Picot J, Hartwell D, Harris P, et al. The effectiveness of interventions to treat severe acute malnutrition in young children: a systematic review. Health Technol Assess 2012;16(19):1–316.

[29] Rytter MJH, Kolte L, Briend A, et al. The immune system in children with malnutrition: a systematic review. PLoS One 2014;9(8):e105017.

[30] Macdougall LG, Moodley G, Eyberg C, et al. Mechanisms of anemia in protein-energy malnutrition in Johannesburg. Am J Clin Nutr 1982;35(2):229–35.

[31] Danaei G, Ding EL, Mozaffarian D, et al. The preventable causes of death in the United States: comparative risk assessment of dietary, lifestyle, and metabolic risk factors. PLoS Med 2009;6:e1000058.

[32] Ma S, Liu X, Jiao B, et al. Low-dose radiation-induced responses: focusing on epigenetic regulation. Int J Radiat Biol 2010;86(7):517–28.

[33] Weber F. Absorption mechanisms for fat-soluble vitamins and the effect of other food constituents. Prog Clin Biol Res 1981;77:119–35.

[34] Lichtenstein AH, Kennedy E, Barrier P, et al. Dietary fat consumption and health. Nutr Rev 1998;56(5 Pt 2):S3–19;discussion S19–28.

[35] Bouillon R, Carmeliet G, Lieben L, et al. Vitamin D and energy homeostasis: of mice and men. Nat Rev Endocrinol 2014;10(2):79–87.

[36] Obregon MJ. Adipose tissues and thyroid hormones. Front Physiol 2014;5:479.

[37] Lu SY, Qi SD, Zhao Y, et al. Type 2 diabetes mellitus non-genetic rhesus monkey model induced by high fat and high sucrose diet. Exp Clin Endocrinol Diabetes 2015;123(1):19–26.

[38] Schwab U, Lauritzen L, Tholstrup T, et al. Effect of the amount and type of dietary fat on cardiometabolic risk factors and risk of developing type 2 diabetes, cardiovascular diseases, and cancer: a systematic review. Food Nutr Res 2014;58:25145.

[39] Judd JT, Kelsay JL, Mertz W. Potential risks from low-fat diets. Semin Oncol 1983;10(3):273–80.

[40] Horvath PJ, Eagen CK, Ryer-Calvin SD, et al. The effects of varying dietary fat on the nutrient intake in male and female runners. J Am Coll Nutr 2000;19(1):42–51.

[41] Yonezawa M, Misonoh J, Hosokawa Y. Two types of X-ray-induced radioresistance in mice: presence of 4 dose ranges with distinct biological effects. Mutat Res 1996;358:237–43.

[42] Schmid M. The micronucleus test. Mutat Res 1975;31:9–15.

[43] Chaubey RC, Bhilwade HN, Joshi BN, et al. Studies on the migration of micronucleated erythrocytes from bone marrow to the peripheral blood in irradiated Swiss mice. Int J Radiat Biol 1993;63:239–45.

[44] Wang B, Tanaka K, Ninomiya Y, et al. Relieved residual damage in the hematopoietic system of mice rescued by radiation-induced adaptive response (Yonezawa effect). J Radiat Res 2013;54:45–51.

[45] Suzuki Y, Nagae Y, Li J, et al. The micronucleus test and erythropoiesis. Effects of erythropoietin and a mutagen on the ratio of polychromatic to normochromatic erythrocytes (P/N ratio). Mutagenesis 1989;4:420–4.

[46] Pospisil M. Hormonal balance and radiation resistance of the mammalian organism. A speculative review. Agressologie 1977;18(2):73–81.

[47] Neta R. Modulation of radiation damage by cytokines. Stem Cells 1997;15(Suppl 2): 87–94.

[48] Singh VK, Yadav VS. Role of cytokines and growth factors in radioprotection. Exp Mol Pathol 2005;78(2):156–69.

[49] Wang B, Sun J, Li X, et al. Resveratrol prevents suppression of regulatory T-cell production, oxidative stress, and inflammation of mice prone or resistant to high-fat diet-induced obesity. Nutr Res 2013;33(11):971–81.

[50] Wang B, Sun J, Ma Y, et al. Resveratrol preserves mitochondrial function, stimulates mitochondrial biogenesis, and attenuates oxidative stress in regulatory T cells of mice fed a high-fat diet. J Food Sci 2014;79(9):H1823–31.

[51] Martin KR, Barrett JC. Reactive oxygen species as double-edged swords in cellular processes: low-dose cell signaling versus high-dose toxicity. Hum Exp Toxicol 2002;21(2):71–5.

[52] Limoli CL, Giedzinski E, Baure J, et al. Altered growth and radiosensitivity in neural precursor cells subjected to oxidative stress. Int J Radiat Biol 2006;82(9):640–7.

[53] Kondo H, Yumoto K, Alwood JS, et al. Oxidative stress and gamma radiation-induced cancellous bone loss with musculoskeletal disuse. J Appl Physiol 2010;108(1): 152–61.

[54] Bladen CL, Kozlowski DJ, Dynan WS. Effects of low-dose ionizing radiation and menadione, an inducer of oxidative stress, alone and in combination in a vertebrate embryo model. Radiat Res 2012;178(5):499–503.

[55] Kieffer TJ, Habener JF. The adipoinsular axis: effects of leptin on pancreatic beta-cells. Am J Physiol Endocrinol Metab 2000;278(1):E1–14.

[56] Collins S, Martin TL, Surwit RS, et al. Genetic vulnerability to diet-induced obesity in the C57BL/6J mouse: physiological and molecular characteristics. Physiol Behav 2004;81(2):243–8.

[57] Supic G, Jagodic M, Magic Z. Epigenetics: a new link between nutrition and cancer. Nutr Cancer 2013;65(6):781–92.

[58] Westenhoefer J. Age and gender dependent profile of food choice. Forum Nutr 2005;(57):44–51.

[59] Stroebele N, De Castro JM. Effect of ambience on food intake and food choice. Nutr 2004;20(9):821–38.

[60] Remick AK, Polivy J, Pliner P. Internal and external moderators of the effect of variety on food intake. Psychol Bull 2009;135(3):434–51.

[61] Oliver G, Wardle J, Gibson EL. Stress and food choice: a laboratory study. Psychosom Med 2000;62(6):853–65.

[62] Habhab S, Sheldon JP, Loeb RC. The relationship between stress, dietary restraint, and food preferences in women. Appetite 2009;52(2):437–44.

[63] Thorne F, Baldwin C. Multimodal interventions including nutrition in the prevention and management of disease-related malnutrition in adults: a systematic review of randomised control trials. Clin Nutr 2014;33(3):375–84.

[64] Langendorf C, Roederer T, de Pee S, et al. Preventing acute malnutrition among young children in crises: a prospective intervention study in Niger. PLoS Med 2014;11(9):e1001714.

[65] Martins VJ, Neves AP, Franco Mdo C, et al. Impact of nutritional recovery with linear growth on the concentrations of adipokines in undernourished children living in Brazilian slums. Br J Nutr 2014;112(6):937–44.

[66] Kiss NK, Krishnasamy M, Isenring EA. The effect of nutrition intervention in lung cancer patients undergoing chemotherapy and/or radiotherapy: a systematic review. Nutr Cancer 2014;66(1):47–56.

[67] Bossola M. Nutritional interventions in head and neck cancer patients undergoing chemoradiotherapy: a narrative review. Nutrients 2015;7(1):265–76.

[68] Bauer J, Jürgens H, Frühwald MC. Important aspects of nutrition in children with cancer. Adv Nutr 2011;2(2):67–77.

[69] Vaquero A, Reinberg D. Calorie restriction and the exercise of chromatin. Genes Dev 2009;23:1849–69.

[70] Longo VD, Fontana L. Calorie restriction and cancer prevention: metabolic and molecular mechanisms. Trends Pharmacol Sci 2010;31(2):89–98.

[71] Trepanowski JF, Canale RE, Marshall KE, et al. Impact of caloric and dietary restriction regimens on markers of health and longevity in humans and animals: a summary of available findings. Nutr J 2011;10:107.

[72] Weindruch R. Effect of caloric restriction on age-associated cancers. Exp Gerontol 1992;27(5–6):575–81.

[73] Lawler DF, Larson BT, Ballam JM, et al. Diet restriction and ageing in the dog: major observations over two decades. Br J Nutr 2008;99(4):793–805.

[74] Colman RJ, Anderson RM, Johnson SC, et al. Caloric restriction delays disease onset and mortality in rhesus monkeys. Science 2009;325(5937):201–4.

[75] Tannenbaum A, Sliverstone H. The influence of the degree of caloric restriction on the formation of skin tumors and hepatomas in mice. Cancer Res 1949;9(12):724–7.

[76] Tucker MJ. The effect of long-term food restriction on tumours in rodents. Int J Cancer 1979;23(6):803–7.

[77] Kritchevsky D, Klurfeld DM. Influence of caloric intake on experimental carcinogenesis: a review. Adv Exp Med Biol 1986;206:55–68.

[78] Gross L. Inhibition of the development of tumors or leukemia in mice and rats after reduction of food intake. Possible implications for humans. Cancer 1988;62(8):1463–5.

[79] Gross L, Dreyfuss Y. Prevention of spontaneous and radiation-induced tumors in rats by reduction of food intake. Proc Natl Acad Sci USA 1990;87(17):6795–7.

[80] Masoro EJ. Aging and proliferative homeostasis: modulation by food restriction in rodents. Lab Anim Sci 1992;42(2):132–7.

[81] Yoshida K, Inoue T, Hirabayashi Y, et al. Radiation-induced myeloid leukemia in mice under calorie restriction. Leukemia 1997;11(Suppl 3):410–2.

[82] Yoshida K, Inoue T, Nojima K, et al. Calorie restriction reduces the incidence of myeloid leukemia induced by a single whole-body radiation in C3H/He mice. Proc Natl Acad Sci USA 1997;94(6):2615–9.

[83] Shang Y, Kakinuma S, Yamauchi K, et al. Cancer prevention by adult-onset calorie restriction after infant exposure to ionizing radiation in B6C3F1 male mice. Int J Cancer 2014;135(5):1038–47.

[84] Burdge GC, Lillycrop KA. Nutrition, epigenetics, and developmental plasticity: implications for understanding human disease. Annu Rev Nutr 2010;30:315–39.

[85] Ingram DK, Roth GS. Calorie restriction mimetics: can you have your cake and eat it, too? Ageing Res Rev 2015;20:46–62.

[86] de Ligt M, Timmers S, Schrauwen P. Resveratrol and obesity: can resveratrol relieve metabolic disturbances? Biochim Biophys Acta 2015;1852(6):1137–1144.

[87] Szkudelski T, Szkudelska K. Resveratrol and diabetes: from animal to human studies. Biochim Biophys Acta 2015;1852(6):1145–1154.

Chapter 4

# The RadioP1 – An Integrative Web Resource for Radioresistant Prokaryotes

Cherif Benhamda, Alia Benkahla,
Slimane Ben Miled, Houria Ouled-Haddar,
María del Carmen Montero-Calasanz, Maher Gtari,
Ameur Cherif, Benjamin Hofner, Kaïs Ghedira and
Haïtham Sghaier

Additional information is available at the end of the chapter

http://dx.doi.org/10.5772/60471

## Abstract

The extremely radioresistant eubacterium *Deinococcus radiodurans* and the phenotypically related prokaryotes, whose genomes have been completely sequenced, are presently used as model species in several laboratories to study the lethal effects of DNA-damaging and protein-oxidizing agents, particularly the effects of ionizing radiation (IR). Unfortunately, providing relevant information about radioresistant prokaryotes (RP) in a neatly centralized and organized manner still remains a need. In this study, we designed RadioP1 Web resource (www.radiop.org.tn) to gather information about RP defined by the published literature with specific emphasis on (i) predicted genes that produce and protect against oxidative stress, (ii) predicted proteins involved in DNA repair mechanisms and (iii) potential uses of RP in biotechnology. RadioP1 allows the complete RP proteogenomes to be queried using various patterns in a user-friendly and interactive manner. The output data can be saved in plain text, Excel or HyperText Markup Language (HTML) formats for subsequent analyses. Moreover, RadioP1 provides for users a tool "START ANALYSIS", including the previously described R-packages "drc" and "lethal", to generate exponential or sigmoid survival curves with $D_{10}$ and $D_{50}$ values. Furthermore, when accessible, links to external databases are provided. Supplementary data will be included in the future when the sequences of other RP genomes will become available.

**Keywords:** Database, Prokaryotes, RadioP1, Radioresistant, Web resource

## 1. Introduction

To be considered as an RP, a microorganism should have a $D_{10}$—the ionizing-radiation (IR) dose necessary to effect a 90 % reduction in colony-forming units (CFU)—threshold that is greater than 1 kilogray (kGy), corresponding to efficient physiological, genetic and proteic protection and repair mechanisms ([1, 2] and references therein). In this context, to our knowledge, even when prokaryotic members belonging to a radioresistant-species-harbouring genus have contrasted optimum temperatures—for example, ranging from 10 to 47 ºC—the least IR-resistant ones do not have $D_{10}$ values inferior to 1 kGy [3]. Furthermore, as suggested from $D_{10}$ and $F_{10}$—the ultraviolet (UV) dose necessary to effect a 90 % reduction in CFU—reported in literature [4-6], an RP is tolerant to both IR (e.g. $\alpha$ and $\beta$ particles, $\gamma$- and x-rays, neutrons) and non-IR (UV light); and correlations were suggested [7]. In this context, it is important to note that UV may cause effects similar to those stimulated by IR [8].

The first RP to be described, designated *Micrococcus radiodurans*, was isolated from irradiated meat [9] and was subsequently reclassified as *D. radiodurans* [10]. IR resistance has been observed in a broad range of prokaryotic groups [11], including halophilic *Archaea* (*Halobacterium* sp.) [12], hyperthermophilic *Archaea* (*Pyrococcus abyssi, Pyrococcus furiosus, Thermococcus gammatolerans, Thermococcus marinus* and *Thermococcus radiotolerans*) [13-16], *Actinobacteria* (*Geodermatophilus dictyosporus, Geodermatophilus obscurus, Geodermatophilus poikilotrophi, Kineococcus radiotolerans, Rubrobacter radiotolerans* and *Rubrobacter xylanophilus*) [17-23], *Cyanobacteria* (*Arthrospira* sp. and *Chroococcidiopsis* sp.) [24, 25], the *Deinococcus-Thermus* group (many *Deinococcus* spp. and *Truepera radiovictrix*) ([3, 26-28] and references therein), *Proteobacteria* (*Acinetobacter radioresistens* and *Methylobacterium radiotolerans*) [29, 30] and *Sphingobacteria* (*Hymenobacter actinosclerus*) [31]. Recently, members of the genera *Knoellia, Lysobacter, Microvirga, Nocardioides, Paracoccus, Pontibacter* and *Rufibacter* were suggested as RP [32]. However, the list of RP with available experimental $D_{10}/F_{10}$ values (see examples in Table 1) and completely sequenced genomes is still limited.

As a complicated multifaceted phenotype, prokaryotic radioresistance is an important subject in radiation microbiology. A focus on just one contributing mechanism is unlikely to yield a complete understanding of the phenomenon [53]. The radioresistance of prokaryotes depends on their ability to protect enzymes including those needed to repair and replicate DNA from inactivation by oxidative protein damage (protein-centric view) and to fully amend their DNA —double-strand breaks (DSBs)—(DNA-centric view). Obviously, much is yet to be discovered from the mesmerizing radioresistance strategies posed by RP. There is an increasing need to compile the entire data about RP in a centralized and organized manner and to mine it regarding prokaryotic radioresistance. RadioP1 is addressing these requests by providing pertinent information as well as diverse analytical tools. This first version of RadioP is a preliminary step towards the establishment of a comprehensive RP database. The increase of the number of side aspects of radioresistance make us keen to collect and to make available for the scientific community the most up-to-date and relevant information.

| Organism | $D_{10}$ value ($^{60}$Co, Gy) | $F_{10}$ value (UV-C, J/m$^2$) | Reference |
|---|---|---|---|
| **Archaea** | | | |
| *Archaeoglobus fulgidus* | 1,087 | 108 | [4] |
| *Ignicoccus hospitalis* with *Nanoarchaeum equitans* | 1,473 | n.d. | [4] |
| *I. hospitalis* | 1,482 | n.d. | [4] |
| *Methanocaldococcus jannaschii* | 1,036 | n.d. | [4] |
| *Pyrococcus furiosus* | 1,018 | 64 | [4] |
| *Halobacterium salinarum* | 5,000 | 280 | [12, 33] |
| *Thermococcus stetteri* | 6,000 | 30 | [11] |
| **Bacteria** | | | |
| *Aquifex pyrophilus* | 2,842 | 63 | [4] |
| *Deinococcus aerius* | 4,900 | n.d. | [34] |
| *Deinococcus aetherius* | >8,000 | n.d. | [35] |
| *Deinococcus alpinitundrae* | 4,000 | 690 | [3] |
| *Deinococcus altitudinis* | 3,800 | 550 | [3] |
| *Deinococcus claudionis* | 3,600 | 310 | [3] |
| *Deinococcus deserti* | >7,500 | n.d. | [36] |
| *Deinococcus ficus* | 11,000 | n.d. | [37] |
| *Deinococcus geothermalis* | 5,100–16,000 | 1800 | [38-40] |
| *Deinococcus gobiensis* | 12,700 | >800 | [41] |
| *Deinococcus grandis* | 7,000–11,000 | n.d. | [37, 42, 43] |
| *Deinococcus guangriensis* | 9,800 | n.d. | [44] |
| *Deinococcus indicus* | 4,200 | n.d. | [37] |
| *Deinococcus mumbaiensis* | 17,000 | n.d. | [45] |
| *Deinococcus murrayi* | 9,100 | n.d. | [40] |
| *Deinococcus piscis* | 7,400 | n.d. | [46] |
| *Deinococcus proteolyticus* | 10,300 | n.d. | [37] |
| *Deinococcus radiodurans* | 6,700–16,000 | 660–2000 | [3, 37-39, 47-50] |
| *Deinococcus radiomollis* | 2,200 | 220 | [3] |
| *Deinococcus radiophilus* | >16,000 | n.d. | [51] |
| *Deinococcus radiopugnans* | 5,300 | n.d. | [52] |
| *Enterococcus faecium* | 2,000 | n.d. | [38] |
| *Geodermatophilus poikilotrophi* | 9,000 | 6300–12600 | [22] |

n.d.: not determined

For further details: www.radiop.org.tn

**Table 1.** $F_{10}$ and $D_{10}$ values for 90 % reduction of selected radioresistant prokaryotes (RP)

## 2. Source of data

Used information was obtained by searching the NCBI database [54]. Clusters of Orthologous Group (COG) [55] were used to classify orthologous gene records in RadioP1. Orthology was calculated with Basic Local Alignment Search Tool (BLAST), the best reciprocal hit approach and InParanoid program.

## 3. Database construction

We built the database on a recent version of Linux operation system. MySQL workbench 5.2.38 was used to handle the database schema and to build the relational database. Perl scripts were developed to retrieve genome data and gene information from GenBank files collected from NCBI using file transfer protocol (FTP) and to store RP information in the appropriate database tables. A frontend user interface was developed using HTML5, JavaScript, Cascading Style Sheet (CSS) and Hypertext Preprocessor (PHP) program languages. Perl CGI (Common Gateway Interface) modules and PHP scripts were developed and used to link the Web interface to the database. These scripts allow all users to send requests via the Web interface to the server, run the jobs on the server and then return and display results on the Web interface (Figure 1).

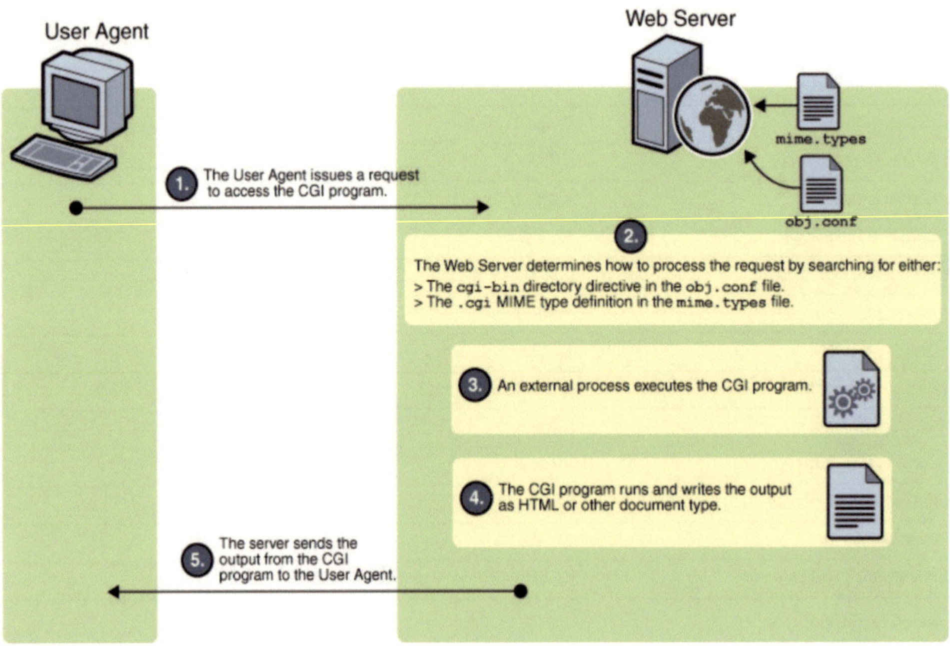


Data source: http://docs.oracle.com/cd/E19146-01/821-1828/6nmpm01g9/index.html

**Figure 1.** Overview of the interactions between the Web interface and the database using CGI modules

http://dx.doi.org/10.5772/60471

The database schema (Figure 2) consists of 13 tables, allowing to search and to retrieve any stored biological data. Among the main tables is the species table (primary information: organism name and taxon ID), which is connected to the taxonomy and chromosome tables. from NCBI, and is linked to the taxonomy and chromosome tables. This later is connected to seqfile tables detailing the different file formats and paths related to each chromosome. The gene table, related to the chromosome table, stores information such as gene name, gene ID, symbol, first position, last position and strand. The gene table is linked to the orthology table.

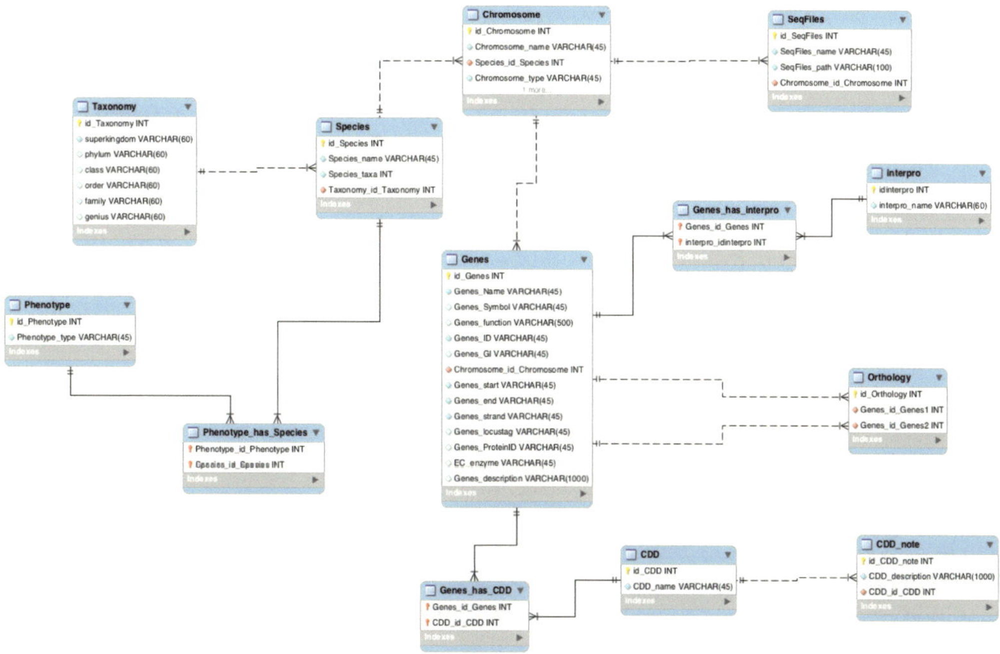


**Figure 2.** RadioP1 database structure

# 4. RadioP1 database user guide

RadioP1 is freely accessible through a Web browser at http://www.radiop.org.tn. There are at least three ways to use the database: browse, search and generate data.

# 5. Browse in the database

In the main page of RadioP1, a clickable list of currently available groups of IRRP—ionizing-radiation-resistant prokaryotes—is organized at the top-left side, allowing users to browse pages for each of the groups, IR-resistant archaea (IRRA) and IR-resistant bacteria (IRRB).

## 6. Search the database

RadioP1 provides a search engine that is able to extract information from the database through: (i) text search, (ii) BLAST search and (iii) function category search..

The text and homology search contains three categories:

1. "SEARCH GENES": This search category allows extracting annotation information—gene symbol, chromosome name, strand, predicted orthologous genes, etc.—using querying gene locus tags.. The querying results are displayed in a table with each hit represented by a row containing a corresponding gene ID and a summary of characteristics—gene name, symbol, strand and product. In addition, each listed row in the output table provides a link to the individual gene pages, which highlight the querying genes found in the page of NCBI [54]. Users can get results in HTML, plain text or Excel formats for further analyses.

2. "RETRIEVE SEQUENCES": This search category enables extracting nucleic or proteic sequences using querying gene locus tags.

3. "HOMOLOGY DATA": This search category enables extracting predicted orthologous gene clusters using querying gene locus tags.

The function category search contains four subclasses:

1. "OXIDATIVE STRESS PRODUCTION":

When the generation of reactive oxygen species (ROS; superoxide ($O_2^{\cdot-}$), hydrogen peroxide ($H_2O_2$) and hydroxyl ($HO^{\cdot}$) radicals) produced by metabolism or irradiation exceeds the capacity of endogenous scavengers to neutralize them, cells become vulnerable to damage, a condition referred to as oxidative stress [56, 57]. Typically, during irradiation, ~80 % of DNA damage is caused indirectly by irradiation-induced ROS and the remaining ~20 % by direct interaction between c-photons and DNA [57]. $HO^{\cdot}$ radicals are the primary product of the radiolysis of water and in the presence of oxygen, can also generate some $O_2^{\cdot-}$ and $H_2O_2$ by dismutation of $O_2^{\cdot-}$ [57]. In contrast, the primary ROS generated by metabolism are $O_2^{\cdot-}$ and $H_2O_2$ [56]. The total intracellular titer of cytochromes and flavins might serve as a marker for the proclivity of cells to survive radiation and other oxidizing conditions [58, 59]. For instance, the total number of c-type cytochromes in *D. radiodurans* and *Shewanella oneidensis* ($D_{10} = 70$ Gy [38]) is 7 and 39, respectively [58]. Searching RadioP1 by this function subcategory provides a way to find out predicted genes involved in ROS production and subsequently estimate cellular radioresistance level.

2. "OXIDATIVE STRESS PROTECTION":

Unlike DNA DSB lesion yields ([6] and references therein), in IR-sensitive cells, yields of IR-induced protein oxidation can be ~100 times greater than in IR-resistant cells [60, 61]. Indeed, presently, it is demonstrated that proteins are major targets of IR damage and that shield against protein oxidation is an important mechanism for survival from IR exposure. IR resistance in some prokaryotes was highly correlated to the accumulation of high intracellular concentration of $Mn^{2+}$, supporting the idea of a common model of $Mn^{2+}$-dependent ROS scavenging in the aerobes ([6, 62] and references therein). For example, the aerobic archaeon *H. salinarum* accumulates high intracellular concentration of $Mn^{2+}$, 155 ng/$10^9$ cells [62, 63]. In contrast, hyperthermophilic anaerobic archaea *T. gammatolerans* and *P. furiosus* do not contain significant amounts of intracellular $Mn^{2+}$, 3 ng/$10^9$ cells and 14 ng/$10^9$ cells, respectively [62]. These low concentrations of $Mn^{2+}$ in anaerobic hyperthermophiles were explained by the low levels of IR-generated ROS under anaerobic conditions combined with efficient detoxification systems [62]. In RadioP1, using the "OXIDATIVE STRESS PROTECTION" function subcategory, a summary table is provided to users giving insights about radioprotectors of each RP.

**3.** "DNA REPAIR GENES":

During irradiation, DNA double-strand breaks (DSBs) are considered as the most lethal damage, although they are the least frequent form of cellular DNA damage—compared to single-strand breaks and DNA base damages [60]. For example, in *D. radiodurans*, PprA protein has an important role in DNA DSBs repair [64]. Exploring RadioP1 by the "DNA REPAIR GENES" function subcategory allows users to generate a list of genes—orthologs of genes in Table 2—for which a functional knockout may change the level of radioresistance of mutant cells.

| Gene name (locus tag, $D_{10}$ of a D. radiodurans strain lacking the gene) | Reference |
|---|---|
| *recA* (DR_2340, 0.1 kGy) | [65] |
| *polA* (DR_1707, 1 kGy) | [66] |
| *recQ* (DR_1289, 6 kGy) | [67] |
| *recD* (DR_1902, 6 kGy) | [68] |
| *crtB* (DR_0862, 9 kGy) | [69] |
| *crtI* (DR_0861, 9 kGy) | [69] |
| *sbcC* (DR_1922, 15 kGy) | [70] |
| *sbcD* (DR_1921, 15 kGy) | [70] |
| *polX* (DR_0467, 15 kGy) | [71] |
| *pprA* (DR_A0346, 2 kGy) | [64] |
| *ddrB* (DR_0070, 8 kGy) | [65] |
| *ddrA* (DR_0423, 12 kGy) | [65] |
| *ddrC* (DR_0003, > 14 kGy) | [65] |
| *ddrD* (DR_0326, > 14 kGy) | [65] |

**Table 2.** Selected important DNA protection and repair genes for ionizing-radiation-resistant prokaryotes (IRRP) based on data of *Deinococcus radiodurans*

**4.** "USE IN BIOTECHNOLOGY":

RP provide inestimable opportunities in therapeutics for multiple diseases [72], biotechnology [73], pharmaceuticals [74] and bioengineering—bioremediation—of toxic and radioactive compounds [75-81]. The function "USE IN BIOTECHNOLOGY" in RadioP1 was designed to present the diversity of IRRP genomes in terms of genes with potential applications in biotechnology.

## 7. Generate survival plots and $D_x$ ($D_{10}$ and $D_{50}$) values

Cell survival models aim to describe the relationship between the absorbed dose and the fraction of surviving cells—cell survival curve. Distinct cell survival models were described [82-86]: the linear—single-hit single-target, the linear-quadratic (LQ) and the repairable-conditionally repairable damage (RCR) models. Other models include those based on target theory first described by Lea [87] and those described by Tobias [88], Curtis [89] and Sontag [90].

For instance, for UV-C-irradiated prokaryotes, as summarized previously [91], the mathematical dose-response models which describe the probability of a specific biological response at a given dose can be represented as follows (Figure 3):

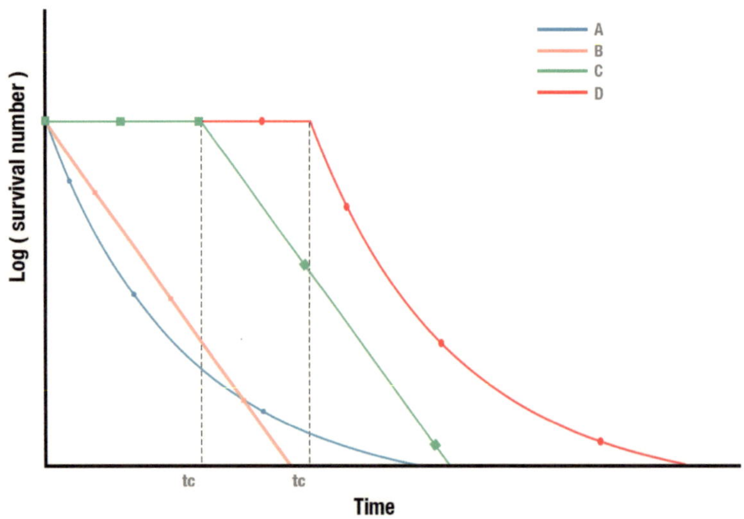


Shoulder curve (A), linear (B), two-stage curve (C) and sigmoid (D)

$N=N_0e^{-\alpha(t-t_c)}$ for $t \geq t_c$ (A)

$N=N_0e^{-\alpha t}$ (B)

$N=N_0[(1-F_0)e^{-\alpha_1 t}+F_0e^{-\alpha_2 t}]$(C)

$N=N_0[(1-F_0)e^{-\alpha_1 (t-t)_c}+F_0e^{\alpha_2 t-t_c}]$ for $t \geq t_c$(D)

Where:

* N and $N_0$ represent the microorganisms surviving at time t and those initially present at time t = 0 respectively.
* $\alpha$ is a parameter proportional to the applied UV-C intensity and depends on the sensitivity of the microorganism to the UV-C ray exposure.
* $t_c$ is the time during which microorganisms are substantially not inactivated.
* $F_0$ represents the most resistant fraction, characterized by a lower sensitivity to the UV-C rays exposure, in a population of microorganisms, compared to the fraction $(1 - F_0)$ less resistant to such exposure.

**Figure 3.** Summary of prokaryotic survival curves

http://dx.doi.org/10.5772/60471

RadioP1 provides a tool "START ANALYSIS" for users to generate exponential survival curves [92, 93]. In addition, it integrated the previously described R-packages "drc" [94] for sigmoid curves and "lethal" [95] that computes lethal doses (LD) with confidence intervals [22]. All curves are supplied with $D_{10}$ and $D_{50}$ values.

## 8. Future directions

RadioP1 is a specialized database aimed at making a comprehensive repository of identified RP with experimentally determined $D_{10}$. It is complemented by data extraction and analysis tools to help further analysis of RP. Researchers are kindly requested and encouraged to invigorate RadioP1 by depositing their new results—$D_{10}$—of RP at RadioP1. Submission might either be performed through the "Submit new RP with a $D_{10}$" form accessible under the IRRP main page or by e-mail to corresponding authors. In the future, we intend to include more detailed information about RP in the area of evolutionary biology, biotechnology and theranostics. Additional data sources like Kyoto Encyclopedia of Genes and Genomes (KEGG) and COGs will be integrated to extract further information about gene functions, clusters and pathways, helping users to categorise genes of interest into functional units and perform more efficient analysis on RP genomes.

## Funding

This work was supported by the Tunisian National Center for Nuclear Sciences and Technology (CNSTN), the Pasteur Institute (Tunis) and a bilateral cooperation project coordinated by Dr Haïtham Sghaier from Tunisia and Dr Houria Ouled-Haddar from Algeria.

## Acknowledgements

Authors would like to thank colleagues who gave their feedbacks after testing the database. Particularly, authors thank Prof. Issay NARUMI from the Radiation Microbiology Laboratory, in Toyo University, Japan, for many helpful comments concerning the content of RadioP1; and InTech team for excellent editorial assistance.

## Author details

Cherif Benhamda[1,2], Alia Benkahla[1,2], Slimane Ben Miled[1,2], Houria Ouled-Haddar[3], María del Carmen Montero-Calasanz[4], Maher Gtari[5], Ameur Cherif[6], Benjamin Hofner[7], Kaïs Ghedira[1,2] and Haïtham Sghaier[6,8*]

*Address all correspondence to: kais.ghedira@pasteur.rns.tn; haitham.sghaier@cnstn.rnrt.tn

1 Institut Pasteur de Tunis, Laboratoire de Parasitologie Médicale Biotechnologie et Biomolécules LR11IPT06. Group of Bioinformatics and Modelling, Tunis, Tunisia

2 Université de Tunis El Manar, Tunis, Tunisia

3 Laboratory of Molecular Toxicology, University of Jijel, Jijel, Algeria

4 Leibniz Institute DSMZ-German Collection of Microorganisms and Cell Cultures, Braunschweig, Germany

5 Laboratoire Microorganismes et Biomolécules Actives, Faculté des Sciences de Tunis, Université Tunis El Manar, Campus Universitaire, Tunis, Tunisia

6 Laboratory BVBGR, ISBST, University of Manouba, La Manouba, Tunisia

7 Institut für Medizininformatik, Biometrie und Epidemiologie, Friedrich-Alexander-Universität Erlangen-Nürnberg, Erlangen, Germany

8 National Center for Nuclear Sciences and Technology (CNSTN), Sidi Thabet Technopark, Ariana, Tunisia

## References

[1] Sghaier H: DNA Repair: Lessons from the Evolution of Ionizing- Radiation-Resistant Prokaryotes – Fact and Theory. In: *Selected Topics in DNA Repair.* Edited by Clark CC, University of California. San Diego: InTech; 2011: 145–156.

[2] Sghaier H, Ghedira K, Benkahla A, Barkallah I: Basal DNA repair machinery is subject to positive selection in ionizing-radiation-resistant bacteria. *BMC Genomics* 2008, 9:297.

[3] Callegan RP, Nobre MF, McTernan PM, Battista JR, Navarro-Gonzalez R, McKay CP, da Costa MS, Rainey FA: Description of four novel psychrophilic, ionizing radiation-sensitive *Deinococcus* species from alpine environments. *Int J Syst Evol Microbiol* 2008, 58(Pt 5):1252–1258.

[4] Beblo K, Douki T, Schmalz G, Rachel R, Wirth R, Huber H, Reitz G, Rettberg P: Survival of thermophilic and hyperthermophilic microorganisms after exposure to UV-C, ionizing radiation and desiccation. *Arch Microbiol* 2011, 193(11):797–809.

[5] Daly MJ: Modulating radiation resistance: insights based on defenses against reactive oxygen species in the radioresistant bacterium *Deinococcus radiodurans. Clin Lab Med* 2006, 26(2):491–504.

[6] Daly MJ: Death by protein damage in irradiated cells. *DNA Repair (Amst)* 2011, 11(1): 12–21.

[7] Arrange AA, Phelps TJ, Benoit RE, Palumbo AV, White DC: Bacterial sensitivity to UV light as a model for ionizing radiation resistance. *J Microbiol Methods* 1993, 18(2): 127–136.

[8] Nikogosyan DN: Two-quantum UV photochemistry of nucleic acids: comparison with conventional low-intensity UV photochemistry and radiation chemistry. *Int J Radiat Biol* 1990, 57(2):233–299.

[9] Anderson AW, Nordan HC, Cain RF, Parrish G, Duggan D: Studies on a radio-resistant *Micrococcus*. I. Isolation, morphology, cultural characteristics, and resistance to gamma radiation. *Food Technol* 1956, 10:575–577.

[10] Brooks BW, Murray RGE: Nomenclature for "*Micrococcus radiodurans*" and other radiation-resistant cocci: *Deinococcaceae* fam. nov. and *Deinococcus* gen. nov., including five species. *Int J Syst Bacteriol* 1981, 31:353–360.

[11] Kopylov VM, Bonch-Osmolovskaya EA, Svetlichnyi VA, Miroshnicheko ML, Skobin VS: Gamma-irradiation resistance and UV sensitivity of extremely thermophilic archaebacteria and eubacteria. *Mikrobiologiya* 1993(62):90–95.

[12] Kottemann M, Kish A, Iloanusi C, Bjork S, DiRuggiero J: Physiological responses of the halophilic archaeon *Halobacterium* sp. strain NRC1 to desiccation and gamma irradiation. *Extremophiles* 2005, 9(3):219–227.

[13] DiRuggiero J, Santangelo N, Nackerdien Z, Ravel J, Robb FT: Repair of extensive ionizing-radiation DNA damage at 95 degrees C in the hyperthermophilic archaeon *Pyrococcus furiosus*. *J Bacteriol* 1997, 179(14):4643–4645.

[14] Jolivet E, Matsunaga F, Ishino Y, Forterre P, Prieur D, Myllykallio H: Physiological responses of the hyperthermophilic archaeon "*Pyrococcus abyssi*" to DNA damage caused by ionizing radiation. *J Bacteriol* 2003, 185(13):3958–3961.

[15] Jolivet E, Corre E, L'Haridon S, Forterre P, Prieur D: *Thermococcus marinus* sp. nov. and *Thermococcus radiotolerans* sp. nov., two hyperthermophilic archaea from deep-sea hydrothermal vents that resist ionizing radiation. *Extremophiles* 2004, 8(3):219–227.

[16] Jolivet E, L'Haridon S, Corre E, Forterre P, Prieur D: *Thermococcus gammatolerans* sp. nov., a hyperthermophilic archaeon from a deep-sea hydrothermal vent that resists ionizing radiation. *Int J Syst Evol Microbiol* 2003, 53(Pt 3):847–851.

[17] Yoshinaka T, Yano K, Yamaguchi H: Isolation of highly radioresistant bacterium, *Arthrobacter radiotolerans* nov. sp. *Agr Biol Chem* 1973, 37(10):2269–2275.

[18] Ferreira AC, Nobre MF, Moore E, Rainey FA, Battista JR, da Costa MS: Characterization and radiation resistance of new isolates of *Rubrobacter radiotolerans* and *Rubrobacter xylanophilus*. *Extremophiles* 1999, 3(4):235–238.

[19] Chen MY, Wu SH, Lin GH, Lu CP, Lin YT, Chang WC, Tsay SS: *Rubrobacter taiwanensis* sp. nov., a novel thermophilic, radiation-resistant species isolated from hot springs. *Int J Syst Evol Microbiol* 2004, 54(Pt 5):1849–1855.

[20] Phillips RW, Wiegel J, Berry CJ, Fliermans C, Peacock AD, White DC, Shimkets LJ: *Kineococcus radiotolerans* sp. nov., a radiation-resistant, gram-positive bacterium. *Int J Syst Evol Microbiol* 2002, 52(Pt 3):933–938.

[21] Gtari M, Essoussi I, Maaoui R, Sghaier H, Boujmil R, Gury J, Pujic P, Brusetti L, Chouaia B, Crotti E *et al*: Contrasted resistance of stone-dwelling *Geodermatophilaceae* species to stresses known to give rise to reactive oxygen species. *FEMS Microbiol Ecol* 2012, 80(3):566–577.

[22] Montero-Calasanz MC, Hofner B, Göker M, Rohde M, Spröer C, Hezbri K, Gtari M, Schumann P, Klenk HP: *Geodermatophilus poikilotrophi* sp. nov.: a multitolerant actinomycete isolated from dolomitic marble. *Biomed Res Int* 2014, 2014:11.

[23] Montero-Calasanz MC, Hezbri K, Göker M, Sghaier H, Rohde M, Spröer C, Schumann P, Klenk HP: Description of gamma radiation-resistant *Geodermatophilus dictyosporus* sp. nov. to accommodate the not validly named *Geodermatophilus obscurus* subsp. *dictyosporus* (Luedemann, 1968). *Extremophiles* 2015, 19(1):77–85.

[24] Billi D, Friedmann EI, Hofer KG, Caiola MG, Ocampo-Friedmann R: Ionizing-radiation resistance in the desiccation-tolerant cyanobacterium *Chroococcidiopsis*. *Appl Environ Microbiol* 2000, 66(4):1489–1492.

[25] Badri H, Monsieurs P, Coninx I, Wattiez R, Leys N: Molecular investigation of the radiation resistance of edible cyanobacterium *Arthrospira* sp. PCC 8005. *Microbiologyopen* 2015:doi: 10.1002/mbo1003.1229.

[26] Albuquerque L, Simoes C, Nobre MF, Pino NM, Battista JR, Silva MT, Rainey FA, da Costa MS: *Truepera radiovictrix* gen. nov., sp. nov., a new radiation resistant species and the proposal of *Trueperaceae* fam. nov. *FEMS Microbiol Lett* 2005, 247(2):161–169.

[27] Dong N, Li HR, Yuan M, Zhang XH, Yu Y: *Deinococcus antarcticus* sp. nov., isolated from soil. *Int J Syst Evol Microbiol* 2015, 65(Pt 2):331–335.

[28] Srinivasan S, Lee JJ, Lim SY, Joe MH, Im SH, Kim MK: *Deinococcus radioresistens* sp. nov., a UV and gamma radiation-resistant bacterium isolated from mountain soil. *Antonie Van Leeuwenhoek* 2015, 107(2):539–545.

[29] Ito H, Iizuka H: Taxonomic studies on a radio-resistant *Pseudomonas*. Part XII. Studies on the microorganisms of cereal grain. *Agric Biol Chem* 1971, 35(10):1566–1571.

[30] Nishimura Y, Uchida K, Tanaka K, Ino T, Ito H: Radiation sensitivities of *Acinetobacter* strains isolated from clinical sources. *J Basic Microbiol* 1994, 34(5):357–360.

[31] Collins MD, Hutson RA, Grant IR, Patterson MF: Phylogenetic characterization of a novel radiation-resistant bacterium from irradiated pork: description of *Hymenobacter actinosclerus* sp. nov. *Int J Syst Evol Microbiol* 2000, 50(Pt 2):731–734.

[32] Yu LZ, Luo XS, Liu M, Huang Q: Diversity of ionizing radiation-resistant bacteria obtained from the Taklimakan Desert. *J Basic Microbiol* 2015, 55(1):135–140.

[33] Shahmohammadi HR, Asgarani E, Terato H, Saito T, Ohyama Y, Gekko K, Yamamoto O, Ide H: Protective roles of bacterioruberin and intracellular KCl in the resistance of *Halobacterium salinarium* against DNA-damaging agents. *J Radiat Res* 1998, 39(4): 251–262.

[34] Yang Y, Itoh T, Yokobori S, Itahashi S, Shimada H, Satoh K, Ohba H, Narumi I, Yamagishi A: *Deinococcus aerius* sp. nov., isolated from the high atmosphere. *Int J Syst Evol Microbiol* 2009, 59(Pt 8):1862–1866.

[35] Yang Y, Itoh T, Yokobori S, Shimada H, Itahashi S, Satoh K, Ohba H, Narumi I, Yamagishi A: *Deinococcus aetherius* sp. nov., isolated from the stratosphere. *Int J Syst Evol Microbiol* 2010, 60(Pt 4):776–779.

[36] de Groot A, Chapon V, Servant P, Christen R, Saux MF, Sommer S, Heulin T: *Deinococcus deserti* sp. nov., a gamma-radiation-tolerant bacterium isolated from the Sahara Desert. *Int J Syst Evol Microbiol* 2005, 55(Pt 6):2441–2446.

[37] Shashidhar R, Kumar SA, Misra HS, Bandekar JR: Evaluation of the role of enzymatic and nonenzymatic antioxidant systems in the radiation resistance of *Deinococcus*. *Can J Microbiol* 2010, 56(3):195–201.

[38] Daly MJ, Gaidamakova EK, Matrosova VY, Vasilenko A, Zhai M, Venkateswaran A, Hess M, Omelchenko MV, Kostandarithes HM, Makarova KS *et al*: Accumulation of Mn(II) in *Deinococcus radiodurans* facilitates gamma-radiation resistance. *Science* 2004, 306(5698):1025–1028.

[39] Makarova KS, Omelchenko MV, Gaidamakova EK, Matrosova VY, Vasilenko A, Zhai M, Lapidus A, Copeland A, Kim E, Land M *et al*: *Deinococcus geothermalis*: the pool of extreme radiation resistance genes shrinks. *PLoS One* 2007, 2(9):e955.

[40] Ferreira AC, Nobre MF, Rainey FA, Silva MT, Wait R, Burghardt J, Chung AP, da Costa MS: *Deinococcus geothermalis* sp. nov. and *Deinococcus murrayi* sp. nov., two extremely radiation-resistant and slightly thermophilic species from hot springs. *Int J Syst Bacteriol* 1997, 47(4):939–947.

[41] Yuan M, Zhang W, Dai S, Wu J, Wang Y, Tao T, Chen M, Lin M: *Deinococcus gobiensis* sp. nov., an extremely radiation-resistant bacterium. *Int J Syst Evol Microbiol* 2009, 59(Pt 6):1513–1517.

[42] Chanal A, Chapon V, Benzerara K, Barakat M, Christen R, Achouak W, Barras F, Heulin T: The desert of Tataouine: an extreme environment that hosts a wide diversity of microorganisms and radiotolerant bacteria. *Environ Microbiol* 2006, 8(3):514–525.

[43] Oyaizu H, Stackebrandt E, Schleifer KH, Ludwig W, Pohla H, Ito H, Hirata A, Oyaizu Y, Komagata K: A radiation-resistant rod-shaped bacterium, *Deinobacter grandis* gen. nov., sp. nov., with peptidoglycan containing ornithine. *Int J Syst Bacteriol* 1987, 37:62–67.

[44] Sun J, Shen P, Chao H, Wu B: Isolation and identification of a new radiation-resistant bacterium *Deinococcus guangriensis* sp.nov. and analysis of its radioresistant character. *Wei Sheng Wu Xue Bao* 2009, 49(7):918–924.

[45] Shashidhar R, Bandekar JR: *Deinococcus mumbaiensis* sp. nov., a radiation-resistant pleomorphic bacterium isolated from Mumbai, India. *FEMS Microbiol Lett* 2006, 254(2):275–280.

[46] Shashidhar R, Bandekar JR: *Deinococcus piscis* sp. nov., a radiation-resistant bacterium isolated from a marine fish. *Int J Syst Evol Microbiol* 2009, 59(Pt 11):2714–2717.

[47] Battista JR, Earl AM, Park MJ: Why is *Deinococcus radiodurans* so resistant to ionizing radiation? *Trends Microbiol* 1999, 7(9):362–365.

[48] Bauermeister A, Bentchikou E, Moeller R, Rettberg P: Roles of PprA, IrrE, and RecA in the resistance of *Deinococcus radiodurans* to germicidal and environmentally relevant UV radiation. *Arch Microbiol* 2009, 191(12):913–918.

[49] Shukla M, Chaturvedi R, Tamhane D, Vyas P, Archana G, Apte S, Bandekar J, Desai A: Multiple-stress tolerance of ionizing radiation-resistant bacterial isolates obtained from various habitats: correlation between stresses. *Curr Microbiol* 2007, 54(2):142–148.

[50] Battista JR: Against all odds: the survival strategies of *Deinococcus radiodurans*. *Annu Rev Microbiol* 1997, 51:203–224.

[51] Lewis NF: Studies on a radio-resistant coccus isolated from Bombay duck (Harpodon nehereus). *J Gen Microbiol* 1971, 66(1):29–35.

[52] Davis NS, Silverman GJ, Keller WH: Combined effects of ultrahigh vacuum and temperature on the viability of some spores and soil organisms. *Appl Microbiol* 1963, 11:202–210.

[53] Harris DR, Pollock SV, Wood EA, Goiffon RJ, Klingele AJ, Cabot EL, Schackwitz W, Martin J, Eggington J, Durfee TJ *et al*: Directed evolution of ionizing radiation resistance in *Escherichia coli*. *J Bacteriol* 2009, 191(16):5240–5252.

[54] Wheeler DL, Chappey C, Lash AE, Leipe DD, Madden TL, Schuler GD, Tatusova TA, Rapp BA: Database resources of the National Center for Biotechnology Information. *Nucleic Acids Res* 2000, 28(1):10–14.

[55] Tatusov RL, Galperin MY, Natale DA, Koonin EV: The COG database: a tool for genome-scale analysis of protein functions and evolution. *Nucleic Acids Res* 2000, 28(1): 33–36.

[56] Imlay JA: Pathways of oxidative damage. *Annu Rev Microbiol* 2003, 57:395–418.

[57] Halliwell B, Gutteridge JMC: Free Radicals in Biology and Medicine, 4 edn. Oxford: Oxford University Press; 2007.

[58] Ghosal D, Omelchenko MV, Gaidamakova EK, Matrosova VY, Vasilenko A, Venkateswaran A, Zhai M, Kostandarithes HM, Brim H, Makarova KS *et al*: How radiation kills cells: survival of *Deinococcus radiodurans* and *Shewanella oneidensis* under oxidative stress. *FEMS Microbiol Rev* 2005, 29(2):361–375.

[59] Messner KR, Imlay JA: Mechanism of superoxide and hydrogen peroxide formation by fumarate reductase, succinate dehydrogenase, and aspartate oxidase. *J Biol Chem* 2002, 277(45):42563–42571.

[60] Daly MJ: A new perspective on radiation resistance based on *Deinococcus radiodurans*. *Nat Rev Microbiol* 2009, 7(3):237–245.

[61] Daly MJ, Gaidamakova EK, Matrosova VY, Kiang JG, Fukumoto R, Lee DY, Wehr NB, Viteri GA, Berlett BS, Levine RL: Small-molecule antioxidant proteome-shields in *Deinococcus radiodurans*. *PLoS One* 2010, 5(9):e12570.

[62] Webb KM, DiRuggiero J: Role of $Mn^{2+}$ and compatible solutes in the radiation resistance of thermophilic Bacteria and Archaea. *Archaea* 2012, 2012:11.

[63] Robinson CK, Webb K, Kaur A, Jaruga P, Dizdaroglu M, Baliga NS, Place A, Diruggiero J: A major role for nonenzymatic antioxidant processes in the radioresistance of *Halobacterium salinarum*. *J Bacteriol* 2011, 193(7):1653–1662.

[64] Narumi I, Satoh K, Cui S, Funayama T, Kitayama S, Watanabe H: PprA: a novel protein from *Deinococcus radiodurans* that stimulates DNA ligation. *Mol Microbiol* 2004, 54(1):278–285.

[65] Tanaka M, Earl AM, Howell HA, Park MJ, Eisen JA, Peterson SN, Battista JR: Analysis of *Deinococcus radiodurans*'s transcriptional response to ionizing radiation and desiccation reveals novel proteins that contribute to extreme radioresistance. *Genetics* 2004, 168(1):21–33.

[66] Gutman PD, Fuchs P, Minton KW: Restoration of the DNA damage resistance of *Deinococcus radiodurans* DNA polymerase mutants by *Escherichia coli* DNA polymerase I and Klenow fragment. *Mutat Res* 1994, 314(1):87–97.

[67] Huang L, Hua X, Lu H, Gao G, Tian B, Shen B, Hua Y: Three tandem HRDC domains have synergistic effect on the RecQ functions in *Deinococcus radiodurans*. *DNA Repair (Amst)* 2007, 6(2):167–176.

[68] Servinsky MD, Julin DA: Effect of a *recD* mutation on DNA damage resistance and transformation in *Deinococcus radiodurans. J Bacteriol* 2007, 189(14):5101–5107.

[69] Zhang L, Yang Q, Luo X, Fang C, Zhang Q, Tang Y: Knockout of *crtB* or *crtI* gene blocks the carotenoid biosynthetic pathway in *Deinococcus radiodurans* $R_1$ and influences its resistance to oxidative DNA-damaging agents due to change of free radicals scavenging ability. *Arch Microbiol* 2007, 188(4):411–419.

[70] Bentchikou E, Servant P, Coste G, Sommer S: Additive effects of SbcCD and PolX deficiencies in the in vivo repair of DNA double-strand breaks in *Deinococcus radiodurans. J Bacteriol* 2007, 189(13):4784–4790.

[71] Lecointe F, Shevelev IV, Bailone A, Sommer S, Hubscher U: Involvement of an X family DNA polymerase in double-stranded break repair in the radioresistant organism *Deinococcus radiodurans. Mol Microbiol* 2004, 53(6):1721–1730.

[72] Singh OV, Gabani P: Extremophiles: radiation resistance microbial reserves and therapeutic implications. *J Appl Microbiol* 2011, 110(4):851–861.

[73] Gabani P, Singh OV: Radiation-resistant extremophiles and their potential in biotechnology and therapeutics. *Appl Microbiol Biotechnol* 2013, 97(3):993–1004.

[74] Gaidamakova EK, Myles IA, McDaniel DP, Fowler CJ, Valdez PA, Naik S, Gayen M, Gupta P, Sharma A, Glass PJ *et al*: Preserving immunogenicity of lethally irradiated viral and bacterial vaccine epitopes using a radio- protective Mn2+-Peptide complex from *Deinococcus. Cell Host Microbe* 2012, 12(1):117–124.

[75] Daly MJ: Engineering radiation-resistant bacteria for environmental biotechnology. *Curr Opin Biotechnol* 2000, 11(3):280–285.

[76] Appukuttan D, Rao AS, Apte SK: Engineering of *Deinococcus radiodurans* R1 for bioprecipitation of uranium from dilute nuclear waste. *Appl Environ Microbiol* 2006, 72(12):7873–7878.

[77] Appukuttan D, Seetharam C, Padma N, Rao AS, Apte SK: PhoN-expressing, lyophilized, recombinant *Deinococcus radiodurans* cells for uranium bioprecipitation. *J Biotechnol* 2011, 154(4):285–290.

[78] Kulkarni S, Ballal A, Apte SK: Bioprecipitation of uranium from alkaline waste solutions using recombinant *Deinococcus radiodurans. J Hazard Mater* 2013, 262:853–861.

[79] Misra CS, Appukuttan D, Kantamreddi VS, Rao AS, Apte SK: Recombinant *D. radiodurans* cells for bioremediation of heavy metals from acidic/neutral aqueous wastes. *Bioeng Bugs* 2012, 3(1):44–48.

[80] Misra CS, Mukhopadhyaya R, Apte SK: Harnessing a radiation inducible promoter of *Deinococcus radiodurans* for enhanced precipitation of uranium. *J Biotechnol* 2014, 189:88–93.

[81] Brim H, Venkateswaran A, Kostandarithes HM, Fredrickson JK, Daly MJ: Engineering *Deinococcus geothermalis* for bioremediation of high-temperature radioactive waste environments. *Appl Environ Microbiol* 2003, 69(8):4575–4582.

[82] Lea D, Catcheside D: The mechanism of the induction by radiation of chromosome aberrations in tradesoantia. *J Genet* 1942, 44:216–245.

[83] Kellerer AM, Rossi HH: A generalized formulation of dual radiation action. *Radiat Res* 1978, 75(3):471–488.

[84] Chadwick KH, Leenhouts HP: A molecular theory of cell survival. *Phys Med Biol* 1973, 18(1):78–87.

[85] Brahme A: Accurate description of the cell survival and biological effect at low and high doses and LET's. *J Radiat Res* 2011, 52(4):389–407.

[86] Lind BK, Persson LM, Edgren MR, Hedlof I, Brahme A: Repairable-conditionally repairable damage model based on dual Poisson processes. *Radiat Res* 2003, 160(3):366–375.

[87] Lea D: Actions of radiations on living cells. Cambridge: University Press; 1946.

[88] Tobias CA: The repair-misrepair model in radiobiology: comparison to other models. *Radiat Res Suppl* 1985, 8:S77–95.

[89] Curtis SB: Lethal and potentially lethal lesions induced by radiation--a unified repair model. *Radiat Res* 1986, 106(2):252–270.

[90] Sontag W: A discrete cell survival model including repair after high dose-rate of ionizing radiation. *Int J Radiat Biol* 1997, 71(2):129–144.

[91] Salata F, D'Orazio A, Fabiani M, D'Alessandro D: Effectiveness of UV radiation for reducing *Aspergillus Niger* and *Actynomices* contamination in air-conditioning systems. In: *Proceedings of Clima 2007 WellBeing Indoors: 10 Jun 2007 – 14 Jun 2007; Helsinki (Finland)*: FINVAC; 2007: 8.

[92] McCullagh P, Nelder JA: Generalized Linear Models. London: Chapman & Hall; 1989.

[93] R Development Core Team: R: a language and environment for statistical computing. *R Foundation for Statistical Computing* http://www.R-project.org 2007.

[94] Knezevic SZ, Streibig JC, Ritz C: Utilizing R software package for dose-response studies: the concept and data analysis. *Weed Technol* 2007, 21:840–848.

[95] Hofner B: Lethal: compute lethal doses (LD) with confidence intervals. *R package*, https://github.com/hofnerb/lethal 2014.

Section 2

# Medical Uses

Chapter 5

# Physical and Radiobiological Evaluation of Radiotherapy Treatment Plan

Suk Lee, Yuan Jie Cao and Chul Yong Kim

Additional information is available at the end of the chapter

http://dx.doi.org/10.5772/60846

**Abstract**

Radiation treatment planning plays an important role in modern radiation therapy; it could simulate to plan the geometric, radiobiological, and dosimetric aspects of the therapy using radiation transport simulations and optimization. In this chapter, we have reviewed several quantitative methods used for evaluating radiation treatment plans and discussed some important considering points. For the purpose of quantitative plan evaluation, we reviewed dosimetrical indexes like PITV, CI, TCI, HI, MHI, CN, COSI, and QF. Furthermore, radiobiological indexes like Niemierko's EUD-based TCP and NTCP were included for the purpose of radiobiological outcome modeling. Additionally, we have reviewed dose tolerance for critical organs including RTOG clinical trial results, QUENTEC data, Emami data, and Milano clinical trial results. For the purpose of clinical evaluation of radiation-induced organ toxicity, we have reviewed RTOG and EORTC toxicity criteria. Several programs could help for the easy calculation and analysis of dosimetrical plan indexes and biological results. We have reviewed the recent trend in this field and proposed further clinical use of such programs. Along this line, we have proposed clinically optimized plan comparison protocols and indicated further directions of such studies.

**Keywords:** Treatment plan evaluation, Dosimetrical indices, Radiobiological indices, Tolerance doses, Radiation toxicity

## 1. Introduction

We have reviewed the methods used for quantitative comparison of different radiation treatment plans, the process of treatment plan comparison protocol, and the further direction of treatment plan evaluation programs. For the purpose of quantitative plan evaluation, we reviewed dosimetrical indexes like prescription isodose to target volume (PITV) ratio, homogeneity index (HI), conformity index (CI), target coverage index (TCI), modified dose homogeneity index (MHI), conformity number (CN), critical organ scoring index (COSI), and quality factor (QF). Furthermore, radiobiological indexes like Niemierko's EUD-based tumor control probability (TCP) and normal tissue complication probability (NTCP) were included for the purpose of radiobiological outcome modeling. Additionally, we have reviewed dose tolerance for critical organs including RTOG clinical trial results, QUENTEC data, Emami data, and Milano clinical trial results. For the purpose of clinical evaluation of radiation-induced organ toxicity, we have reviewed RTOG and EORTC toxicity criteria. Several programs could help for the easy calculation and analysis of dosimetrical plan indexes and biological results. We have reviewed the recent trend in this field and proposed further clinical use of such programs. It is well known that plan comparison study still remain many controversies. The major issue is that plan evaluation methods are used in plan comparison and plan optimization. We have reviewed well-known dosimetric and biological plan indexes and several commercial and non-commercial plan evaluation programs. Along this line, we have proposed clinically optimized plan comparison protocols and indicated the further directions of such studies.

## 2. Background: Radiotherapy, radiation treatment planning, and planning decision support program

### 2.1. Radiotherapy

Over the past few decades, radiation treatment has become a technologically advanced field in modern medicine, especially with the advent of intensity-modulated radiation therapy (IMRT) [1]. Traditional radiation therapy planning is a manual, iterative, and simple process in which treatment fields are placed and beam modifiers are inserted.

Modifications are then made after manual inspection of the dose distribution calculated after each iteration [2]. In IMRT, the dose calculation engine specified dose distribution over the target volume and surrounding normal structures. Furthermore, dose calculation engine displayed a 2D dose intensity map by using its optimization algorithms [3]. Moreover, the inverse planning algorithm required users to set a dose/volume criteria for the specific organ/structure, and the computer calculated to find out a final solution to satisfy the criteria. [4]. Another breakthrough of modern radiation treatment is image-guided radiotherapy (IGRT). With the adoption and integration of imaging information in treatment designs, IGRT is the most innovative area in advanced radiotherapy [5]. IGRT has increased knowledge of exact tumor targets and their movements during the treatment process [6]. Despite improvements in target coverage and normal tissue sparing, the implementation of IMRT and IGRT remains a labor-intensive trial and error process. The creation of optimized treatment plans for personalized therapy still requires significant time and effort. Radiation treatment includes CT simulation, organ contouring, treatment planning, quality assurance, and dose delivery (Figure 1) [7].

http://dx.doi.org/10.5772/60846

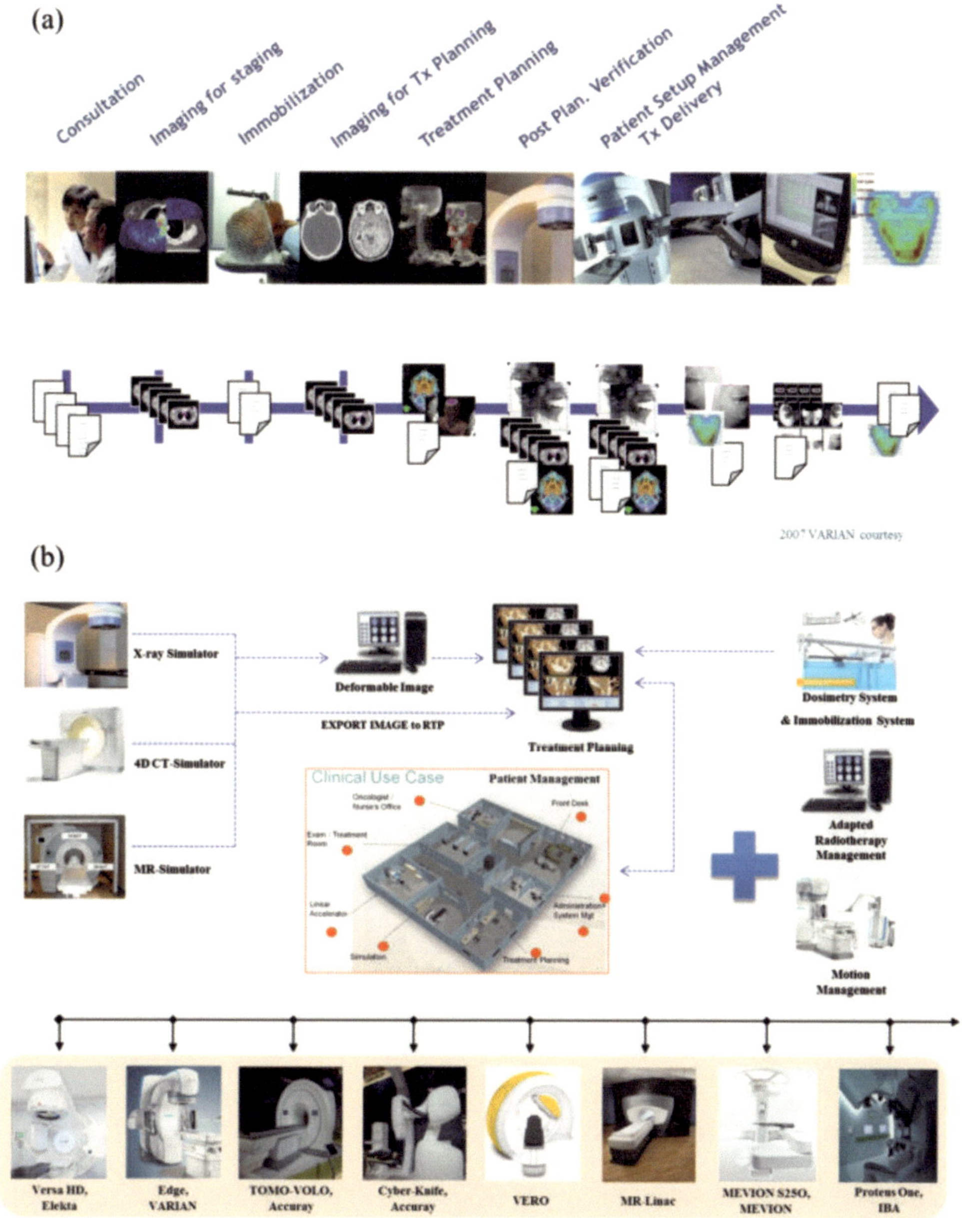


**Figure 1.** Clinical workflow of radiation treatment plan (a); radiation treatment includes CT simulation, organ contouring, treatment planning, quality assurance, and dose delivery. (b); configuration of radiotherapy equipment.

## 2.2. Radiation treatment planning

For radiation treatment, a team of radiation oncologists, radiation therapists, medical physicists, and medical dosimetrists plan the appropriate external beam radiotherapy treatment technique for a patient with cancer [8]. There are generally two different types of planning algorithms, forward planning and inverse planning. The forward planning technique is mostly used in external-beam radiotherapy treatment planning process. For example, a medical physicist determines the beam angles in the treatment planning systems to maximize tumor dose when sparing the healthy tissues. This type of planning is used for the majority of external-beam radiotherapy treatments, but is only useful for relatively uncomplicated cases in which the tumor has a simple shape and is not near any critical organs. Inverse planning is a technique used to inversely design radiotherapy treatment plans (Figure 2). The radiation oncologist defines a patient's critical organs and tumor. Then, the dosimetrist provides target doses for each. An optimization program is then run to find the treatment plan that best matches all input criteria. This type of trial-and-error planning process is time and labor intensive.

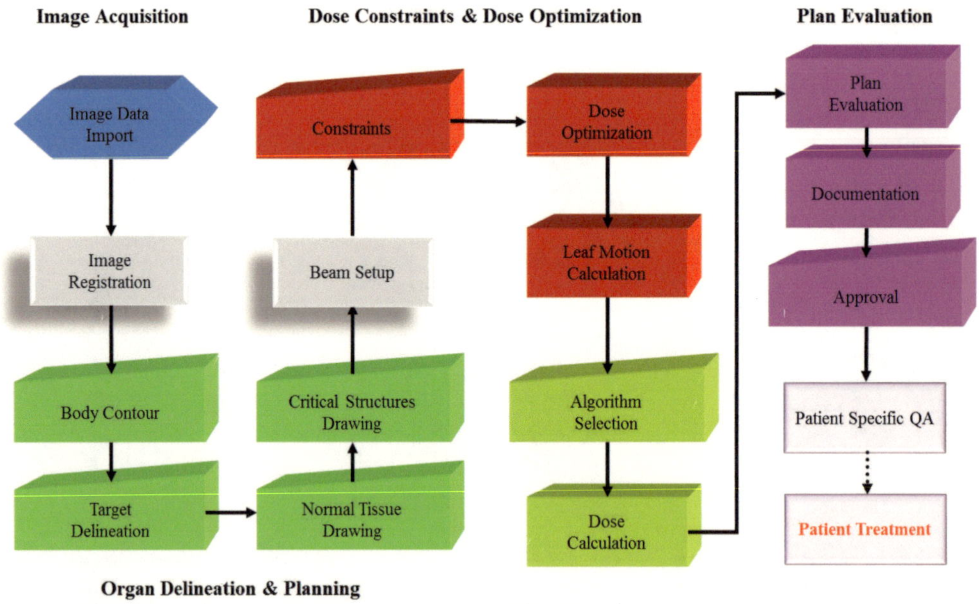


**Figure 2.** Workflow of inverse radiation treatment planning.

There are several commercial treatment planning systems (TPS) available nowadays. Table 1 summarizes information about commercial TPS [9].

### 2.3. Planning decision support program

Dose volume histogram (DVH) provides dose volume coverage information. However, it fails to provide more information like hot spot and dose homogeneity. Dosimetrical indices were widely used for plan evaluation for a specific purpose. For example, a homogeneity index refers to the intensity of dose distributions in target volume, those plans with both "hot" spot and "cold" spot could be distinguished by this index. Additionally, some indices consider dose conformity in the target volume. Conformity index was an example of such indices. Another method to review and evaluate treatment plan quality was biological index. A tumor control probability could indirectly estimate a tumor could be controlled by a certain dose. Furthermore, normal tissue complication probability could estimate the probability of a surrounding critical structure becomes some radiation-induced complications. Many programs have been designed and developed to calculate both dosimetrical and biological indices since the 2000s [10-29]. This is shown in Figure 3.

| Treatment planning system | Company | Website |
|---|---|---|
| ScandiPlan | Scanditronix | http://www.scanditronix-magnet.se |
| Pinnacle3 | Philips Healthcare | http://www.healthcare.philips.com |
| ISOgray | DOSIsoft | http://www.dosisoft.com |
| iPlan | Brainlab | https://www.brainlab.com |
| XiO | Electa | http://www.elekta.com |
| Monaco | Electa | http://www.elekta.com |
| Theraplan Plus | Electa | http://www.elekta.com |
| Oncentra MasterPlan | Electa | http://www.elekta.com |
| Oncentra Prostate | Electa | http://www.elekta.com |
| Oncentra GYN | Electa | http://www.elekta.com |
| Pinnacle | Philips Healthcare | http://www.healthcare.philips.com |
| Plato RTS | Electa | http://www.elekta.com |
| Plato BPS | Electa | http://www.elekta.com |
| Cad Plan | Varian Medical Systems | http://www.varian.com |
| Corvus | nomos | http://www.nomos.com |
| KL-Medical Electron Linear Accelerator treatment system | KLZ Healthcare | http://klz.comedb.com |
| Prowess 3-D | Prowess | http://www.prowess.com/ |
| Brachyvision | Varian | http://www.varian.com |
| Leksell GammaPlan® | Electa | http://www.elekta.com |
| Eclipse | Varian Medical Systems | http://www.varian.com |
| VariSeed | Varian Medical Systems | http://www.varian.com |
| RayStation | RaySearch Laboratories | http://www.raysearchlabs.com |

**Table 1.** Commercial RTP lists

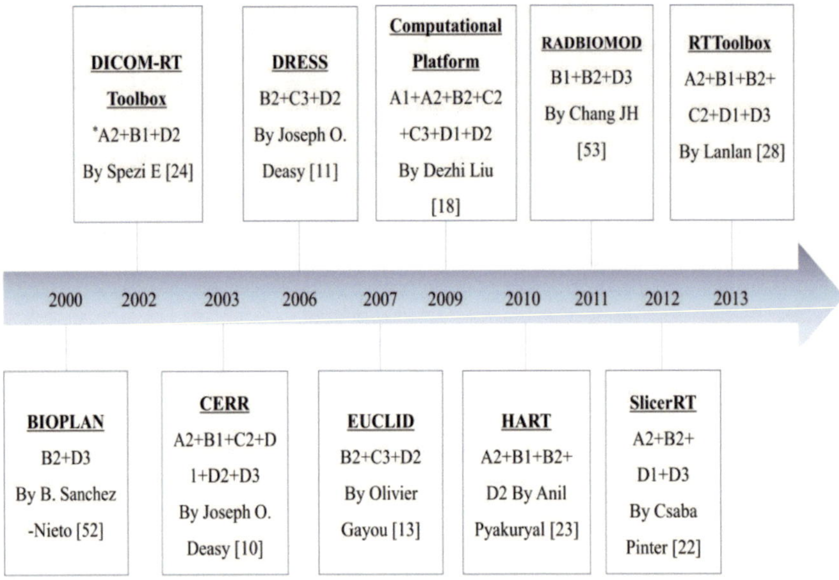


| *Legend | | 1 | 2 | 3 |
|---|---|---|---|---|
| A | DICOM RT Platform | PACS compatible | 3D Image module | - |
| B | Plan Comparison | Physical | Biological | Overall |
| C | Plan Analysis | Multi-RTP | Analysis DB | Statistical Analysis |
| D | Program | GUI Independence | Matlab | C, VB, Java, xls etc. |

**Figure 3.** Timeline of plan analysis programs [10-11, 13, 17-18, 22-24, 28, 52-53].

# 3. Plan evaluation

## 3.1. Plan evaluation methods

### *3.1.1. Qualitative analysis*

In conventional radiation therapy, an isodose distribution is used for plan analysis and evaluation. Figure 4 shows the typical isodose distribution of 3D conformal treatment plans and IMRT plans.

### *3.1.2. Quantitative analysis*

DVH is the relationship between the dose distribution of a certain organ and 100% normalized volume of such organ. It was calculated and generated based on 3D reconstructed images in the treatment planning systems [9]. DVH could simplify 3D information of dose distribution

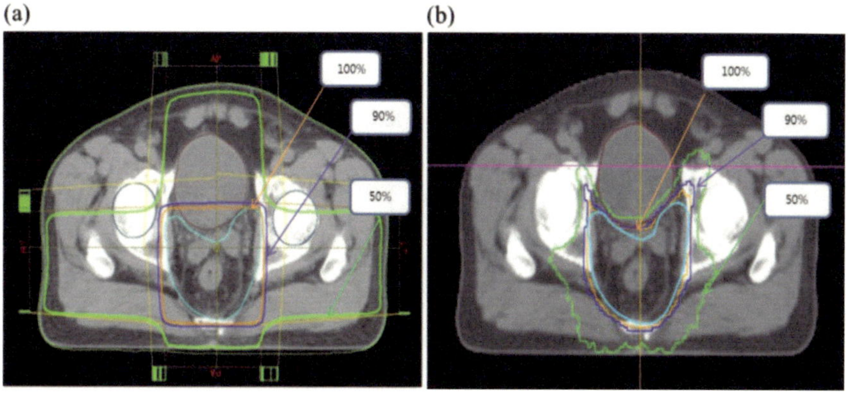


**Figure 4.** Typical isodose distribution of (a) 3D conformal treatment plan and (b) IMRT plan.

into a 2D graph or quantitative values [30-34]. Figure 5 shows a typical DVH for helical tomotherapy (HT) and intensity modulated proton therapy (IMPT) plans for prostate cancer.

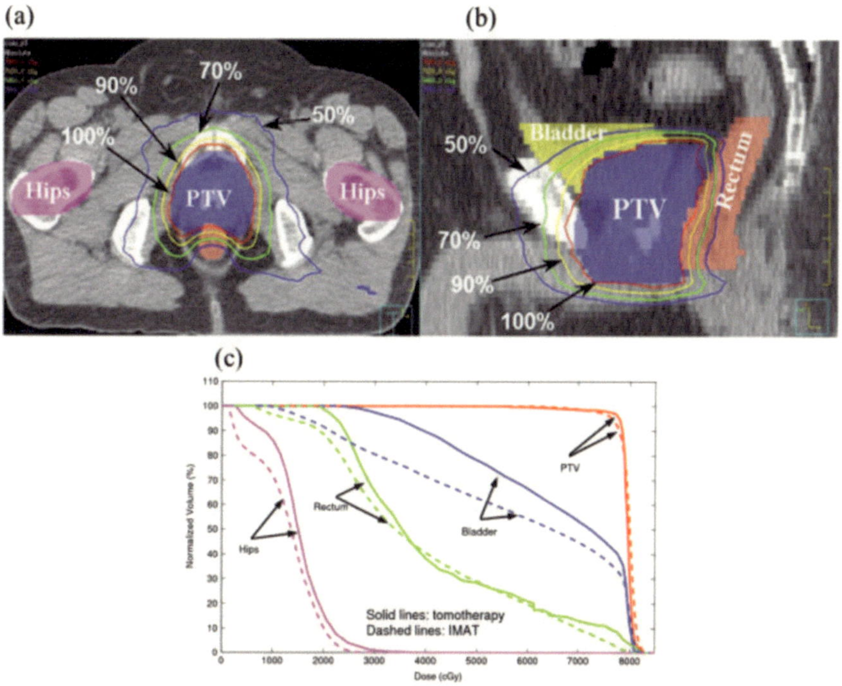


**Figure 5.** Typical DVH for helical tomotherapy (HT) treatment plan and intensity modulated arc therapy (IMAT) plan of prostate cancer: (a) axial slice, (b) sagittal slice. Planning target volume (PTV), critical structures, and four different isodose lines shown. (c) Dose-volume histogram comparison for prostate case. Solid lines, tomotherapy plan; dashed lines, intensity modulated arc therapy (IMAT) plan (International Journal of Radiation Oncology Biology Physics, 69(1), 2007).

## 4. Plan analysis

Isodose distribution and DVH analysis were insufficient compared to complicated and advanced planning techniques. As the femoral head DVHs in Figure 4 show, it was difficult to distinguish whether IMPT (continuous red line) or HT (dashed red line) plans were superior. For low dose volume ($V_0$ to $V_{20}$), IMPT was more favorable than HT. However, this relationship reversed for high dose volume ($V_{20}$ to $V_{50}$). As a result, there are several indexes that may represent target conformity and dose homogeneity [31, 35-38].

### 4.1. Dosimetrical analysis

#### *4.1.1. Index*

Several quantitative evaluation tools were reviewed in this paper. These included the prescription isodose to target volume (PITV) ratio, homogeneity index (HI), conformity index (CI), target coverage index (TCI), modified dose homogeneity index (MHI), conformity number (CN), quality factor (QF) for PTV, maximum dose, mean dose, dose volume histogram (DVH), and critical organ scoring index (COSI) for the OAR (Figure 6).

#### *4.1.2. PTV index*

The PITV ratio, obtained by dividing prescription isodose surface volume by target volume, is expressed as:

$$PITV = \frac{PIV}{TV} \tag{1}$$

In the above equation, PIV represents prescription isodose surface volume and TV refers to target volume [39]. The PITV ratio is a conformity measure, and a value of 1.0 indicates that the volume of the prescription isodose surface equals that of the PTV. A PITV ratio of 1.0 does not necessarily imply that both volumes are similar. To ensure adequate PTV coverage, this measure should always be used in conjunction with a PTV-DVH [39]. The CI and HI indices for targets were computed to assess the quality of IMRT plans. CI is defined as the ratio of target volume and the volume inside the isodose surface that corresponds to the prescription dose. CI is generally used to indicate the portion of a prescription dose that is delivered inside the PTV [40].

CI is expressed as:

$$CI = \frac{PTV_{PD}}{PIV} \tag{2}$$

In the above equation, PIV represents prescription isodose surface volume and $PTV_{PD}$ represents PTV coverage at the prescription dose. CI of 1 indicates that 100% of a prescription dose is delivered to the PTV, and no dose is delivered to any adjacent tissue [40]. The CI is less than 1 for most clinical cases. Higher CI values indicate poorer dose conformity to the PTV. HI is defined as the ratio of maximum dose delivered to the PTV divided by the prescription dose delivered to the PTV [41].

HI is expressed as:

$$HI = \frac{D_{max}}{PD} \tag{3}$$

In the above equation, $D_{max}$ represents PTV maximum dose. An HI of 1 represents the ideal uniform dose within a target. Higher HI values indicate greater dose heterogeneity in the PTV [39].

TCI refers to the exact coverage of PTV in a treatment plan for a given prescription dose.

TCI is expressed as:

$$TCI = \frac{PTV_{PD}}{PTV} \tag{4}$$

In the above equation, $PTV_{PD}$ represents PTV coverage at the prescription dose.

MHI is similar to HI, and is expressed as [41]:

$$MHI = \frac{D_{95}}{D_5} \tag{5}$$

In the above equation, $D_{95}$ and $D_5$ represent doses received at 95% and 5% of the volume coverage, respectively.

Conformity number (CN) is a relative measurement of dosimetric target coverage and sparing of normal tissues in a treatment plan [42]. The CN is expressed as:

$$CN = TCI \times CI = \frac{PTV_{PD}}{PTV} \times \frac{PTV_{PD}}{PIV} \tag{6}$$

In the above equation, $PTV_{PD}$ refers to PTV coverage at the prescription dose and PIV represents prescription isodose surface volume [42].

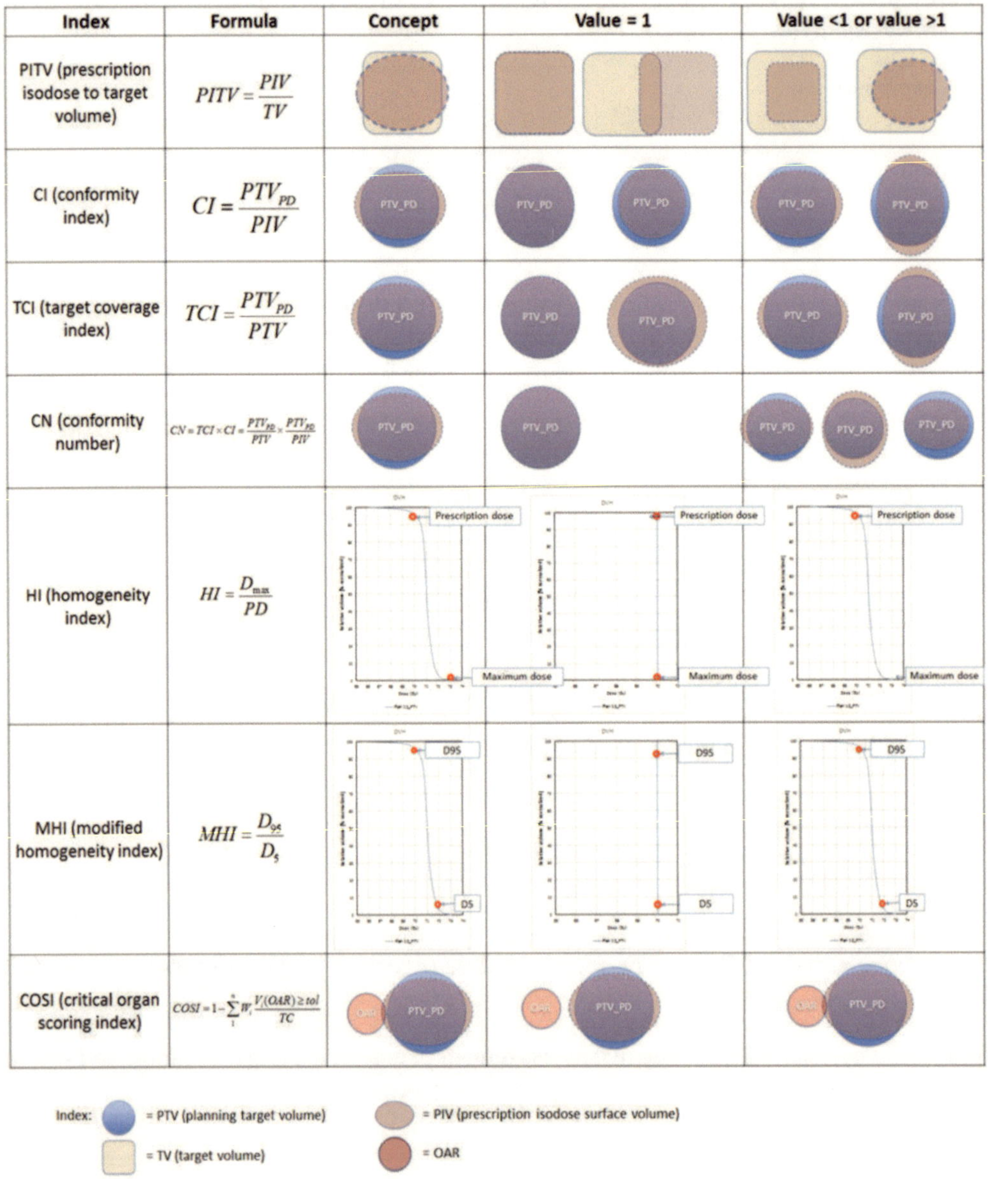


**Figure 6.** Comparison of the various dosimetrical indices in various clinical cases.

## 4.2 Biological analysis

### *4.2.1. Overview of biological models*

For radiobiological model-based plan evaluation, Niemierko's equivalent uniform dose (EUD)-based NTCP and TCP model were reviewed [12, 19]. First, the DVHs from each plan were exported from the appropriate treatment planning system (TPS) for each modality. The DVHs were then imported into MATLAB version R2012a (The Math Works, Inc., Natick, MA, USA) for TCP and NTCP modeling analysis. According to Neimierko's phenomenological model, EUD is defined as:

$$EUD = \left[ \sum_{i=1} \left( V_i EQD_i^a \right) \right]^{\frac{1}{a}} \tag{7}$$

where $a$ is a unitless model parameter that is specific to the nominal tumor structure of interest, and $V_i$ is a unitless parameter that represents the $i^{th}$ partial volume receiving dose $D_i$ in Gy [12]. Since the relative volume of the whole structure of interest corresponds to 1, the sum of all partial volumes $V_i$ will equal 1. In equation [5], the EQD is a biologically equivalent physical dose of 2 Gy defined as:

$$EQD = D \times \frac{\left( \frac{\alpha}{\beta} + \frac{D}{n_f} \right)}{\left( \frac{\alpha}{\beta} + 2 \right)} \tag{8}$$

where $n_f$ and $d_f = D/n_f$ are the number of fractions and the dose per fraction size of the treatment course, respectively. In this equation, $\alpha/\beta$ is the tissue-specific linear quadratic (LQ) parameter of the organ being exposed. Niemierko's TCP [12] is defined as:

$$TCP = \frac{1}{1 + \left( \frac{TCD_{50}}{EUD} \right)^{\gamma 50}} \tag{9}$$

where $TCD_{50}$ is the tumor dose required to control 50% of cancer cells when a tumor is homogeneously irradiated and $\gamma_{50}$ is a unitless model parameter that is specific to the tumor of interest. The slope of the dose response curve is described by $\gamma_{50}$. Niemierko's NTCP [19] is defined as:

$$TCP = \frac{1}{1 + \left( \frac{TCD_{50}}{EUD} \right)^{\gamma 50}} \tag{10}$$

where $TD_{50}$ is the tolerance dose of a 50% complication rate at a specific time (e.g. 5 years in the Emami et al. normal tissue tolerance data [43]) for an entire organ of interest. This parameter also describes the slope of the dose response curve.

### 4.3. Overall plan index

#### 4.3.1. Overall plan index

A comprehensive quality index (CQI) including surrounding OARs were introduced to evaluate the individual difference between OARs and PTV and the small volume of critical structures. CQI is expressed as [44]:

$$CQI = \frac{1}{N}\sum_{i=1}^{N} QI_i = \frac{1}{N}\sum_{i=1}^{N} \frac{\left(D_{\max}^{plan1}\right)}{\left(D_{\max}^{plan2}\right)} \tag{11}$$

In this equation, I is the index of the critical organs, which are several critical structures in certain plan. CQI was designed to compare the ability of avoiding these organs around the PTV given the same weighting to all organs. Although CQI may overweight certain organs that are below tolerance, we chose this index as it represents a global measure of the capability of avoiding sensitive structures. Individual Qis are shown for direct comparison of each OAR. A CQI less than one indicates that HT provides a better plan for the surrounding OARs, and vice versa.

#### 4.3.2. COSI

The COSI index accounts for both target coverage and critical organ irradiation [45]. The main advantage of this index is its ability to distinguish between different critical organs. COSI is expressed as:

$$COSI = 1 - \sum_{1}^{n} w_i \frac{V_i\left(OAR\right)_{>tol}}{TCI} \tag{12}$$

where $V_i(OAR)_{>tol}$ is the volume fraction of OAR that receives more than a predefined tolerance dose. TCV is the volumetric target coverage, which is defined as the fractional volume of PTV covered by the prescribed isodose. Modified COSI is expressed as:

$$mCOSI = \sum_{i=1} W_i \left( \frac{COSI_{10} + COSI_{20} + \cdots + COSI_{80}}{8} \right) \tag{13}$$

Although the COSI index focuses only on OARs that receive high dose region volumes, the modified COSI considers both high dose and low dose regions.

#### 4.3.3. *Quality factor*

The quality factor (QF) introduced in this study is a dosimetrical index that can evaluate the quality of an entire plan [23]. The QF of a plan is analytically expressed as:

$$QF = \left[ 2.718 \exp\left( -\sum_{i=1}^{N} W_i X_i \right) \right] \tag{14}$$

In the above equation, $X_i$ represents all PTV indices, including PITV, CI, HI, TCI, MHI, CN, and COSI. The weighting factor ($W_i$) values can be adjusted between 0 and 1 for all relatively weighted indices for a user-defined number of indices (N). A weighting factor of 1 was used for all separate indices. Thus, the QF was mainly used to compare the conformity of plans throughout various trials of a treatment.

## 5. Radiation tolerance dose and toxicity

The dose to critical structures plays an important role in treatment plan evaluation and is a challenging parameter in radiotherapy treatment planning. Here, Emami data [43], QUENTEC data [46], RTOG data, and the Milano study were reviewed. Doses based on tumor location in the body related to critical organs are as follows (Table 2-4).

### 5.1. Radiation toxicities

The assessment and reporting of toxicity plays a central role in oncology [47-50]. The foundation of toxicity reporting is the toxicity criteria system. Multiple systems have been developed in the last 30 years, and they have evolved substantially since their first introduction. The wide adoption of standardized criteria will facilitate comparison between institutions and clinical trials.

The Radiation Therapy Oncology Group (RTOG) acute radiation morbidity scoring criteria developed in 1984 consists of 13 scales that cover most body regions [51]. This system was used by the RTOG and in other clinical trials for over 30 years. The inclusion of acute radiation criteria into a multimodality grading system facilitated toxicity grading in all oncologic disciplines. This system also allows radiation oncologists to recognize and grade toxicities that were not available in the previous RTOG system. Tables 5 and 6 summarize acute toxicity categorized by body region.

The RTOG/EORTC (European Organization for Research and Treatment of Cancer) system for scoring late effects was developed in 1984 alongside the RTOG acute criteria. It contains 16 organ categories (Tables 7, 8) and has been used widely. However, its shortcomings have prompted the development of other systems.

| RTOG data | | | | | | | QUANTEC data | | | | | | | Emami Data | | | | | | Milano Data | | |
|---|---|---|---|---|---|---|---|---|---|---|---|---|---|---|---|---|---|---|---|---|---|---|
| Critical Structure | Dose/ fx | Vol. | Dose | Max. Dose | Protocol | Treated organ | Critical Structure | Vol. | Dose/ Vol. | Max. Dose | Toxicity Rate | Toxicity Endpoint | Organ | TD 5/5 | | | TD 50/5 | | | Organ | Dose tolerance | Endpoint |
| | | | | | | | | | | | | | | Whole | 2/3 | 1/3 | Whole | 2/3 | 1/3 | | | |
| Brachial Plexus | 2 Gy | 5% | 60 Gy | | 619 | Postop H&N | Brain | | | <60 Gy | <3% | Symptomatic necrosis | | | | | | | | | | |
| | 2 Gy | | | 60 Gy | 522 | Definitive H&N | | | | 72 Gy | 5% | Symptomatic necrosis | Brain | 4500 | 5000 | 6000 | 6000 | 6500 | 7500 | | | |
| | 2 Gy | | | 66 Gy | 0619, 0617 | Postop H&N, lung, nasopharynx | | | | 90 Gy | 10% | Symptomatic necrosis | | | | | | | | | | |
| | 3 Gy | | | 36 Gy | 937 | Lung | | | | | | | Brachial plexus | 6000 | 6100 | 6200 | 7500 | 7600 | 7700 | | | |
| | 4 Gy | | | 30 Gy | 937 | Lung | | | | | | | | | | | | | | | | |
| Brainstem | 1.8-2Gy | 0.03cc | | 55 Gy (0.03 cc) | 539 | Intermediate risk meningioma | Brain stem | | | <54 Gy | <5% | Neuropathy or necrosis | Brain stem | 5000 | 5300 | 6000 | 6500 | – | – | Brain stem | V60 < 0.9 mL | <5% grade >= 1 toxicity |
| | 33 fxs | | | 54 Gy | 615 | Nasopharynx | | D1-10 cc | <= 59 Gy | | <5% | Neuropathy or necrosis | | | | | | | | | | |
| | 1.8-2Gy | | | 60 Gy (0.03 cc) | 0539, 0825 | High risk meningioma, glioblastoma | | | | <64 Gy | <5% | Neuropathy or necrosis | | | | | | | | | | |
| | 2 Gy | | | 52 Gy (0.03 cc | 1016 | Oropharynx | | | | | | | | | | | | | | | | |
| Cochlea | 33 fxs | 5% | 55 Gy | | 615 | Nasopharynx | Cochlea | Mean | <=45 Gy | | <30% | Sensory-neural hearing loss | Ear | 5500 | 5500 | 5500 | 6500 | 6500 | 6500 | | | |
| Larynx, glottis | Mean | 20 Gy | | | 1016 | Oropharynx | Larynx | | | <66 Gy | <20% | Vocal dysfunction | Larynx (necrosis) | 7000 | 7000 | 7900 | 8000 | 8000 | 9000 | | | |
| | 2 Gy | | | 45 Gy | 0619, 0615 | Postop H&N, definitive H&N, nasopharynx | | Mean | <50 Gy | | <30% | Aspiration | Larynx (edema) | 4500 | 4500 | – | 8000 | – | – | | | |
| | | | | | | | | Mean | <44 Gy | | <20% | Edema | | | | | | | | | | |
| | | | | | | | | V50 | <27% | | <20% | Edema | | | | | | | | | | |

| RTOG data | | | | | | | QUANTEC data | | | | | | Emami Data | | | | | | | Milano Data | | |
|---|---|---|---|---|---|---|---|---|---|---|---|---|---|---|---|---|---|---|---|---|---|---|
| Critical Structure | Dose/ fx | Vol. | Dose | Max. Dose | Protocol | Treated organ | Critical Structure | Vol. | Dose/ Vol. | Max. Dose | Toxicity Rate | Toxicity Endpoint | Organ | TD 5/5 | | | TD 50/5 | | | Organ | Dose tolerance | Endpoint |
| Lens | 1.8-2Gy | | | 5 Gy (0.03 cc) | 539 | Intermediate risk meningioma | | | | | | | Lens | 1000 | – | – | 1800 | – | – | | | |
| | 1.8-2Gy | | | 7 Gy (0.03 cc) | 0539, 0825 | High risk meningioma, glioblastoma | | | | | | | | | | | | | | | | |
| | 33 fxs | | | 25 Gy | 615 | Nasopharynx | | | | | | | | | | | | | | | | |
| Lips | 2 Gy | Mean | <20 Gy | | 1016 | Oropharynx | | | | | | | | | | | | | | | | |
| Mandible/ TM joint | 2 Gy | | | 66 Gy | 1016 | Oropharynx | | | | | | | | | | | | | | | | |
| | 33 fxs | 1 cc | 75 Gy | | 615 | Nasopharynx | | | | | | | TMJ mandi-ble | 6000 | 6000 | 6500 | 7200 | 7200 | 7700 | | | |
| | 33 fxs | | | 50 Gy | 615 | Nasopharynx | | | | | | | | | | | | | | | | |
| Optic chiasm | 33 fxs | | | 50 Gy | 615 | Nasopharynx | Optic nerve/ chiasm | | | <55 Gy | <3% | Optic neuropathy | Optic chiasm | 5000 | – | – | 6500 | – | – | | | |
| | 1.8-2 Gy | | | 54 Gy (0.03 cc) | 539 | Intermediate risk meningioma | | | | 55-60 Gy | 3-7% | Optic neuropathy | | | | | | | | | | |
| | 1.8-2 Gy | | | 56 Gy (0.03 cc) | 0539, 0825 | High risk meningioma, glioblastoma | | | | >60 Gy | >7-20% | Optic neuropathy | | | | | | | | | | |
| Optic nerve | 1.8-2 Gy | | | 50 Gy (0.03 cc) | 0539, 0615 | Intermediate risk meningioma, nasopharynx | | | | | | | | | | | | | | | | |
| | 1.8-2 Gy | | | 55 Gy (0.03 cc) | 0539, 0825 | High risk meningioma, glioblastoma | | | | | | | Optic nerve | 5000 | – | – | 6500 | – | – | | | |
| Oral cavity | 33 fxs | Mean | <40 Gy | | 615 | Nasopharynx | | | | | | | | | | | | | | | | |
| Oral cavity (non-involved) | 2 Gy | Mean | <30 Gy | 60 Gy | 1016 | Oropharynx | | | | | | | | | | | | | | | | |

| RTOG data | | | | | | | QUANTEC data | | | | | | Emami Data | | | | | | | Milano Data | | |
|---|---|---|---|---|---|---|---|---|---|---|---|---|---|---|---|---|---|---|---|---|---|---|
| Critical Structure | Dose/ fx | Vol. | Dose | Max. Dose | Protocol | Treated organ | Critical Structure | Vol. | Dose/ Vol. | Max. Dose | Toxicity Rate | Toxicity Endpoint | Organ | TD 5/5 | | | TD 50/5 | | | Organ | Dose tolerance | Endpoint |
| | 2 Gy | Mean one gland | <26 Gy | | 0619, 0522, 1016 | Postop H&N, definitive H&N, oropharynx | Parotid, bilateral | Mean | <=25 Gy | | <20% | Long-term salivary function <25% | | | | | | | | | | |
| Parotid Glands | 2 Gy | V50 one gland | <30 Gy | | 0619, 0522 | Postop H&N, definitive H&N | | Mean | <=39 Gy | | <50% | Long-term salivary function <25% | Parotid gland | 3200 | 3200 | – | 4600 | 4600 | | Parotid | Mean dose < 26 Gy | Late grade 2 xerostomia, >75% functional loss |
| | 2 Gy | Combined 20 cc | <20 Gy | | 0619, 0522 | Postop H&N, definitive H&N | Parotid, unilateral | Mean | <=20 Gy | | <20% | Long-term salivary function <25% | | | | | | | | | | |
| Pharynx, postcricoid | 33 fxs | | | 45 Gy | 615 | Nasopharynx | | | | | | | | | | | | | | | | |
| Pharynx, posterior wall | 2 Gy | 33% | 50Gy | | 1016 | Oropharynx | Pharyngeal constrictors | Mean | <=50 Gy | | <20% | Symptomatic dysphagia and aspiration | | | | | | | | | | |
| | 2 Gy | 15% | 60Gy | | 1016 | Oropharynx | | | | | | | | | | | | | | | | |
| | 2 Gy | Mean | 45Gy | | 1016 | Oropharynx | | | | | | | | | | | | | | | | |
| Retina | 1.8-2 Gy | | | 45 Gy (0.03 cc) | 539 | Intermediate risk meningioma | | | | | | | | | | | | | | | | |
| | 1.8-2 Gy | | | 50 Gy (0.03 cc) | 0539, 0825, 0615 | High risk meningioma, glioblastoma, nasopharynx | | | | | | | Retina | 4500 | – | – | 6500 | – | – | | | |
| Spinal Cord | 1.8 Gy | | | 45 Gy | 0623, 0615 | Lung, Nasopharynx | | | | 50 Gy | 0.20% | Myelopathy | Spinal cord | (20cm) | (10cm) | (5 cm) | – | (10cm) | (5 cm) | Spinal cord | Max < 50 Gy | <5% grade >= 3 toxicity |
| | 2 Gy | | | 48 Gy (0.03 cc) | 0619, 0522 | Postop H&N, definitive H&N | Spinal cord | | | 60 Gy | 6% | Myelopathy | | 4700 | 5000 | 5000 | | 7000 | 7000 | Cervical spinal cord | EUD < 52 Gy, Max < 55 Gy | <5% grade >= 3 toxicity |
| Submandibular Gland | 2 Gy | Mean | <39 Gy | | 1016 | Oropharynx | | | | 69 Gy | 50% | Myelopathy | | | | | | | | | | |

**Table 2.** Radiation tolerance dose in head and neck

| RTOG data | | | | | | | QUANTEC data | | | | | | Emami Data | | | | | | | Milano Data | | |
|---|---|---|---|---|---|---|---|---|---|---|---|---|---|---|---|---|---|---|---|---|---|---|
| Critical Structure | Dose/ fx | Vol. | Dose | Max. Dose | Protocol | Treated organ | Critical Structure | Vol. | Dose/Vol. | Max. Dose | Toxicity Rate | Toxicity Endpoint | Organ | TD 5/5 | | | TD 50/5 | | | Organ | Dose tolerance | Endpoint |
| | | | | | | | | | | | | | | Whole | 2/3 | 1/3 | Whole | 2/3 | 1/3 | | | |
| Esopha- gus | 1.8 Gy | Mean | 34 Gy | | 0623, 0617 | Lung | Esophagus | Mean | <34 Gy | | 5-20% | Grade 3+ esophagitis | | | | | | | | Esopha- gus | V50 and S50 < 30% | 5% risk of late toxicity |
| | 1.8 Gy | 10 cm | 60 Gy | | 623 | Lung | | V35 | <50% | | <30% | Grade 2+ esophagitis | Esophagus (stricture, perforation) | 5500 | 5800 | 6000 | 6800 | 7000 | 7200 | | | |
| | 2 Gy | Mean | 30 Gy | | 1016 | Oropharynx | | V50 | <40% | | <30% | Grade 2+ esophagitis | | | | | | | | | | |
| | 3 Gy | | | 47 Gy | 937 | Lung | | V70 | <20% | | <30% | Grade 2+ esophagitis | Thyroid | 4500 | | | 8000 | | | | | |
| Heart | 1.8 Gy | 33% | 60 Gy | | 0623, 0617 | Lung | Heart ,Pericardiu m | Mean | <26 Gy | | <15% | Pericarditis | Heart (pericarditis) | 4000 | 4500 | 6000 | 5000 | 5500 | 7000 | Heart | V33 < 60% | 5% excess cardiac mortality |
| | 1.8 Gy | 33% | 50 Gy | | 436 | Esophagus | | V30 | <46% | | <15% | Pericarditis | | | | | | | | | V38 < 33% | |
| | 1.8 Gy | 67% | 45 Gy | | 0623, 0617 0436 | Lung, esophagus | Heart | V25 | <10% | | <1% | Long term cardiac mortality | | | | | | | | | V42 < 20% | |
| | 1.8 Gy | 100% | 40 Gy | | 0623, 0617 0436 | Lung, esophagus | | | | | | | | | | | | | | | | |
| | 3 Gy | | | 47 Gy | 937 | Lung | | | | | | | | | | | | | | | | |
| | 3 Gy | V45 | <30% | | 937 | Lung | | | | | | | | | | | | | | | | |

| RTOG data | | | | | | | QUANTEC data | | | | | | Emami Data | | | | | | | Milano Data | | |
|---|---|---|---|---|---|---|---|---|---|---|---|---|---|---|---|---|---|---|---|---|---|---|
| Critical Structure | Dose/ fx | Vol. | Dose | Max. Dose | Protocol | Treated organ | Critical Structure | Vol. | Dose/Vol. | Max. Dose | Toxicity Rate | Toxicity Endpoint | Organ | TD 5/5 | | | TD 50/5 | | | Organ | Dose tolerance | Endpoint |
| Lung, single | 2 Gy | 3 cm CW to field | | | 413 | Breast | | V20 | <=30% | | <20% | Symptomatic pneumonitis | Lung | 1750 | 3000 | 4500 | 2450 | 4000 | 6500 | | V13 < 40% | Late grade 2 in <10-20% |
| | 2 Gy | V20 | 20% | | 630 | Sarcoma | | Mean | 7 Gy | | 5% | Symptomatic pneumonitis | Rib cage | – | – | 5000 | – | – | 6500 | Lung | V20 < 25-30% | Late grade 3 in <5-10% |
| | 2 Gy | V20 | 37% | | 0617, 0623 | Lung | | Mean | 13 Gy | | 10% | Symptomatic pneumonitis | | | | | | | | | V30 < 10-15% | |
| Lungs, total | 2 Gy | Mean | 20 Gy | | 617 | Lung | Lung | Mean | 20 Gy | | 20% | Symptomatic pneumonitis | | | | | | | | | MLD < 10-20 Gy | |
| | 3 Gy | Mean | 20 Gy | | 937 | Lung | | Mean | 24 Gy | | 30% | Symptomatic pneumonitis | | | | | | | | | | |
| | 3 Gy | V20 | <= 30% | | 937 | Lung | | Mean | 27 Gy | | 40% | Symptomatic pneumonitis | | | | | | | | | | |
| | 3 Gy | 150 cc | 30 Gy | | 937 | Lung | Small bowel (individual loops) | V15 | <120 cc | | <10% | Grade 3+ toxicity | Small intestine (obstruction, perforation) | 4000 | – | 5000 | 5500 | – | 6000 | | | |
| | 3 Gy | 100 cc | 35 Gy | | 937 | Lung | Small bowel (peritoneal cavity) | V45 | <195 cc | | <10% | Grade 3+ toxicity | | | | | | | | | | |
| | 3 Gy | 50 cc | 40 Gy | | 937 | Lung | | | | | | | | | | | | | | | | |
| Small Bowel | 3 Gy | 1 cc | 45 Gy | | 937 | Lung | | | | | | | | | | | | | | | | |
| | 4 Gy | 100 cc | 30 Gy | | 937 | Lung | | | | | | | | | | | | | | | | |
| | 4 Gy | 50 cc | 35 Gy | | 937 | Lung | | | | | | | Stomach (ulceration,perforation) | 5000 | 5500 | 6000 | 6500 | 6700 | 7000 | | | |
| | 4 Gy | 1 cc | 40 Gy | | 937 | Lung | Stomach | D100 | <45 Gy | | <7% | Ulceration | | | | | | | | | ▯ | |

| RTOG data | | | | | | | QUANTEC data | | | | | | Emami Data | | | | | | | Milano Data | | |
|---|---|---|---|---|---|---|---|---|---|---|---|---|---|---|---|---|---|---|---|---|---|---|
| Critical Structure | Dose/ fx | Vol. | Dose | Max. Dose | Protocol | Treated organ | Critical Structure | Vol. | Dose/Vol. | Max. Dose | Toxicity Rate | Toxicity Endpoint | Organ | TD 5/5 | | | TD 50/5 | | | Organ | Dose tolerance | Endpoint |
| Spinal Cord | 1.8 Gy | | | 45 Gy | 0623, 0615 | Lung, Nasopharynx | Spinal cord | | | 50 Gy | 0.20% | Myelopathy | Spinal cord | (20 cm) | (10 cm) | (5 cm) | – | (10 cm) | (5 cm) | Spinal cord | Max < 50 Gy | <5% grade >= 3 toxicity |
| | 2 Gy | | | 50.5 | 617 | Lung | | | | 60 Gy | 6% | Myelopathy | | 4700 | 5000 | 5000 | | 7000 | 7000 | Cervical spinal cord | EUD < 52 Gy | <5% grade >= 3 toxicity |
| | 1.8 Gy | 10 cm | 50 Gy | | 436 | Esophagus | | | | 69 Gy | 50% | Myelopathy | | | | | | | | | Max < 55 Gy | |
| | 1.8 Gy | 20 cm | 47 Gy | | 436 | Esophagus | | | | | | | | | | | | | | | | |
| | 3 Gy | | | 36 Gy | 937 | Lung | | | | | | | | | | | | | | | | |
| | 4 Gy | | | 30 Gy | 937 | Lung | | | | | | | | | | | | | | | | |
| Kidney | 1.8 Gy | 100% | 23 Gy | | 436 | Esophagus | Kidney, bilateral | Mean | <15-18 Gy | | <5% | Clinical dysfunction | Kidney | 2300 | 3000 | 5000 | 2800 | 4000 | – | Kidney | Median dose < 17.5 Gy | Anemia,, azotemia, HTN, edema |
| | 1.8 Gy | 67% | 30 Gy | | 436 | Esophagus | | Mean | <28 Gy | | <50% | Clinical dysfunction | | | | | | | | | | |
| | 1.8 Gy | 33% | 50 Gy | | 436 | Esophagus | | V12 | <55% | | <5% | Clinical dysfunction | | | | | | | | | | |
| | 2 Gy | 50% | 14 Gy | | 630 | Sarcoma | | V20 | <32% | | <5% | Clinical dysfunction | | | | | | | | | | |
| | 3 Gy | V18 | < 25% | | 937 | Lung | | V23 | <30% | | <5% | Clinical dysfunction | | | | | | | | | | |

| RTOG data | | | | | | | QUANTEC data | | | | | | Emami Data | | | | | | | Milano Data | | |
|---|---|---|---|---|---|---|---|---|---|---|---|---|---|---|---|---|---|---|---|---|---|---|
| Critical Structure | Dose/ fx | Vol. | Dose | Max. Dose | Protocol | Treated organ | Critical Structure | Vol. | Dose/Vol. | Max. Dose | Toxicity Rate | Toxicity Endpoint | Organ | TD 5/5 | | | TD 50/5 | | | Organ | Dose tolerance | Endpoint |
| Liver | 1.8 Gy | 50% | 35 Gy | | 436 | Esophagus | Liver | Mean | <30-32 Gy | | <5% | RILD (in normal liver function) | Liver | 3000 | 3500 | 5000 | 4000 | 4500 | 5500 | Liver | 1/3: 40-80 Gy | Late grade 3-4 liver toxicity < 5% |
| | 3 Gy | >700 cc | <18 Gy | | 937 | Lung | | Mean | <42 Gy | | <50% | RILD (in normal liver function) | | | | | | | | | 2/3: 30-50 Gy | |
| | | | | | | | | Mean | <28 Gy | | <5% | RILD (in Child-Pugh A or HCC) | | | | | | | | | 3/3: 25-35% | |
| | | | | | | | | Mean | <36 Gy | | <50% | RILD (in Child-Pugh A or HCC) | | | | | | | | | | |

**Table 3.** Radiation tolerance dose in abdomen

| | RTOG data | | | | | | Handbook | | | | QUANTEC data | | | | | | Emami Data | | | | | | | Milano Data | | |
|---|---|---|---|---|---|---|---|---|---|---|---|---|---|---|---|---|---|---|---|---|---|---|---|---|---|---|
| Critical Structure | Dose/fx | Vol. | Dose | Max. Dose | Protocol | Treated organ | Organ | Partial Organ | Tolerance (1.8 – 2.0 Gy/fx) | Critical Structure | Vol. | Dose/ Vol. | Max. Dose | Toxicity Rate | Toxicity Endpoint | Organ | TD 5/5 | | | TD 50/5 | | | Organ | Dose tolerance | End point |
| Bladder | 1.8 Gy | 60% | 50 Gy | | 621 | Prostate | Bladder | Whole | 50 Gy (5-10% late) | Bladder (bladder cancer) | | | <65 | <6% | Grade 3+ toxicity | | Whole | 2/3 | 1/3 | Whole | 2/3 | 1/3 | | | |
| | 1.8 Gy | 60% | 40 Gy | | 534 | Postop prostate | | Whole | 60 Gy (10-40% late) | | V65 | <50% | | | Grade 3+ toxicity | Bladder | 6500 | 8000 | N/A | 8000 | 8500 | N/A | | | |
| | 1.8 Gy | 55% | 50 Gy | | PMID 18947938 | RTOG Prostate Group Consensus 2009 | | 1/3 | 60 Gy (5-10% late) | Bladder (prostate cancer) | V70 | <35% | | | Grade 3+ toxicity | | | | | | | | | | |
| | 1.8 Gy | 50% | 35 Gy | | 529 | Anus | | 1/3 | 70 Gy (20% late) | | V75 | <25% | | | Grade 3+ toxicity | | | | | | | | | | |
| | 1.8 Gy | 50% | 65 Gy | | 415 | Prostate | | GYN HDR | <70% Point A | | V80 | <15% | | | Grade 3+ toxicity | | | | | | | | | | |
| | 1.8 Gy | 40% | 40 Gy | | 822 | Rectum | Urethra | | 60-70 Gy | | | | | | | | | | | | | | | | |
| | 1.8 Gy | 40% | 65 Gy | | 534 | Postop prostate | Testis | Transient azosperima | 0.5 Gy | | | | | | | | | | | | | | | | |
| | 1.8 Gy | 40% | 66.6 Gy | | 621 | Prostate | | Total azospermia | 0.8-1 Gy | | | | | | | | | | | | | | | | |
| | 1.8 Gy | 35% | 40 Gy | | 529 | Anus | | Sterilization | 2-3 Gy | | | | | | | | | | | | | | | | |
| | 1.8 Gy | 35% | 45 Gy | | 418 | Endometrial | Ovary | Sterilization | 2-3 Gy | | | | | | | | | | | | | | | | |
| | 1.8 Gy | 35% | 70 Gy | | 415 | Prostate | | Ovarian failure | 5-10 Gy | | | | | | | | | | | | | | | | |
| | 1.8 Gy | 30% | 70 Gy | | PMID 18947938 | RTOG Prostate Group Consensus 2009 | Vagina | Upper mucosa | 120 Gy | | | | | | | | | | | | | | | | |
| | 1.8 Gy | 25% | 75 Gy | | 415 | Prostate | | Mid mucosa | 80-90 Gy | | | | | | | | | | | | | | | | |
| | 1.8 Gy | 15% | 45 Gy | | 822 | Rectum | | Lower mucosa | 60-70 Gy | | | | | | | | | | | | | | | | |
| | 1.8 Gy | 15% | 80 Gy | | 415 | Prostate | | Fibrosis/stenosis | >50-60 Gy | | | | | | | | | | | | | | | | |
| | 1.8 Gy | 5% | 50 Gy | | 529 | Anus | Ureter | | <75 Gy | | | | | | | | | | | | | | | | |
| | 1.8 Gy | | | 50 Gy | 822 | Rectum | | | | | | | | | | | | | | | | | | | |

| RTOG data | | | | | | | Handbook | | | QUANTEC data | | | | | | Emami Data | | | Milano Data | | |
|---|---|---|---|---|---|---|---|---|---|---|---|---|---|---|---|---|---|---|---|---|---|
| Critical Structure | Dose/fx | Vol. | Dose | Max. Dose | Protocol | Treated organ | Organ | Partial Organ | Tolerance (1.8 – 2.0 Gy/fx) | Critical Structure | Vol. | Dose/ Vol. | Max. Dose | Toxicity Rate | Toxicity Endpoint | Organ | TD 5/5 | TD 50/5 | Organ | Dose tolerance | End point |
| External genitalia | 1.8 Gy | 50% | 20 Gy | | 529 | Anus | | | | | | | | | | | | | | | |
| | 1.8 Gy | 35% | 30 Gy | | 529 | Anus | | | | | | | | | | | | | | | |
| | 1.8 Gy | 5% | 40 Gy | | 529 | Anus | | | | | | | | | | | | | | | |
| Femoral Head | 1.8 Gy | 50% | 30 Gy | | 529 | Anus | Femoral head | Adult | 42-50 Gy | | | | | | | | | | | | |
| | 1.8 Gy | 15% | 30 Gy | | 418 | Endometrial | | Slipped epiphysis | <25 Gy | | | | | | | | | | | | |
| | 1.8 Gy | 40% | 40 Gy | | 822 | Rectum | | Avascular necrosis | 30-40 Gy | | | | | | | | | | | | |
| | 1.8 Gy | 35% | 40 Gy | | 529 | Anus | | | | | | | | | | | | | | | |
| | 1.8 Gy | 25% | 45 Gy | | 822 | Rectum | | | | | | | | | | | | | | | |
| | 1.8 Gy | 10% | 50 Gy | | 534 | Postop prostate | | | | | | | | | | | | | | | |
| | 1.8 Gy | 5% | 44 Gy | | 529 | Anus | | | | | | | | | | | | | | | |
| | 1.8 Gy | 5% | 50 Gy | | PMID 18947938 | RTOG Prostate Group Consensus 2009 | | | | | | | | | | | | | | | |
| | 2 Gy | 5% | 60 Gy | | 630 | Sarcoma | | | | | | | | | | | | | | | |
| | 1.8 Gy | | | 50 Gy | 822 | Rectum | | | | | | | | | | | | | | | |
| | 1.8 Gy | | | 45 Gy | 712 | Bladder | | | | | | | | | | | | | | | |
| Iliac crests | 1.8 Gy | 50% | 30 Gy | | 529 | Anus | | | | | | | | | | | | | | | |
| | 1.8 Gy | 35% | 40 Gy | | 529 | Anus | | | | | | | | | | | | | | | |
| | 1.8 Gy | 5% | 50 Gy | | 529 | Anus | | | | | | | | | | | | | | | |
| Large Bowel | 1.8 Gy | 50% | 35 Gy | | 529 | Anus | | | | | | | | | | Colon | | | | | |
| | 1.8 Gy | 35% | 40 Gy | | 529 | Anus | | | | Penile bulb | Mean dose to 95% gland | <50 Gy | | <35% | Severe erectile dysfunction | obstruction, perforation, ulceration | 4500 – 5500 | 5500 – 6500 | | | |
| | 1.8 Gy | 5% | 50 Gy | | 529 | Anus | | | | | D90 | <50 Gy | | <35% | Severe erectile dysfunction | | | | | | |
| Penile Bulb | 1.8 Gy | Mean | 52.5 Gy | | 415 | Prostate | | | | | D60-70 | <70 Gy | | <55% | Severe erectile dysfunction | | | | | | |

| RTOG data | | | | | | | Handbook | | | QUANTEC data | | | | | | Emami Data | | | | | | | Milano Data | | |
|---|---|---|---|---|---|---|---|---|---|---|---|---|---|---|---|---|---|---|---|---|---|---|---|---|---|
| Critical Structure | Dose/fx | Vol. | Dose | Max. Dose | Protocol | Treated organ | Organ | Partial Organ | Tolerance (1.8 – 2.0 Gy/fx) | Critical Structure | Vol. | Dose/ Vol. | Max. Dose | Toxicity Rate | Toxicity Endpoint | Organ | TD 5/5 | | | TD 50/5 | | | Organ | Dose tolerance | End point |
| Rectum | 1.8 Gy | 60% | 30 Gy | | 418 | Endometrial | Rectum | Whole | 60 Gy | Rectum | V50 | <50% | | <10% | Grade 3+ toxicity | Rectum | | | | | | | Rectum | V70-80 <= 15 cc | Late grade 2 in < 5-10% |
| | 1.8-2 Gy | 50% | 55 Gy | | 712 | Bladder | | GYN HDR | <70% Point A | | V60 | <35% | | <10% | Grade 3+ toxicity | severe proctitis, necrosis, fistula, stenosis | 6000 | – | – | 8000 | – | – | | V70 <= 20-25% | |
| | 1.8 Gy | 50% | 50 Gy | | 0621, PM ID 18947938 | Prostate | | | | | V65 | <25% | | <10% | Grade 3+ toxicity | | | | | | | | | | |
| | 1.8 Gy | 50% | 60 Gy | | 415 | Prostate | | | | | V70 | <20% | | <10% | Grade 3+ toxicity | | | | | | | | | | |
| | 1.8 Gy | 35% | 65 Gy | | 415 | Prostate | | | | | V75 | <15% | | <10% | Grade 3+ toxicity | | | | | | | | | | |
| | 1.8 Gy | 25% | 66.6 Gy | | 621 | Prostate | | | | | | | | | | | | | | | | | | | |
| | 1.8 Gy | 25% | 70 Gy | | 415 | Prostate | | | | | | | | | | | | | | | | | | | |
| | 1.8 Gy | 15% | 75 Gy | | 415 | Prostate | | | | | | | | | | | | | | | | | | | |
| | 1.8 Gy | 45% | 40 Gy | | 534 | Postop prostate | | | | | | | | | | | | | | | | | | | |
| | 1.8 Gy | 25% | 65 Gy | | 534 | Postop prostate | | | | | | | | | | | | | | | | | | | |
| | 1.8 Gy | 20% | 70 Gy | | PMID 18947938 | RTOG Prostate Group Consensus 2009 | | | | | | | | | | | | | | | | | | | |

| RTOG data | | | | | | | Handbook | | | QUANTEC data | | | | | | | Emami Data | | Milano Data | | |
|---|---|---|---|---|---|---|---|---|---|---|---|---|---|---|---|---|---|---|---|---|---|
| Critical Structure | Dose/fx | Vol. | Dose | Max. Dose | Protocol | Treated organ | Organ | Partial Organ | Tolerance (1.8 – 2.0 Gy/fx) | Critical Structure | Vol. | Dose/ Vol. | Max. Dose | Toxicity Rate | Toxicity Endpoint | Organ | TD 5/5 | TD 50/5 | Organ | Dose tolerance | End point |
| Small Bowel | 1.8 Gy | 200 cc | 30 Gy | | 529 | Anus | Small bowel | Small volume | 50 Gy | Small bowel (individual loops) | V15 | <120 cc | | <10% | Grade 3+ toxicity | Small intestine | 4000 – 5000 | 5500 – 6000 | | | |
| | 1.8 Gy | 150 cc | 35 Gy | | 529 | Anus | | Whole | <40 Gy | Small bowel (peritoneal cavity) | V45 | <195 cc | | <10% | Grade 3+ toxicity | obstruction, perforation | | | | | |
| | 1.8 Gy | 180 cc | 35 Gy | | 822 | Rectum | | | | | | | | | | | | | | | |
| | 1.8 Gy | 100 cc | 40 Gy | | 822 | Rectum | | | | | | | | | | | | | | | |
| | 1.8 Gy | 20 cc | 45 Gy | | 529 | Anus | | | | | | | | | | | | | | | |
| | 1.8 Gy | 65 cc | 45 Gy | | 822 | Rectum | | | | | | | | | | | | | | | |
| | 1.8 Gy | | | 50 Gy | 0822, 0529 | Rectum, anus | | | | | | | | | | | | | | | |
| | 1.8 Gy | | | 52 Gy | PMID 18947938 | RTOG Prostate Group Consensus 2009 | | | | | | | | | | | | | | | |
| | 1.8 Gy | 30% | 40 Gy | | 418 | Endometrial | | | | | | | | | | | | | | | |
| Skin, longitudinal | 2 Gy | 50% | 20 Gy | | 630 | Sarcoma | | | | | | | | | | | | | | | |
| Testis | 2 Gy | 50% | 3 Gy | | 630 | Sarcoma | Extremity | Circumferential | 20-30 Gy | | | | | | | | | | | | |
| Vulva | 2 Gy | 50% | 30 Gy | | 630 | Sarcoma | Bone marrow | Whole abdomen | <30 Gy | | | | | | | | | | | | |
| Anus | 2 Gy | 50% | 30 Gy | | 630 | Sarcoma | | Ablation | >40 Gy | | | | | | | | | | | | |
| Bone, weight-bearing | 2 Gy | 50% | 50 Gy | | 630 | Sarcoma | Bone | Cortex | 50 Gy | | | | | | | | | | | | |
| Joints | 2 Gy | 50% | 50 Gy | | 630 | Sarcoma | Joint space | Fibrotic constriction | 40-45 Gy | | | | | | | | | | | | |

**Table 4.** Radiation tolerance dose in pelvis

| Tissue | Grade 0 | Grade 1 | Grade 2 | Grade 3 | Grade 4 |
|---|---|---|---|---|---|
| Skin | No change over baseline | Follicular, faint, or dull erythema/epilation/dry desquamation/decreased sweating | Tender or bright erythema, patchy moist desquamation/moderate edema | Confluent, moist desquamation other than skin folds, pitting edema | Ulceration, hemorrhage, necrosis |
| Mucosal membrane | No change over baseline | Injection/may experience mild pain not requiring analgesic | Patchy mucositis which may produce an inflammatory serosanguinitis discharge/may experience moderate pain requiring analgesia | Confluent fibrinous mucositis/may include severe pain requiring narcotic | Ulceration, hemorrhage, or necrosis |
| Eye | No change | Mild conjunctivitis with or without scleral injection/increased tearing | Moderate conjunctivitis with or without keratitis requiring steroids and/or antibiotics/dry eye requiring artificial tears/iritis with photophobia | Severe keratitis with corneal ulceration/objective decrease in visual acuity or in visual fields/acute glaucoma/panophthalmitis | Loss of vision (unilateral or bilateral) |
| Ear | No change over baseline | Mild external otitis with erythema, pruritus, secondary to dry desquamation not requiring medication. Audiogram unchanged from baseline | Moderate external otitis requiring topical medication/serious otitis medius/hypoacusis on testing only | Severe external otitis with discharge or moist desquamation/symptomatic hypoacusis/tinnitus, not drug related | Deafness |
| Salivary gland | No change over baseline | Mild mouth dryness/slightly thickened saliva/may have slightly altered taste such as metallic taste/these changes not reflected in alteration in baseline feeding behavior, such as increased use of liquids with meals | Moderate to complete dryness/thick, sticky saliva/markedly altered taste | | Acute salivary gland necrosis |
| Pharynx and esophagus | No change over baseline | Mild dysphagia or odynophagia/may require topical anesthetic or non-narcotic analgesics/may require soft diet | Moderate dysphagia or odynophagia/may require narcotic analgesics/may require puree or liquid diet | Severe dysphagia or odynophagia with dehydration or weight loss (>15% from pre-treatment baseline) requiring N-G feeding tube, I.V. fluids, or hyperalimentation | Complete obstruction, ulceration, perforation, fistula |
| Larynx | No change over baseline | Mild or intermittent hoarseness/cough not requiring antitussive/erythema of mucosa | Persistent hoarseness but able to vocalize/referred ear pain, sore throat, patchy fibrinous exudate or mild arytenoid edema not requiring narcotic/cough requiring antitussive | Whispered speech, throat pain, or referred ear pain requiring narcotic/confluent fibrinous exudate, marked arytenoid edema | Marked dyspnea, stridor, or hemoptysis with tracheostomy or intubation necessary |
| CNS | No change | Fully functional status (i.e., able to work) with minor neurologic findings, no medication needed | Neurologic findings present sufficient to require home case/nursing assistance may be required/medications including steroids/anti-seizure agents may be required | Neurologic findings requiring hospitalization for initial management | Serious neurologic impairment which includes paralysis, coma, or seizures >3 per week despite medication/hospitalization required |

**Table 5.** Summary of RTOG acute toxicity criteria for head and neck region.

| Organ/Tissue | Grade 0 | Grade 1 | Grade 2 | Grade 3 | Grade 4 |
|---|---|---|---|---|---|
| Upper G.I | No change | Anorexia with <=5% weight loss from pretreatment baseline/nausea not requiring antiemetics/abdominal discomfort not requiring parasympatholytic drugs or analgesics | Anorexia with <=15% weight loss from pretreatment baseline/nausea and/or vomiting requiring antiemetics/abdominal pain requiring analgesics | Anorexia with >15% weight loss from pretreatment baseline or requiring N-G tube or parenteral support. Nausea and/or vomiting requiring tube or parenteral support/abdominal pain, severe despite medication/hematemesis or melena/abdominal distention (flat plate radiograph demonstrates distended bowel loops) | Ileus, subacute or acute obstruction, performation, GI bleeding requiring transfusion/abdominal pain requiring tube decompression or bowel diversion |
| Lower G.I | No change | Increased frequency or change in quality of bowel habits not requiring medication/rectal discomfort not requiring analgesics | Diarrhea requiring parasympatholytic drugs (e.g., Lomotil)/mucous discharge not necessitating sanitary pads/rectal or abdominal pain requiring analgesics | Diarrhea requiring parenteral support/severe mucous or blood discharge necessitating sanitary pads/abdominal distention (flat plate radiograph demonstrates distended bowel loops) | Acute or subacute obstruction, fistula or perforation; GI bleeding requiring transfusion; abdominal pain or tenesmus requiring tube decompression or bowel diversion |
| Lung | No change | Mild symptoms of dry cough or dyspnea on exertion | Persistent cough requiring narcotic, antitussive agents/dyspnea with minimal effort but not at rest | Severe cough unresponsive to narcotic antitussive agent or dyspnea at rest/clinical or radiologic evidence of acute pneumonitis/intermittent oxygen or steroids may be required | Severe respiratory insufficiency/continuous oxygen or assisted ventilation |
| Genitourinary | No change | Frequency of urination or nocturia twice pretreatment habit/dysuria, urgency not requiring medication | Frequency of urination or nocturia that is less frequent than every hour. Dysuria, urgency, bladder spasm requiring local anesthetic (e.g., Pyridium) | Frequency with urgency and nocturia hourly or more frequently/dysuria, pelvis pain, or bladder spasm requiring regular, frequent narcotic/gross hematuria with/without clot passage | Hematuria requiring transfusion/acute bladder obstruction not secondary to clot passage, ulceration, or necrosis |
| Heart | No change over baseline | Asymptomatic but objective evidence of EKG changes or pericardial abnormalities without evidence of other heart disease | Symptomatic with EKG changes and radiologic findings of congestive heart failure or pericardial disease/no specific treatment required | Congestive heart failure, angina pectoris, pericardial disease responding to therapy | Congestive heart failure, angina pectoris, pericardial disease, arrhythmias not responsive to non-surgical measures |

**Table 6.** Summary of RTOG acute toxicity criteria for body region.

http://dx.doi.org/10.5772/60846

| Organ/Tissue | Grade 0 | Grade 1 | Grade 2 | Grade 3 | Grade 4 | Organ/Tissue |
|---|---|---|---|---|---|---|
| Subcutaneous tissue | None | Slight atrophy; pigmentation change; some hair loss | Patch atrophy; moderate telangiectasia; total hair loss | Marked atrophy; gross telangiectasia | Ulceration | Death related to radiation effects |
| Mucosis membrane | None | slight induration (fibrosis), and loss of subcutaneous fat | Moderate fibrosis but asymptomatic; slight field contracture; <10% linear reduction | Severe induration and loss of subcutaneous tissue; field contracture > 10% linear measurement | Necrosis | Death related to radiation effects |
| Mucosis membrane | None | Slight atrophy and dryness | Moderate atrophy and telangiectasia; little mucous | Marked atrophy with complete dryness; severe telangiectasia | Ulceration | Death related to radiation effects |
| Salivary gland | None | Slight dryness of mouth; good response on stimulation | Moderate dryness of mouth; poor response on stimulation | Complete dryness of mouth; no response on stimulation | Fibrosis | Death related to radiation effects |
| Spinal cord | None | Mild L'Hermitte's syndrome | Severe L'Hermitte's syndrome | Objective neurological findings at or below cord level treated | Mono, para quadriplegia | Death related to radiation effects |
| Brain | None | Mild headache; slight lethargy | Moderate headache; great lethargy | Severe headaches; severe CNS dysfunction (partial loss of power or dyskinesia) | Seizures or paralysis; coma | Death related to radiation effects |
| Eye | None | Asymptomatic cataract; minor corneal ulceration or keratitis | Symptomatic cataract; moderate corneal ulceration; minor retinopathy or glaucoma | Severe keratitis; severe retinopathy or detachment severe glaucoma | Panophthalmitis/blindness | Death related to radiation effects |
| Larynx | None | Hoarseness; slight arytenoid edema | Moderate arytenoid edema; chondritis | Severe edema; severe chondritis | Necrosis | Death related to radiation effects |

**Table 7.** Summary of RTOG late toxicity criteria for head and neck regions.

| Organ/Tissue | Grade 0 | Grade 1 | Grade 2 | Grade 3 | Grade 4 | Grade 5 |
|---|---|---|---|---|---|---|
| Lung | None | Asymptomatic or mild symptoms (dry cough); slight radiographic appearances | Moderate symptomatic fibrosis or pneumonitis (severe cough); low grade fever; patchy radiographic appearances | Severe symptomatic fibrosis or pneumonitis; dense radiographic changes | Severe respiratory insufficiency/continuous O2/assisted ventilation | Death related to radiation effects |
| Heart | None | Asymptomatic or mild symptoms; transient T wave inversion and ST changes; sinus tachycardia >110 | Moderate angina on effort; mild pericarditis; normal heart size; persistent abnormal T wave and ST changes; low ORS | Severe angina; pericardial effusion; constrictive pericarditis; moderate heart failure; cardiac enlargement; EKG abnormalities | Tamponade/severe heart failure/severe constrictive pericarditis | Death related to radiation effects |
| Esophagus | None | Mild fibrosis; slight difficulty in swallowing solids; no pain on swallowing | Unable to take solid food normally; swallowing semi-solid food; dilation may be indicated | Severe fibrosis; able to swallow only liquids; may have pain on swallowing; dilation required | Necrosis/perforation fistula | Death related to radiation effects |
| Small/large intestine | None | Mild diarrhea; mild cramping; bowel movement 5 times daily; slight rectal discharge or bleeding | Moderate diarrhea and colic; bowel movement >5 times daily; excessive rectal mucus or intermittent bleeding | Obstruction or bleeding, requiring surgery | Necrosis/perforation fistula | Death related to radiation effects |
| Liver | None | Mild lassitude; nausea, dyspepsia; slightly abnormal liver function | Moderate symptoms; some abnormal liver function tests; serum albumin normal | Disabling epatitis insufficiency; liver function tests grossly abnormal; low albumin; edema or ascites | Necrosis/hepatic coma or encephalopathy | Death related to radiation effects |
| Kidney | None | Transient albuminuria; no hypertension; mild impairment of renal function; urea 25–35 mg%; creatinine 1.5–2.0 mg%; Creatinine clearance > 75% | Persistent moderate albuminuria (2+); mild hypertension; no related anemia; moderate impairment of renal function; urea > 36–60mg%; creatinine clearance (50–74%) | Severe albuminuria; severe hypertension persistent anemia (< 10%); severe renal failure; urea > 60 mg%; creatinine > 4.0 mg%; creatinine clearance < 50% | Malignant hypotension; uremic coma/urea > 100% | Death related to radiation effects |
| Bladder | None | Slight epithelial atrophy; minor telangiectasia (microscopic hematuria) | Moderate frequency; generalized telangiectasia; intermittent macroscopic hematuria | Severe frequency and dysuria; severe generalized telangiectasia (often with petechiae); frequent hematuria; reduction in bladder capacity (< 150 cc) | Necrosis/contracted bladder (capacity < 100 cc); severe hemorrhagic cystitis | Death related to radiation effects |
| Bone | None | Asymptomatic; no growth retardation; reduced bone density | Moderate pain or tenderness; growth retardation; irregular bone sclerosis | Severe pain or tenderness; complete arrest of bone growth; dense bone sclerosis | Necrosis/spontaneous fracture | Death related to radiation effects |
| Joint | None | Mild joint stiffness; slight limitation of movement | moderate stiffness; intermittent or moderate joint pain; moderate limitation of movement | Severe joint stiffness; pain with severe limitation of movement | Necrosis/complete fixation | Death related to radiation effects |

**Table 8.** Summary of RTOG late toxicity criteria by body region.

# 6. Radiation treatment plan analysis programs

In modern radiation therapy, physical dose indices, such as mean doses, dose-volume histograms (DVHs), and isodose distribution charts, are often used for treatment plan evaluation. DVHs provide dose volume coverage information. However, they fail to provide information regarding hot spots and dose homogeneity. When reviewing physical dose indices, the resulting biological objectives, such as tumor control rate and normal tissue complication probability, must be indirectly estimated based on clinical experience and knowledge. In some competing plans, it is possible that a similar mean dose, maximum dose, or minimum dose might have significantly different radiobiological outcomes. To facilitate the direct and accurate comparison and ranking of treatment plans, radiobiological models for treatment plan evaluation have been introduced. These radiobiological models are based on the idea that the radio-sensitivity of different organs should be taken into account. As a result, the physical dose delivered to an organ is directly associated with the dose–response probability of inducing complications in normal tissues. Many programs have been designed and developed to calculate both dosimetrical and biological indices, as shown in Table 9 [10-29].

# 7. Multidisciplinary strategies: Planning decision support concept

## 7.1. Methods could be used for planning a decision support system

In this section, we highlight dosimetrical and biological models in radiation oncology treatment planning, with focus on the methodological aspects of prediction model development. In radiation treatment planning analysis, dose volume histograms were the most widely used quantitative results. To comprehensively evaluate a certain DVH, we proposed several dosimetrical and biological models in the earlier sections. For dosimetrical models, there were PTTV, CI, and TCI for target coverage index, and MHI, HI for homogeneity index and COSI, QF, and CQI for overall index. For radiobiological models, there were TCP and NTCP for tumor or critical structures, representatively. There were still other factors like treatment time, planning time, or overall moniter unites irradiated in patients could be helpful for making more reasonable decision. Some characteristic prognostic and predictive factors like radiation-induced organ toxicities were discussed in earlier sections. We also enumerate the normal tissue tolerance criteria including QUENTEC and EMAMI database.

## 7.2. The need of plan decision support concept in RT

With the emergence of individualized medicine and the increasing amount and complexity of available medical data, a growing need exists for the development of planning decision-support systems based on prediction models of treatment outcome [55-57]. In radiation oncology, these models combine both predictive and prognostic data factors from dosimetrical, biological, imaging, and other sources to achieve the highest accuracy to predict tumor response and follow-up event rates. The central challenge, however, is how to integrate diverse, multimodal information (imaging, dosimetrical, biological, and other data) in a quantitative manner to provide specific clinical predictions that accurately and robustly estimate patient outcomes as a function of the possible decisions. Currently, many prediction models are being published that consider factors related to disease and treatment, but without standardized assessments of their robustness, reproducibility, or clinical utility [58]. Consequently, these prediction models might not be suitable for clinical decision-support systems for routine care.

**Review of previous programs**

| | Input system | | | Dicom RT platform | | Plan comparison | | | | Plan analysis | | | | Program features | | Paper publication | | | |
|---|---|---|---|---|---|---|---|---|---|---|---|---|---|---|---|---|---|---|---|
| | | | | | | Physical | | Biological | | | | Statistical analysis | | | | | | | |
| Program | Patient information | Data format | Data compatibility | Compatible with PACS | 3D image module | DVH calculator | Physical index | TCP/NTCP | Overall | Multi-RTP | Analysis database | Normal statistic | Survival statistic | Independence from GUI | Platform | Author | Paper | Year | Others |
| HART | × | AAPM/ RTOG, DicomRT | Pinnacle | × | √ | √ | √ | √ | × | × | × | × | × | × | MatLab | Anil Pyakuryal | (23) | 2010 | http://www2.uic.edu/~apyaku1/ |
| CERR | × | AAPM/ RTOG, DicomRT (toolbox) | Pinnacle, expand function with DicomRT toolbox | × | √ | √ | × | × | × | × | √ | × | × | √ | MatLab, Fortran, C/C++, Java | Joseph O. Deasy | (10) | 2003 | http://www.cerr.info/about.php |
| DREES | √ | Matlab's human-readable data structures | No | × | × | × | × | √ | × | × | × | × | √ | × | MatLab | Joseph O. Deasy | (11) | 2006 | http://cerr.info/drees/about.php |
| EUD-based mathematical model | × | DVH file | specialized format | × | × | × | × | √ | × | × | × | × | × | × | MatLab | Andrzej Niemierko | (12) | 2007 | - |
| EUCLID | √ | AAPM/ RTOG, DicomRT | LANTIS and IMPAC based on DRESS | × | × | × | × | √ | × | × | × | × | √ | × | MatLab | Olivier Gayou | (13) | 2007 | - |
| Dose Volume Histogram Analyzer | × | Eclipse, Pinnacle, Tomo, DVH files | Eclipse, Pinnacle, Tomo | × | × | × | × | √ | × | × | × | × | × | × | MatLab | Jin Sung Kim | (17) | 2008 | http://mpjinsung.tistory.com/entry/DVH-Analyzer-v10 |

**Review of previous programs**

| Program | Input system | | | Dicom RT platform | | Plan comparison | | | | | Plan analysis | | | Program features | | Paper publication | | | |
|---|---|---|---|---|---|---|---|---|---|---|---|---|---|---|---|---|---|---|---|
| | | | | | | Physical | | Biological | | | | Statistical analysis | | | | | | | |
| | Patient information | Data format | Data compatibility | Compatible with PACS | 3D image module | DVH calculator | Physical index | TCP/NTCP | Overall | Multi-RTP | Analysis database | Normal statistic | Survival statistic | Independence from GUI | Platform | Author | Paper | Year | Others |
| computational platform | √ | AAPM/ RTOG, DicomRT | compatible with ARIA, different RTP | √ | √ | × | × | √ | × | × | √ | × | √ | √ | MatLab, Web, ARIA | Dezhi Liu | (18) | 2009 | - |
| BIOPLAN | × | DVH file | DVH file | × | × | × | × | √ | × | × | × | × | × | × | Microsoft Visual Basic | B. SANCHEZ -NIETO | (52) | 2000 | - |
| Anonymous | × | DVH file | DVH file | × | × | × | × | √ | × | × | × | × | × | × | MatLab | Arun S. Oinam | (21) | 2011 | - |
| SlicerRT | √ | Dicom RT | compatible with cormercial RTP | × | √ | × | × | √ | × | × | × | × | × | √ | C++ | Csaba Pinter | (22) | 2012 | https://www.assembla.com/spaces/slicerrt/wiki |
| MERT | √ | Dicom RT | This was RTP | × | √ | √ | × | √ | × | × | × | × | × | √ | Multi format(MC) | Murat Surucu | (26) | 2010 | - |
| DIRART | × | Dicom RT | use CERR import engine | × | √ | × | × | √ | × | × | × | × | × | √ | MatLab | Deshan Yang | (27) | 2010 | http://code.google.com/p/dirart/ |
| SABER | × | Dicom RT | Eclipse | × | × | × | × | √ | × | × | × | × | × | × | MatLab | Jay Burmeister | (29) | 2010 | - |
| DICOM RT toolbox | √ | Dicom RT | Helax TMS | × | √ | √ | × | × | × | × | × | × | × | × | MatLab | Spezi E | (24) | 2002 | - |

| Review of previous programs | | | | | | | | | | | | | | | | | | | |
|---|---|---|---|---|---|---|---|---|---|---|---|---|---|---|---|---|---|---|---|
| | Input system | | | Dicom RT platform | | Plan comparison | | | | | Plan analysis | | | Program features | | Paper publication | | | |
| | | | | | | Physical | | Biological | | | | Statistical analysis | | | | | | | |
| Program | Patient information | Data format | Data compatibility | Compatible with PACS | 3D image module | DVH calculator | Physical index | TCP/NTCP | Overall | Multi-RTP | Analysis database | Normal statistic | Survival statistic | Independence from GUI | Platform | Author | Paper | Year | Others |
| BEUDcal | × | DVH file | DVH file | × | × | √ | × | √ | × | × | × | × | × | × | MatLab | Su FC | (25) | 2010 | - |
| Comp Plan | × | DVH file | DVH file in Excel | × | × | × | × | √ | × | × | × | × | × | × | MatLab | Holloway LC | (15) | 2012 | - |
| CalcNTCP | × | Manual input | Manual input | × | × | × | × | √ | × | × | × | × | × | × | Visual Basic | Khan HA | (16) | 2007 | - |
| RADBIOMOD | × | DVH file | Manual input | × | × | × | √ | √ | × | × | × | × | × | × | Microsoft Ecel | Chang JH | (53) | 2011 | https://sites.google.com/site/radbiomod/home |
| BioSuite | × | DVH file | Pinnacle, Eclups | × | × | √ | × | √ | × | × | × | × | × | √ | C++ | J Uzan | (54) | 2012 | - |
| RTToolbox | √ | Dicom RT | Virtuos, our in-house developed planning system | × | √ | √ | × | √ | × | × | √ | × | × | √ | C++ | Lanlan, Zhang | (28) | 2013 | - |

**Table 9.** Review of previous programs

Decision making in radiotherapy is mainly based on clinical features, such as the patient performance status, organ function, and grade and extent of the tumor (e.g., as defined by the TNM system). In almost all studies, such features have been found to be prognostic for survival and development of toxicity [59, 60]. Consequently, these features should be evaluated in building robust and clinically acceptable radiotherapy prognostic and predictive models. Moreover, measurement of some clinical variables, such as performance status, can be captured with minimal effort.

Toxicity measurements and scoring should also build on validated scoring systems, such as the Common Terminology Criteria for Adverse Events (CTCAE), which can be scored by the physician or patient [50, 61]. Indeed, a meta-analysis showed that high-quality toxicity assessments from observational trials are similar to those of randomized trials. [45, 46] However, a prospective protocol must clarify which scoring system was used and how changes in toxicity score were dealt with over time with respect to treatment. Finally, to ensure a standardized interpretation, the reporting of clinical and toxicity data and their analyses should be performed in line with the STROBE (Strengthening the Reporting of Observational Studies in Epidemiology) statement for observational studies and genetic-association studies, which is represented as checklists of items that should be addressed in reports to facilitate the critical appraisal and interpretation of these types of studies (Figure 7).

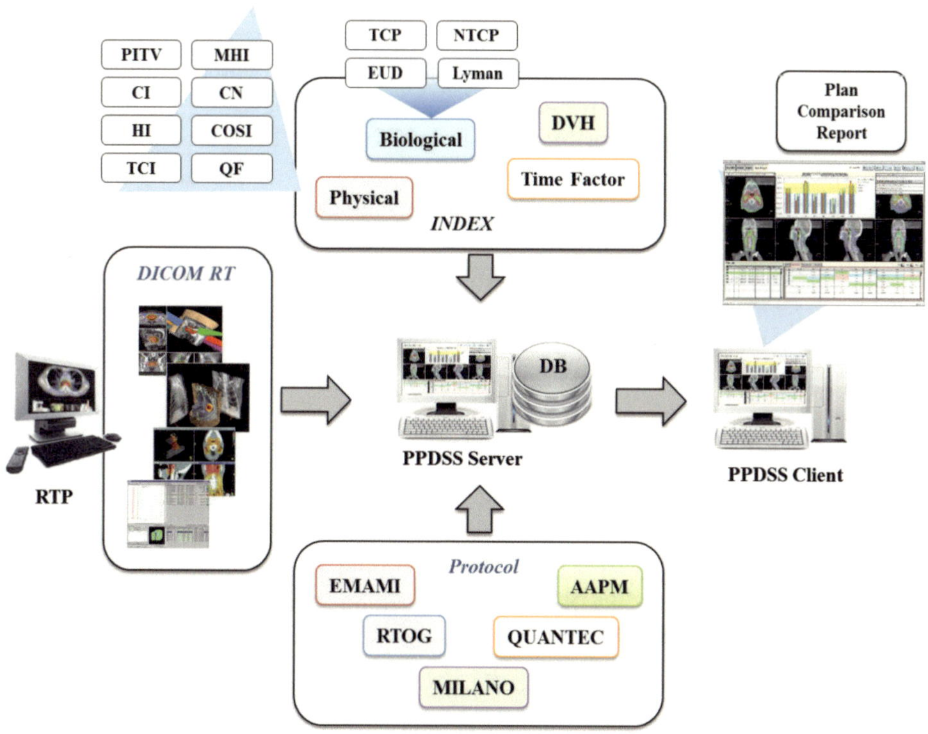


**Figure 7.** Design of planning decision support concept in radiotherapy treatment planning.

Despite the challenges that remain, the vision of predictive models leading to plan decision support concept that are continuously updated via rapid learning on large datasets is clear, and numerous steps have already been taken. These include universal data-quality assurance programs and semantic interoperability issues. However, we believe that this truly innovative journey will lead to necessary improvement of healthcare effectiveness and efficiency. Indeed, investments are being made in research and innovation for health-informatics systems, with an emphasis on interoperability and standards for secured data transfer, which shows that "eHealth" will be among the largest health-care innovations of the coming decade. Accurate, externally validated prediction models are being rapidly developed, whereby multiple features related to the patient's disease are combined into an integrated prediction. The key, however, is standardization—mainly in data acquisition across all areas, including dosimetrical-based and biological-based models, patient preferences, and possible treatments. These crucial features are the basis of validating a plan decision support system, which, in turn, will stimulate developments in rapid-learning health care and will enable the next major advances in shared decision making.

## 8. Conclusion

Plan comparison studies still remain controversial. The main reason for this is because plan parameters, optimization methods, and OAR constraints are difficult to clearly define. Many researchers have focused on the influence of planning parameters on the results of treatment plans [62-64]. For instance, Gutiérrez et al. [65] reported that the use of a field width of 1 cm resulted in dosimetrically superior plans for brain irradiation compared to plans that use a field width of 2.5 cm. More recently, Skorska and Piotrowski studied the influence of treatment-planning parameters on plan qualities for prostate cancer patients using helical tomotherapy [66]. This study revealed that using a field width of 1 cm, instead of 5 cm, leads to decreases in the D20%, D40%, D60%, and D80% of the small intestine by 2.45%, 8.48%, 6.36%, and 5%. This results in 1.22Gy, 4.24Gy, 3.18Gy, and 2.50Gy, respectively, for the prescribed dose of 50 Gy. Another bias of plan comparison studies is that the quality of a planner's abilities and planning techniques may vary. Performing repeat planning processes and using multiple planners to cross check would minimize such bias. The use of OAR dose tolerance guidelines, such as RTOG or QUENTEC protocols, would minimize human error.

Other major issues among plan comparison studies are the method of plan analysis and evaluation. Many studies have focused on developing a simple index that represents the overall quality of plans [14, 19, 41, 42, 67]. However, none of these plans are easily used in a clinic. There is a need for programs that can easily calculate dosimetrical and biological indices [10, 12, 13, 15, 16, 22-25, 28, 68, 78-82].

There is a growing trend of studying the relationships between treatment plan results and clinical outcomes, such as toxicities, survival, and patterns of failure [69-77]. Such studies may help physicians and physicists learn more about the influence of plan results and plan quality on patient treatment.

## Acknowledgements

This chapter was developed by a special working group of the Korea University Medical Physics Lab from the department of radiation oncology, college of medicine, Korea University, Seoul, 136-705, Korea. Members of the planning index study working group include Kwang Hyeon Kim, M.S., Kyung Hwan Chang, Ph.D., and Jang Bo Shim, M.S.

## Author details

Suk Lee*, Yuan Jie Cao and Chul Yong Kim

*Address all correspondence to: sukmp@korea.ac.kr

Department of Radiation Oncology, College of Medicine, Korea University, Seoul, Korea

## References

[1] Bernier J, Hall EJ, Giaccia A. Radiation oncology: a century of achievements. Nature Reviews Cancer. 2004;4(9):737–47.

[2] Thariat J, Hannoun-Levi J-M, Sun Myint A, Vuong T, Gerard J-P. Past, present, and future of radiotherapy for the benefit of patients. Nature Reviews Clinical Oncology. 2013;10(1):52–60.

[3] Rembielak A, Woo TCS. Intensity-modulated radiation therapy for the treatment of pediatric cancer patients. Nature Clinical Practical Oncology. 2005;2(4):211–7.

[4] Mittauer K, Lu B, Yan G, Kahler D, Gopal A, Amdur R, et al. A study of IMRT planning parameters on planning efficiency, delivery efficiency, and plan quality. Medical Physics. 2013;40(6):061704.

[5] Verellen D, Ridder MD, Linthout N, Tournel K, Soete G, Storme G. Innovations in image-guided radiotherapy. Nature Reviews Cancer. 2007;7(12):949–60.

[6] Jaffray DA. Image-guided radiotherapy: from current concept to future perspectives. Nature Reviews Clinical Oncology. 2012;9(12):688–99.

[7] Lo SS, Fakiris AJ, Chang EL, Mayr NA, Wang JZ, Papiez L, et al. Stereotactic body radiation therapy: a novel treatment modality. Nature Reviews Clinical Oncology. 2010;7(1):44–54.

[8] Galvin JM, Ezzell G, Eisbrauch A, Yu C, Butler B, Xiao Y, et al. Implementing IMRT in clinical practice: a joint document of the American Society for Therapeutic Radiol-

ogy and Oncology and the American Association of Physicists in Medicine. International Journal of Radiation Oncology, Biology, Physics. 2004;58(5):1616–34.

[9] Cadman P, Bassalow R, Sidhu NP, Ibbott G, Nelson A. Dosimetric considerations for validation of a sequential IMRT process with a commercial treatment planning system. Physics in Medicine and Biology. 2002;47(16):3001–10.

[10] Deasy JO, Blanco AI, Clark VH. CERR: a computational environment for radiotherapy research. Medical Physics. 2003;30(5):979.

[11] El Naqa I, Suneja G, Lindsay PE, Hope AJ, Alaly JR, Vicic M, et al. Dose response explorer: an integrated open-source tool for exploring and modelling radiotherapy dose-volume outcome relationships. Physics in Medicine and Biology. 2006;51(22): 5719–35.

[12] Gay HA, Niemierko A. A free program for calculating EUD-based NTCP and TCP in external beam radiotherapy. Physica Medica: PM: An International Journal Devoted to the Applications of Physics to Medicine and Biology: Official Journal of the Italian Association of Biomedical Physics. 2007;23(3–4):115–25.

[13] Gayou O, Parda DS, Miften M. EUCLID: an outcome analysis tool for high-dimensional clinical studies. Physics in Medicine and Biology. 2007;52(6):1705–19.

[14] Henriquez FC, Castrillon SV. A quality index for equivalent uniform dose. Journal of Medical Physics/Association of Medical Physicists of India. 2011;36(3):126–32.

[15] Holloway LC, Miller JA, Kumar S, Whelan BM, Vinod SK. Comp Plan: a computer program to generate dose and radiobiological metrics from dose-volume histogram files. Medical Dosimetry: Official Journal of the American Association of Medical Dosimetrists. 2012;37(3):305–9.

[16] Khan HA. CalcNTCP: a simple tool for computation of normal tissue complication probability (NTCP) associated with cancer radiotherapy. International Journal of Radiation Biology. 2007;83(10):717–20.

[17] Kim JS. A dose volume histogram analyzer program for external beam radiotherapy. Journal of the Korean Society of Radiology. 2008.

[18] Liu D, Ajlouni M, Jin J-Y, Ryu S, Siddiqui F, Patel A, et al. Analysis of outcomes in radiation oncology: an integrated computational platform. Medical Physics. 2009;36(5):1680.

[19] Luxton G, Keall PJ, King CR. A new formula for normal tissue complication probability (NTCP) as a function of equivalent uniform dose (EUD). Physics in Medicine and Biology. 2008;53(1):23–36.

[20] NAHUM BS-NaAE. BIOPLAN: software for the biological evaluation of radiation therapy. Medical Dosimetry. 2000.

[21] Oinam AS, Singh L, Shukla A, Ghoshal S, Kapoor R, Sharma SC. Dose volume histogram analysis and comparison of different radiobiological models using in-house de-

veloped software. Journal of Medical Physics/Association of Medical Physicists of India. 2011;36(4):220–9.

[22] Pinter C, Lasso A, Wang A, Jaffray D, Fichtinger G. SlicerRT: radiation therapy research toolkit for 3D Slicer. Medical Physics. 2012;39(10):6332–8.

[23] Pyakuryal A. A computational tool for the efficient analysis of dose-volume histograms for radiation therapy treatment plans. Journal of Applied Clinical Medical Physics. 2010.

[24] Spezi E, Lewis DG, Smith CW. A DICOM-RT-based toolbox for the evaluation and verification of radiotherapy plans. Physics in Medicine and Biology. 2002;47(23): 4223–32.

[25] Su FC, Mavroidis P, Shi C, Ferreira BC, Papanikolaou N. A graphic user interface toolkit for specification, report and comparison of dose-response relations and treatment plans using the biologically effective uniform dose. Computer Methods and Programs in Biomedicine. 2010;100(1):69–78.

[26] Surucu M, Klein EE, Mamalui-Hunter M, Mansur DB, Low DA. Planning tools for modulated electron radiotherapy. Medical Physics. 2010;37(5):2215.

[27] Yang D. Technical note: DIRART – a software suite for deformable image registration and adaptive radiotherapy research. Medical Physics. 2010.

[28] Zhang L, Hub M, Mang S, Thieke C, Nix O, Karger CP, et al. Software for quantitative analysis of radiotherapy: overview, requirement analysis and design solutions. Computer Methods and Programs in Biomedicine. 2013;110(3):528–37.

[29] Zhao B, Joiner MC, Orton CG, Burmeister J. "SABER": a new software tool for radiotherapy treatment plan evaluation. Medical Physics. 2010;37(11):5586.

[30] Oliver M, Ansbacher W, Beckham WA. Comparing planning time, delivery time and plan quality for IMRT, RapidArc and tomotherapy. J Appl Clin Med Phys. 2009;10(4):3068.

[31] Rao M, Yang W, Chen F, Sheng K, Ye J, Mehta V, et al. Comparison of Elekta VMAT with helical tomotherapy and fixed field IMRT: plan quality, delivery efficiency and accuracy. Medical Physics. 2010;37(3):1350–9.

[32] Reynders T, Tournel K, De Coninck P, Heymann S, Vinh-Hung V, Van Parijs H, et al. Dosimetric assessment of static and helical TomoTherapy in the clinical implementation of breast cancer treatments. Radiotherapy and Oncology. 2009;93(1):71–9.

[33] Trofimov A, Nguyen PL, Coen JJ, Doppke KP, Schneider RJ, Adams JA, et al. Radiotherapy treatment of early-stage prostate cancer with IMRT and protons: a treatment planning comparison. International Journal of Radiation Oncology, Biology, Physics. 2007;69(2):444–53.

[34] Vargas C, Fryer A, Mahajan C, Indelicato D, Horne D, Chellini A, et al. Dose-volume comparison of proton therapy and intensity-modulated radiotherapy for prostate

cancer. International Journal of Radiation Oncology, Biology, Physics. 2008;70(3):744–51.

[35] Ceylan C, Kucuk N, Bas Ayata H, Guden M, Engin K. Dosimetric and physical comparison of IMRT and CyberKnife plans in the treatment of localized prostate cancer. Reports of Practical Oncology and Radiotherapy: Journal of Greatpoland Cancer Center in Poznan and Polish Society of Radiation Oncology. 2010;15(6):181–9.

[36] Kumar SA, Holla R, Sukumar P, Padmanaban S, Vivekanandan N. Treatment planning and dosimetric comparison study on two different volumetric modulated arc therapy delivery techniques. Reports of Practical Oncology and Radiotherapy: Journal of Greatpoland Cancer Center in Poznan and Polish Society of Radiation Oncology. 2012;18(2):87–94.

[37] Poon DM, Kam M, Leung CM, Chau R, Wong S, Lee WY, et al. Dosimetric advantages and superior treatment delivery efficiency of RapidArc over conventional intensity-modulated radiotherapy in high-risk prostate cancer involving seminal vesicles and pelvic nodes. Clinical oncology (Royal College of Radiologists (Great Britain)). 2013;25(12):706–12.

[38] Widesott L, Pierelli A, Fiorino C, Lomax AJ, Amichetti M, Cozzarini C, et al. Helical tomotherapy vs. intensity-modulated proton therapy for whole pelvis irradiation in high-risk prostate cancer patients: dosimetric, normal tissue complication probability, and generalized equivalent uniform dose analysis. International Journal of Radiation Oncology, Biology, Physics. 2011;80(5):1589–600.

[39] Shaw E, Kline R, Gillin M, Souhami L, Hirschfeld A, Dinapoli R, et al. Radiation Therapy Oncology Group: radiosurgery quality assurance guidelines. International Journal of Radiation Oncology, Biology, Physics. 1993;27(5):1231–9.

[40] Knoos T, Kristensen I, Nilsson P. Volumetric and dosimetric evaluation of radiation treatment plans: radiation conformity index. International Journal of Radiation Oncology, Biology, Physics. 1998;42(5):1169–76.

[41] Yoon M, Park SY, Shin D, Lee SB, Pyo HR, Kim DY, et al. A new homogeneity index based on statistical analysis of the dose-volume histogram. Journal of Applied Clinical Medical Physics. 2007;8(2):9–17.

[42] van't Riet A, Mak AC, Moerland MA, Elders LH, van der Zee W. A conformation number to quantify the degree of conformality in brachytherapy and external beam irradiation: application to the prostate. International Journal of Radiation Oncology, Biology, Physics. 1997;37(3):731–6.

[43] Emami B, Lyman J, Brown A, Coia L, Goitein M, Munzenrider JE, et al. Tolerance of normal tissue to therapeutic irradiation. International Journal of Radiation Oncology, Biology, Physics. 1991;21(1):109–22.

[44] Leung LH, Kan MW, Cheng AC, Wong WK, Yau CC. A new dose-volume-based plan quality index for IMRT plan comparison. Radiotherapy and Oncology. 2007;85(3):407–17.

[45] Menhel J, Levin D, Alezra D, Symon Z, Pfeffer R. Assessing the quality of conformal treatment planning: a new tool for quantitative comparison. Physics in Medicine and Biology. 2006;51(20):5363–75.

[46] Bentzen SM, Constine LS, Deasy JO, Eisbruch A, Jackson A, Marks LB, et al. Quantitative analyses of normal tissue effects in the clinic (QUANTEC): an introduction to the scientific issues. International Journal of Radiation Oncology, Biology, Physics. 2010;76(3 Suppl):S3–9.

[47] Kirwan JM, Symonds P, Green JA, Tierney J, Collingwood M, Williams CJ. A systematic review of acute and late toxicity of concomitant chemoradiation for cervical cancer. Radiotherapy and Oncology. 2003;68(3):217–26.

[48] Maduro JH, Pras E, Willemse PH, de Vries EG. Acute and long-term toxicity following radiotherapy alone or in combination with chemotherapy for locally advanced cervical cancer. Cancer Treatment Reviews. 2003;29(6):471–88.

[49] Trotti A. The evolution and application of toxicity criteria. Seminars in Radiation Oncology. 2002;12(1 Suppl 1):1–3.

[50] Trotti A, Colevas AD, Setser A, Rusch V, Jaques D, Budach V, et al. CTCAE v3.0: development of a comprehensive grading system for the adverse effects of cancer treatment. Seminars in Radiation Oncology. 2003;13(3):176–81.

[51] Cox JD, Stetz J, Pajak TF. Toxicity criteria of the Radiation Therapy Oncology Group (RTOG) and the European Organization for Research and Treatment of Cancer (EORTC). International Journal of Radiation Oncology, Biology, Physics. 1995;31(5): 1341–6.

[52] Sanchez-Nieto B, Nahum AE. BIOPLAN: software for the biological evaluation of. Radiotherapy treatment plans. Medical Dosimetry: Official Journal of the American Association of Medical Dosimetrists. 2000;25(2):71–6.

[53] Chang JH LJD, Hamilton C, Khoo V. An Excel-based radiation biological modelling tool for the evaluation of radiotherapy treatment plans. Radiotherapy and Oncology 2011;99:S564–S5.

[54] Uzan J, Nahum AE. Radiobiologically guided optimisation of the prescription dose and fractionation scheme in radiotherapy using BioSuite. The British Journal of Radiology. 2012;85(1017):1279–86.

[55] Abernethy AP, Etheredge LM, Ganz PA, Wallace P, German RR, Neti C, et al. Rapid-learning system for cancer care. Journal of Clinical Oncology: Official Journal of the American Society of Clinical Oncology. 2010;28(27):4268–74.

[56] Fraass BA, Moran JM. Quality, technology and outcomes: evolution and evaluation of new treatments and/or new technology. Seminars in Radiation Oncology. 2012;22(1):3–10.

[57] Vogelzang NJ, Benowitz SI, Adams S, Aghajanian C, Chang SM, Dreyer ZE, et al. Clinical cancer advances 2011: Annual Report on Progress Against Cancer from the American Society of Clinical Oncology. Journal of Clinical Oncology: Official Journal of the American Society of Clinical Oncology. 2012;30(1):88–109.

[58] Vickers AJ. Prediction models: revolutionary in principle, but do they do more good than harm? Journal of Clinical Oncology: Official Journal of the American Society of Clinical Oncology. 2011;29(22):2951–2.

[59] Dehing-Oberije C, Yu S, De Ruysscher D, Meersschout S, Van Beek K, Lievens Y, et al. Development and external validation of prognostic model for 2-year survival of non-small-cell lung cancer patients treated with chemoradiotherapy. International Journal of Radiation Oncology, Biology, Physics. 2009;74(2):355–62.

[60] Klopp AH, Eifel PJ. Biological predictors of cervical cancer response to radiation therapy. Seminars in Radiation Oncology. 2012;22(2):143–50.

[61] Trotti A, Colevas AD, Setser A, Basch E. Patient-reported outcomes and the evolution of adverse event reporting in oncology. Journal of Clinical Oncology: Official Journal of the American Society of Clinical Oncology. 2007;25(32):5121–7.

[62] Moldovan M, Fontenot JD, Gibbons JP, Lee TK, Rosen, II, Fields RS, et al. Investigation of pitch and jaw width to decrease delivery time of helical tomotherapy treatments for head and neck cancer. Medical Dosimetry: Official Journal of the American Association of Medical Dosimetrists. 2011;36(4):397–403.

[63] Westerly DC, Soisson E, Chen Q, Woch K, Schubert L, Olivera G, et al. Treatment planning to improve delivery accuracy and patient throughput in helical tomotherapy. International Journal of Radiation Oncology, Biology, Physics. 2009;74(4):1290–7.

[64] Woch K SE, Westerly D, Olivera G, Kaptoes J.. Acceleration of tomotherapy treatment delivery by increasing pitch and decreasing modulation. Medical Physics. 2008;35:2636–7.

[65] Gutierrez AN, Westerly DC, Tome WA, Jaradat HA, Mackie TR, Bentzen SM, et al. Whole brain radiotherapy with hippocampal avoidance and simultaneously integrated brain metastases boost: a planning study. International Journal of Radiation Oncology, Biology, Physics. 2007;69(2):589–97.

[66] Skorska M, Piotrowski T. Optimization of treatment planning parameters used in tomotherapy for prostate cancer patients. Physica Medica: PM : An International Journal Devoted to the Applications of Physics to Medicine and Biology: Official Journal of the Italian Association of Biomedical Physics. 2013;29(3):273–85.

[67] Feuvret L, Noel G, Mazeron JJ, Bey P. Conformity index: a review. International Journal of Radiation Oncology, Biology, Physics. 2006;64(2):333–42.

[68] Ebert MA, Haworth A, Kearvell R, Hooton B, Hug B, Spry NA, et al. Comparison of DVH data from multiple radiotherapy treatment planning systems. Physics in Medicine and Biology. 2010;55(11):N337–46.

[69] Semenenko VA, Tarima SS, Devisetty K, Pelizzari CA, Liauw SL. Validation of normal tissue complication probability predictions in individual patient: late rectal toxicity. International Journal of Radiation Oncology, Biology, Physics. 2012;85(4):1103–9.

[70] Taniguchi CM, Murphy JD, Eclov N, Atwood TF, Kielar KN, Christman-Skieller C, et al. Dosimetric analysis of organs at risk during expiratory gating in stereotactic body radiation therapy for pancreatic cancer. International Journal of Radiation Oncology, Biology, Physics. 2012;85(4):1090–5.

[71] Vanasek J, Odrazka K, Dolezel M, Kolarova I, Jarkovsky J, Pavlik T, et al. Statistical analysis of dose–volume profiles and its implication for radiation therapy planning in prostate carcinoma. International Journal of Radiation Oncology, Biology, Physics. 2013;86(4):769–76.

[72] Carillo V, Cozzarini C, Rancati T, Avuzzi B, Botti A, Borca VC, et al. Relationships between bladder dose–volume/surface histograms and acute urinary toxicity after radiotherapy for prostate cancer. Radiotherapy and Oncology. 2014;111(1):100–5.

[73] Hoskin PJ, Rojas AM, Ostler PJ, Hughes R, Bryant L, Lowe GJ. Dosimetric predictors of biochemical control of prostate cancer in patients randomised to external beam radiotherapy with a boost of high dose rate brachytherapy. Radiotherapy and Oncology. 2014;110(1):110–3.

[74] Kestin L, Grills I, Guckenberger M, Belderbos J, Hope AJ, Werner-Wasik M, et al. Dose–response relationship with clinical outcome for lung stereotactic body radiotherapy (SBRT) delivered via online image guidance. Radiotherapy and Oncology. 2014;110(3):499–504.

[75] Nomden CN, de Leeuw AAC, Roesink JM, Tersteeg RJHA, Moerland MA, Witteveen PO, et al. Clinical outcome and dosimetric parameters of chemo-radiation including MRI guided adaptive brachytherapy with tandem-ovoid applicators for cervical cancer patients: a single institution experience. Radiotherapy and Oncology. 2013;107(1): 69–74.

[76] Strom TJ, Wilder RB, Fernandez DC, Mellon EA, Saini AS, Hunt DC, et al. A dosimetric study of polyethylene glycol hydrogel in 200 prostate cancer patients treated with high-dose rate brachytherapy±intensity modulated radiation therapy. Radiotherapy and Oncology. 2014;111(1):126–31.

[77] Thor M, Apte A, Deasy JO, Karlsdóttir À, Moiseenko V, Liu M, et al. Dose/volume–response relations for rectal morbidity using planned and simulated motion-inclusive dose distributions. Radiotherapy and Oncology. 2013; 109(3):388–93.

[78] Comparison of linac-based fractionated stereotactic radiotherapy and tomotherapy treatment plans for intra-cranial tumors, Jang Bo Shim, Suk Lee, Sam Ju Cho, Sang Hoon Lee, Juree Kim, Kwang Hwan Cho, ChulKee Min, Hyun Do Huh, Rena Lee, DaeSik Yang, Young Je Park, Won Seob Yoon, Chul Yong Kim, Soo Il Kwon, Chinese Physics C, 34(11): 1768-1774, Nov., 2010

[79] Optimization of beam orientation and virtual organ delineation for lung IMRT, Kyung Hwan Chang, Suk Lee, Yuan Jie Cao, Jang Bo Shim, Ji Eun Lee, Nam Kwon Lee, Jung Ae Lee, DaeSik Yang, Young Je Park, Won Sup Yoon, and Chul Yong Kim, Sam Ju Cho, Sang Hoon Lee, Woo Chul Kim, ChulKee Min, Kwang Hwan Cho, Hyun Do Huh, Journal of the Korean Physical Society, Volume 64, Issue 7, 1047-1054, 2014 (SCI, 0.476)

[80] Patient performance-based plan parameter optimization for prostate cancer in tomotherapy, Yuan Jie Cao, Suk Lee, Kyung Hwan Chang, Jang Bo Shim, KwangHyeon Kim, et al., Medical Dosimetry, 2015, article in press.

[81] Optimized planning target volume margin in helical tomotherapy for prostate cancer: is there a preferred method?, Yuan Jie Cao, Suk Lee, Kyung Hwan Chang, Jang Bo Shim, KwangHyeon Kim, et al., Journal of the Korean Physical Society, 67(1), 26-32, 2015

[82] Dosimetrical and radiobiological comparison of intensity modulated planning techniques for prostate radiotherapy: a multi-institutional study, Suk Lee, Yuan Jie Cao, Kyung Hwan Chang, Jang Bo Shim, KwangHyeon Kim, et al., Journal of the Korean Physical Society, 2015, article in press.

Chapter 6

# Quality Control of Ionizing Radiation in Radiotherapy

Ernesto Lamanna, Bianco Cataldo, Giulia Marvaso,
Marco D'Andrea and Lidia Strigari

Additional information is available at the end of the chapter

http://dx.doi.org/10.5772/60421

**Abstract**

This work includes the results of our research on the measurement of the dose delivered by an external beam in radiotherapy. The use of scintillating fibers in high-energy experiments produced rapid and reliable results and allows new dosimeters to be built and extends their use to measure the dose of an external beam of electrons, photons, and hadrons in radiotherapy.The chapter starts from the description of the radiation used in radiotherapy, presents the new approaches and then the tools used to perform the quality control of therapeutic beams, and finally shows the characteristics and differences compared to the traditional quality controls by using our results on the scintillating fibers used as a dosimeter. Some care should be taken into account during the collection and processing of data, for the treatment of some systematic errors in the method. In this chapter, we describe the procedure to be followed.

**Keywords:** radiotherapy, IMRT, dosimeter, scintillating fibers

## 1. Introduction

The use of ionizing radiation for the treatment of cancer can be traced back to the discovery of X-rays and radioactive isotopes by WC Roentgen (1895), H. Becquerel (1896), and M. and P. Curie (1898). X-rays have been used in clinical medicine since the early years of the twentieth century [1]. Initially, X-rays used energies required for diagnostic purposes and then higher energies (180–200 kV) of the photons produced by X-ray tubes developed to treat tumors. The ability to treat tumors located in deeper tissues was guaranteed later by the development in the 1950s of linear accelerators of electrons with energies of 4–20 MeV and the production of intense beams of photons with delivery of higher doses [2]. The most important limitation to their use was the excessive radiation in normal tissue surrounding the tumors. To overcome this, the approach of multifield treatment plans has been developed to guarantee the dose necessary for the tumor, thus reducing the high dose to normal tissues. In the same years, the technology developed to produce accelerators of massive particles allowed therapeutic trials of protons and ions to kill cancer cells to be started [3]. The development of more powerful computers allowed treatment plans to be set up with the assistance of the computer, and from 1984, excellent treatment systems with beams of protons and ions for clinical use were realized. The method of treatment with protons and ions is advantageous for its characteristic of maximum energy transfer near the stopping point of the particle. Using electrons or photons the dose delivered is continuously reduced by increasing its depth. For this reason, hadron beams are preferable for deep tumors. Currently, the systems required for the use of hadrons are expensive, and there are few centers in the world in which they are available [4]. The systems that produce beams of electrons and photons are cheaper and have spread widely in health facilities. They are referred to as "conventional radiotherapy." Research for the improvement of their use has produced an impact on the treatment method by introducing new ways to deliver the dose: the intensity-modulated radiation therapy (IMRT) and the volumetric-modulated arc therapy (VMAT). The new methodologies require more accurate detectors for measuring the dose delivered and for assuring quality controls of the system used, with high spatial and temporal resolutions.

## 2. Ionizing radiation used in radiotherapy

Radiotherapy, also called radiation therapy (RT), indicates the treatment of cancer and other diseases with ionizing radiation, which are used to deposit energy in tumor cells and directly or indirectly damage the genetic material (DNA) in the individual cells, making it impossible for them to continue to grow.

One type of radiation therapy commonly used is with photons, "packets" of energy, or particles, which, depending on the amount of energy they possess, destroy cancer cells on the surface of an area or penetrate to tissues deeper in the body. The higher the energy of the photon beam, the deeper the distance at which a given dose is delivered into the target tissue.

Linear accelerators are generally adopted in Radiotherapy departments to focus ionizing radiation on a cancer site, and this modality is called external beam radiotherapy (EBR). With modern radiation equipment, the radiation is focused on the target thanks to a multileaf collimator (MLC) and a complex sequence of their movements aimed at delivering conformal dose distribution using static or dynamic position of leaves at different static or dynamic gantry positions.

All the above delivery modalities are potentially associated to an on board imaging device in order to improve the treatment setup reproducibility, thanks to 2D/3D imaging based on kV or MV-based imaging.

In fact, in radiation therapy, a sharply defined dose distribution minimizes the side effects of treatment because only small amounts of radiation travel to the surrounding tissues. Gamma rays are produced spontaneously as certain elements (such as cobalt 60) release radiation as they decompose, or decay and are another form of photons used in radiotherapy. Each element decays at a specific rate and gives off energy in the form of gamma rays and other particles.

Another technique for delivering radiation to cancer cells relies on the possibility of placing radioactive implants directly into a tumor or body cavity and are denoted as brachytherapy, interstitial, or intracavity irradiation. This technique is also called internal radiotherapy, and it is appreciated because it is able to deliver high doses concentrated in a small area, generating a high-dose gradient. Internal radiotherapy is sometimes used for cancers of the tongue, uterus, prostate, and cervix but in some cases is considered an invasive procedure.

Other proposed approaches include intraoperative irradiation, in which a large dose of external radiation is directed on to the tumor or tumor bed during surgery, thanks to conventional use of dedicated mobile linear accelerators. Other investigational techniques include the particle beam radiation therapy based on the acceleration of proton or ions to treat localized cancers. The acceleration of these particles requires very sophisticated machines generating modulated or conformal beams to damage tumor cells. Several of these particles, depending on their energy, could produce damage radiobiologically more efficacious than conventional radiotherapy on both tumor and normal tissues, the latter spared by using very conformal dose distributions.

Other recent radiotherapy modalities are based on radiolabeled antibodies to deliver doses of radiation directly to the cancer site due the presence of tumor-specific antibodies (radioimmunotherapy) or in general radionuclides, which thank to their chemical features or their direct injection in the target tissue/tumor bed can target more precisely. The success of this technique will depend upon both the identification of appropriate radioactive substances and the determination of the safe and effective dose of radiation that can be delivered in this way.

Scientists are also looking for ways to increase the effectiveness of existing radiation therapy techniques, based on investigational drugs including radiosensitizers, which make the tumor cells more likely to be damaged, radioprotectors, which protect normal tissues from the effects of radiation, or anti-angiogenic drug, interfering with the neo-angiogenic process. Hyperthermia, or the use of heat, is also being studied for its effectiveness in sensitizing tissues to radiation.

## 3. New methodology in radiotherapy

Radiation therapy (RT) is an integrated part of the modern comprehensive cancer management. Radiotherapy has seen continuous technological improvements since the discovery of

X-rays in 1895 [5]. The main focus in radiation therapy has always been to increase the level of precision and accuracy of dose delivery to the tumor target volume while sparing normal tissue. Remarkable progress has been made in this area, which is especially based on the new delivery systems, new imaging modalities, and more powerful computers and software. These include, for external beam radiotherapy (EBR), the development of advanced linear accelerators, with higher energies and better dose characteristics and skin sparing, as well as smaller sources for reduced lateral penumbra, which date back to the 1950s [6]. One of the major advances in radiation oncology in the early 1990s was the application of computer-graphics technology to CT scanning, when 2D RT was largely replaced by 3D RT, based on CT imaging. Furthermore, radiation dosimetry based on 3D conformal therapy has been studied more accurately, and Monte Carlo methodology [7] has been also introduced into the current calculation for patient dosimetry. Based on EBTR approach, in the first few years of this century, the next technical step forward consisted of the great interest generated in another form of conformal treatment planning, that is, intensity-modulated radiation therapy (IMRT). IMRT allowed better conformation around the tumor and surrounding normal tissue, involving the delivery of optimized, nonuniform irradiation beam intensities. A uniform dose distribution can be created around the tumor by either modulating the intensity of the beam during its journey through the linear accelerator or by using multileaf collimators (MLCs). IMRT is now available in many clinical departments and can be delivered by linear accelerators with smaller segments of differing MLC shape, such as in the case of static IMRT, or modulated by continuously moving MLC, such as in the case of dynamic IMRT [8]. Other derived techniques include tomotherapy, which uses a dedicated CT scan unit and is well suited to treat large volumes [9]. This has led to improvements in the therapeutic ratio for several tumor sites, such as head and neck [10], prostate [11], and gynecological cancers [12]. Volumetric-modulated arc therapy (VMAT) is an advanced form of IMRT that can be delivered using conventional linear accelerators with conventional MLC. VMAT can provide highly conformal dose distribution and improve the IMRT delivery efficiency significantly. The basic concept of arc therapy is the delivery of radiation by means of a continuous rotation of the radiation source and allows the patient to be treated from a full 360 beam angle. However, a major advantage over fixed gantry IMRT is the improvement in treatment delivery efficiency, as a result of the reduction in treatment delivery time and the reduction in monitor unit (MU) usage, with subsequent reduction of integral radiation dose to the rest of the body [13, 14]. In the last decade, there has been a very rapid growth in the clinical application of stereotactic radiosurgery (SRS) and stereotactic body radiation therapy (SBRT). Although the first stereotactic devices were designed by Leksell [15] for the treatment of intracranial benign or malignant lesions, and this was the primary indication, technical improvements in SBRT planning have allowed its use for extracranial lesions [16]. Stereotactic treatment requires strict immobilization, advanced image guidance, and sophisticated treatment planning and delivery systems, resulting in highly conformal dose distributions that allow decreases in the size of treatment volumes relative to conventional radiotherapy. This, in turn, allows for delivery of large doses of radiation per fraction and increased biologically effective doses (BED) beyond those possible with conventional treatments.

## 4. Quality control and dosimetry of the beam

IMRT and VMAT treatments are extremely complex and require patients' specific quality assurance be performed to ensure the dose calculated by treatment planning systems to be the actual dose delivered to the patient at the treatment unit. Guidelines for IMRT commissioning have been published by AAPM Task Group (TG) 119 [17]. The guidelines established test cases to benchmark the overall accuracy of IMRT planning and delivery. The AAPM TG 119 relies on two preliminary tests to evaluate a dose calculating module and four commissioning cases: test prostate, head and neck (H&N), C-shaped target, and multitarget [17].

Typically, IMRT patient-specific QA is performed on a linac using a homogeneous phantom and a dose-measuring device to measure the absolute dose in representative points in the phantom or planar doses. This method requires time on a linac and increases the workload for medical physicist staff. Unfortunately, it is difficult to replicate patient geometries and heterogeneity using a phantom-based QA method.

Treatment plan QA software has been proposed to act as an independent plan evaluation and dosimetry check, thus removing the need to carry out measurements. Unfortunately, this method does not take into account potential failure during delivery that could affect the expected fluence flow map generated by linac.

VMAT [14] is an arc-based IMRT to be delivered on a conventional linac provided of MLC. During arc beam delivery, the dose rate, the speed of the gantry, and the position of the MLC leaves can be adjusted dynamically. RapidArc and SmartArc are examples of VMAT. For most of the commercial planning solutions, no more than two arcs are needed to significantly improve the dose conformity. Rotating the MLC by 45° for VMAT can improve monitor unit efficiency [18].

Generally, VMAT deliver doses faster than IMRT, generating plans with higher or at least equivalent quality, with a very few exceptions. Due to necessary synchronization of both dose rate and gantry motion with MLC movement, it is clear that VMAT involves new and different QA steps relative to conventional IMRT. This should be reflected in acceptance testing (AT), commissioning (COM), and routine QA for VMAT. Finally, VMAT uses fewer monitor units resulting in lower patient total body dose. Plan quality is determined by the number of independent aperture variations generated by a conversion algorithm to produce the calculated fluence maps. Specific controls should be used for this purpose [19].

Testing tools and Devices for VMAT Commissioning include dedicated phantoms, electronic portal imager or films, dedicated programmed MLC files (provided by vendors), and software for analysis. Testing protocols should be based on few parameters, defined method, and appropriate tolerance, supported by documentation or specific QA baselines.

In the Ling et al. paper based on a Varian accelerator for testing, proposed procedures were based on a good knowledge of the use of log files and relatively simple equipment.

In the Bedford et al. paper an Elekta accelerator was used for testing based on complex and expensive equipment [20, 21].

To validate VMAT, the understanding of the limits of planning optimization, gantry rotation, beam blocking, couch rotation, leaf speed, and collimator settings is a prerequisite.

Machine-specific QA should include the following:

- Accuracy of the MLC leaf positions during VMAT delivery
- Ability of the system to vary accurately the dose rate and gantry speed during VMAT delivery
- Ability of the system to vary accurately the MLC leaf speed during VMAT delivery
- Tolerances: baselines from commissioning.

The AAPM TG 142 report represents an update on the TG-40 report, specifying new tests and their tolerance, and added recommendations for not only the new ancillary delivery technologies but also for imaging devices that are part of the linear accelerator. In particular, the imaging devices include X-ray imaging, photon portal imaging, and cone-beam CT.

Deviation from the baseline values could result in suboptimal treatment of patients. Machine parameters can deviate from their baseline values as a result of many reasons, such as unexpected changes in machine performance due to machine malfunctioning, mechanical breakdowns, physical accidents, or component failure. Major component replacement (waveguide, bending magnet, etc.) or degradation of components due to the aging of machines may also alter machine performance from the original parameters. These patterns of failure must be considered when establishing a periodic QA program.

Machines used for SRS/SBRT treatments, TBI, VMAT, or IMRT require different tests and/or tolerance. Specific tests could be adopted for dedicated machines such as tomotherapy, VERO, and Cyberknife. For these devices, vendors in some cases provide phantoms and software to be used, such as black box.

The patient-based QA for irradiation techniques involving spatially dishomogenous fluences needs 2D/3D arrays of dosimeters or matrixes (such as Mapcheck/Arccheck, seven-29/Octavius, delta4, MatriXX/Compass, and Gafcromich), EPID-based dosimetric systems (EPIQA, dosimetry check, and DISO) and online systems attached to linac collimator (compass AP and DAVID). To be noted, a dosimetry check has been proposed as useful for tomotherapy, based on measurements from detectors integrated in the accelerators [22–26].

Additional dosimetric issues requiring novel devices/correction factors include the following: (1) small field dosimetry for which ongoing research suggests diamond-based detectors, microchambers, and scintillators; and (2) high-fluence irradiations chambers with appropriate correction factors, alanine, and Gafcromich [27–30].

## 5. Use of scintillating fibers

Scintillating fibers have been used extensively in experiments of high energy particles. They were used in detector tracers but primarily in calorimetric detectors to measure the corre-

http://dx.doi.org/10.5772/60421

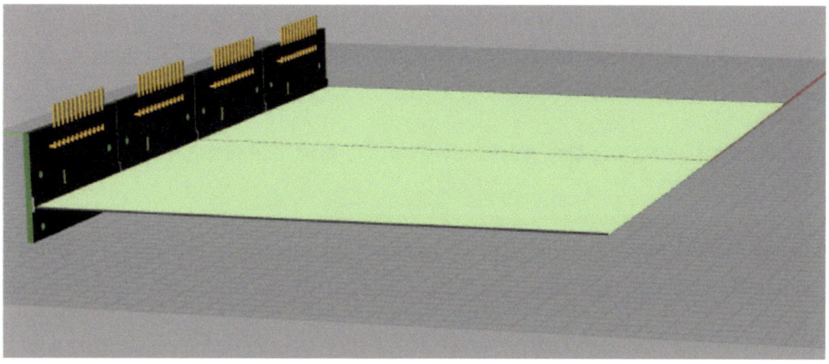

**Figure 1.** Scintillating fibers, glued in one ribbon, connected to the four arrays of photodiodes.

sponding features of hadronic particles and electromagnetic primaries. The measure of the energy absorbed and the topology of the shower were related to the properties of the incident particle. In recent years, scintillating fibers have been used to measure the density of energy absorbed in the fibers directly and hence the dose delivered by a beam of particles used in radiotherapy. Scintillating fibers have been used in some tests [31–34] in small volumes coupled to light guides for connection to the electronics of conversion and reading; in other tests, they were placed in a water phantom or water equivalent to guarantee an accurate measurement 3D [35–37].

Both the scintillating fibers and the light guides are subjected to the radiation beam in all the tests mentioned. The production of light for Cerenkov effect is negligible in the scintillating fibers compared to the scintillation light [38, 39]; it cannot be ignored in the light guides [40]. Systems with subtraction of this contribution were then required [40].

We experimented a different approach using scintillating fibers, performing tests of therapeutic beams of electrons and photons. The detailed description of the material used and some results have been included in the papers [41–43]. In this chapter, we present the main issues raised and described in previous works and insert recent results obtained using the detector in a vertical configuration.

We used a homogeneous plane of scintillating fibers (380 BCF-60 square scintillating optical fibers produced by Saint Gobain Crystals, each fiber of 0.5 × 0.5 mm$^2$) directly coupled to the conversion system and reading as shown in Fig. 1.

The fibers are square (0.5 mm per side) and glued in a single ribbon as shown in Fig. 1. To eliminate optical crosstalk among nearby fibers, each fiber is covered with a white EMA (Extra Mural Absorber).

The detector has been made using a ribbon of 20 × 20.5 cm$^2$ but with a useful sensitive area of 17 × 20.5 cm$^2$ because 3 cm of the fibers is covered of lead to shield the electronics.

The readout system is composed of photodiodes arranged along four arrays optically coupled with the fiber bundles.

We selected four arrays S8865-128 (Hamamatsu, Shizuoka, Japan), each with 128 photodiodes with a sensitive area of 0.3 × 0.6 mm$^2$ and a pitch of 0.4 mm assembled as shown in Fig. 1, and four drivers C9118 (Hamamatsu) for the readout in sequence of the photodiodes.

The linear connection between the response in volts and the energy absorbed in the fibers is shown in [41].

The detector was tested using a 6-MV photon beam generated through a Varian Clinac 2100 DHX (Varian Medica Systems, Palo Alto, CA, USA) at the Cosenza Hospital (Cosenza, Italy).

The modalities used for the exposition were the same described by Lamanna et al. [43]. The integration time of 0.3 s was chosen to ensure a dynamic range large enough to have a linear response from the detector.

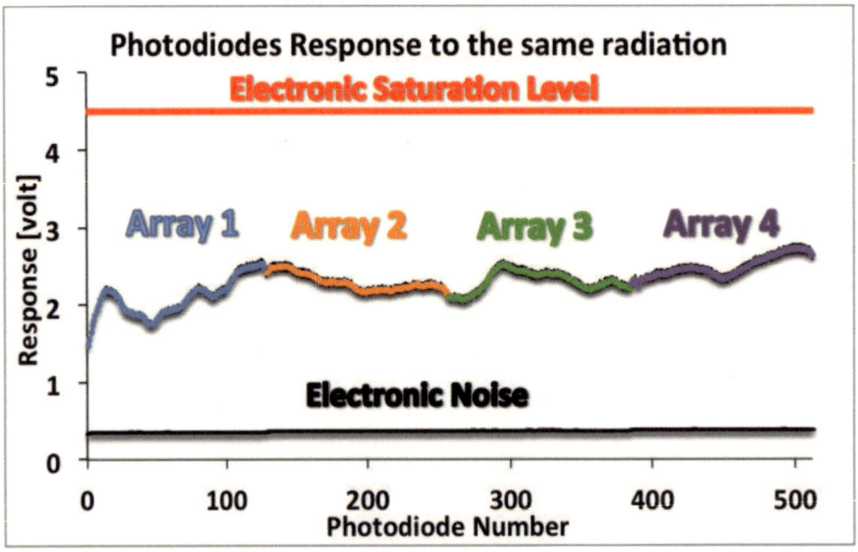


**Figure 2.** Detector response delivering the same dose to the fibers. The channels of the four arrays of photodiode are shown together with the electronic noise.

The response of the detector to the energy deposited in the scintillating fiber is determined by various contributions such as the production and the propagation of light in the fibers, the electronic gain, the coupling between the fibers and the photodiode array, and geometrical and optical. All these effects have been taken into account measuring a calibration factor for each photodiode response by exposing the fibers to the same dose, as explained in detail by Lamanna et al. [43]. We selected a beam of photons with field of view (FOV) 17 × 28 cm$^2$, large enough to cover half the size of the ribbon along the fiber axis.

The response of the detector is shown in Fig. 2. The choice of an integration time of 0.3 s ensures a dynamic range wide enough to have a voltage response in the linear region of the electronics readout. This region is limited between the response of the electronic noise and the saturation voltage. A weight for each channel was estimated to normalize the response of the photodiodes. All the measurements that follow were calibrated using Equation (1):

$$D_i = W_i \times (V_i - N_i) \text{ with } W_i = \frac{1}{R_i - N_{Ri}} \tag{1}$$

The index $i$ is related to the $i$th photodiodes. $D_i$ is the relative calibrated response of the photodiode. $W_i$ is the calibration factor (weight). $V_i$ is the measure in volts. $N_i$ is the electronic noise. $R_i$ is the response when all the fibers are exposed to same dose (as shown in Fig. 2). $N_{Ri}$ is the electronic noise associated with the response $R_i$.

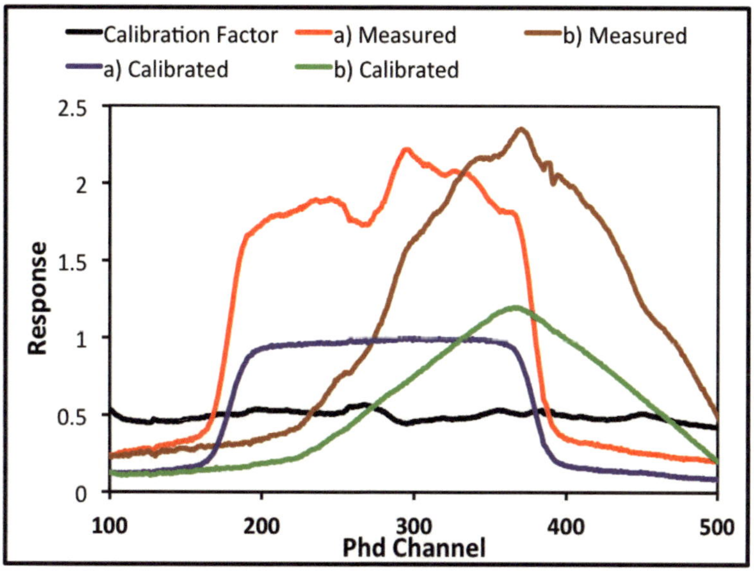


**Figure 3.** Data taken in two different conditions, (a) red and (b) brown; the same data in blue and green after the calibration.

In Fig. 3, two distributions of data taken in different conditions and the same data after calibration are shown. The procedure described was used for the series of measurements that follow.

## 6. Dose profile at fixed depth

The dose profile at a fixed depth is reconstructed using an approach very similar to that adopted in X-ray computed tomography [38], reading 1D radon transformation of the dose absorbed.

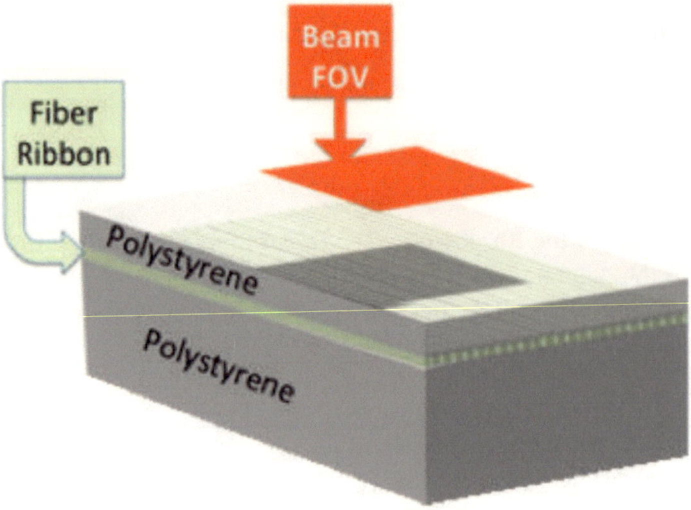


**Figure 4.** Setup configuration for the horizontal exposition; the red area describes the beam field of view (FOV). The fiber ribbon is enclosed between two sheets of polystyrene to ensure the homogeneity top down.

The response of the detector was tested through exposure to a beam orthogonal to the ribbon surface as shown in Fig. 4. Two passive layers were inserted on the top and bottom of the fibers of thickness, respectively, of 1.6 and 5 cm. The polystyrene of density 1.05 g/cm$^3$, transparent to the light as the core of the fibers, was selected to build a homogeneous phantom.

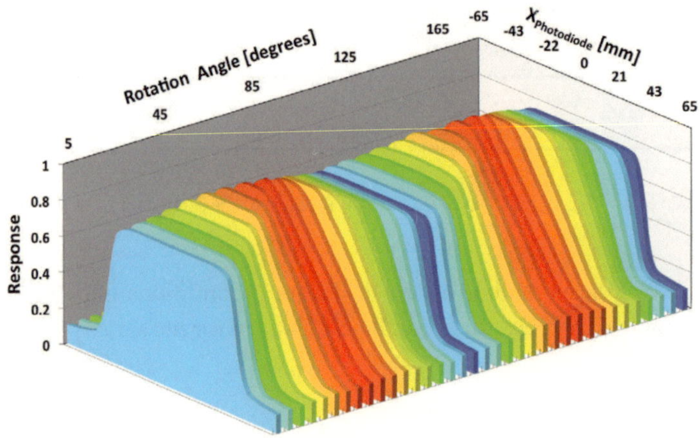


**Figure 5.** Projected dose at 37 rotating angles around the beam axis. The data are normalized to the maximum value measured.

The field of view (FOV) of 8 × 8 cm$^2$ was selected. The system was rotated manually around the beam axis, and the data were collected every 5° for a total of 36 positions from 5° to 180°. The distribution of doses, on the axis orthogonal to that of the fibers, is shown for each angle of rotation in Fig. 5. We observe a variation of the dose and of the width of the distributions, by varying the angle. Angles with maximum dose and minimum width are 45° and 135°, corresponding to a fiber orientation in the direction of the diagonals of rectangular FOV of the beam.

The 36 projections were used to reconstruct the transverse profile of the dose. Using a back-projection approach without filter, we obtained a poor profile, while the reconstruction, shown in Fig. 6, top left, seems to be more precise with the introduction of a ramp filter. A better reconstruction is obtained using the Hann filter, shown in Fig. 6, top right.

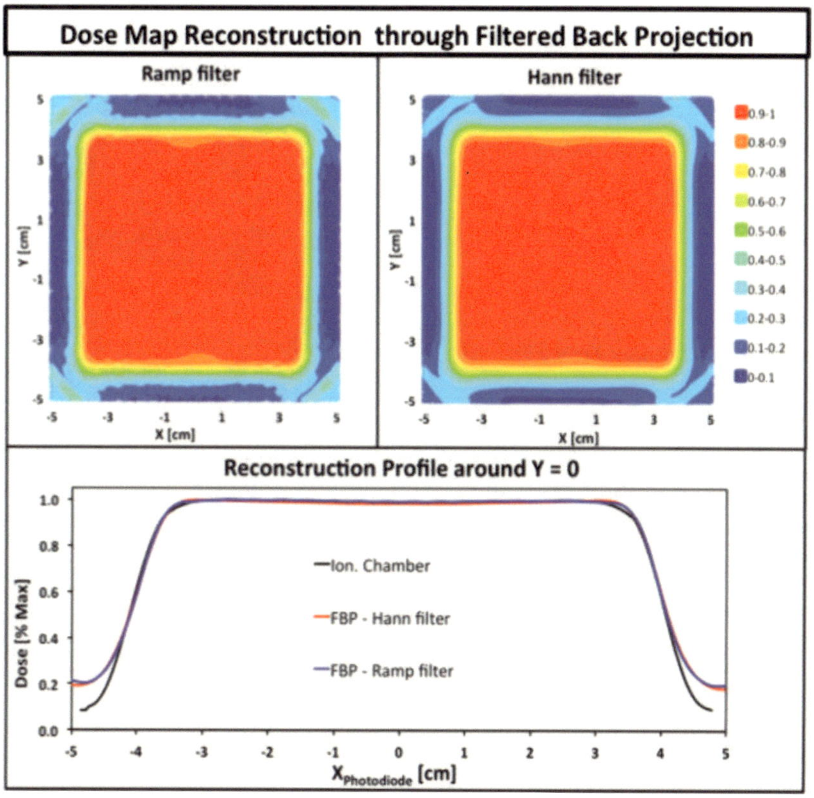


**Figure 6.** Top: dose map reconstruction using a filtered back projection with ramp filter (left) and Hann filter (right). Bottom: comparison dose profile around $Y = 0$ measured using an ionization chamber superimposed on the reconstructed dose profiles using the back projection with and without filters.

A more accurate comparison is shown at the bottom of Fig. 6 where the reconstructed dose profiles, around the center of the beam, are superimposed on the profile obtained using the PTW Freiburg TM31010 ionization chamber. The reconstruction through the back projection approach with ramp or Hann filter, described in [44, 45], reproduces quite accurately the dose profile. A more detailed description of the method and some comments are provided by Lamanna et al. [43].

## 7. Percentage depth dose

The delivered dose in depth was measured by rotating the detector in a vertical exposition of the layer, as shown in Fig. 7.

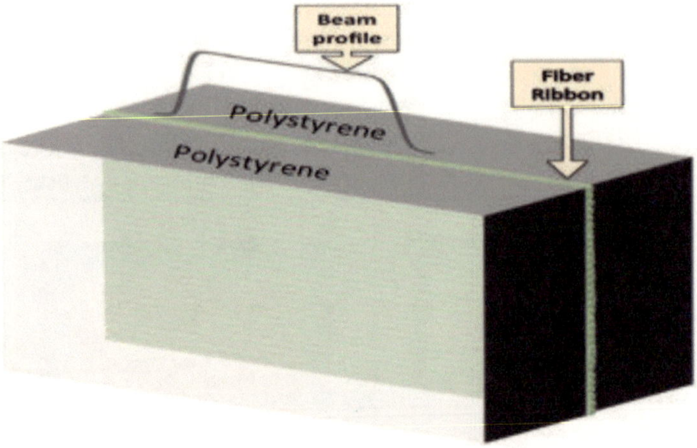


**Figure 7.** Configuration of the detector for the vertical exposure.

The photon beam was delivered to the sensitive layer placed between two polystyrene sheets of 2.5 cm of thickness. An FOV of 6 × 6 $cm^2$ was selected to contain the beam completely all along the fiber. In this configuration, the dose is measured along a line in the FOV area and a granularity of 0.5 mm in depth.

The percentage dose in depth (PDD) measured through the fiber detector (b) is superimposed on those obtained using the PTW TM31010 ionization chamber and the water phantom PTW MP3 (PTW, Freiburg, Germany) (a) in Fig. 8. The response of the fibers after the calibration is shown in curve c.

Two features emerge from the comparison of curves a and c:

- Shift of the point of maximum dose
- Different slopes of the curves

The fiber response is in agreement with the ionization chamber after two corrections of the data. The first correction is reported in Equation (2):

$$\mathrm{PDD}_i^d = W_i * \mathrm{PDD}_i^c \text{ with } W_i = \frac{\mathrm{width}_0}{\mathrm{width}_i}$$
$$\mathrm{width}_0 = 6\,\mathrm{cm}\ ; \mathrm{width}_i = 0.08 * \mathrm{width}_0 + 6\,cm \tag{2}$$

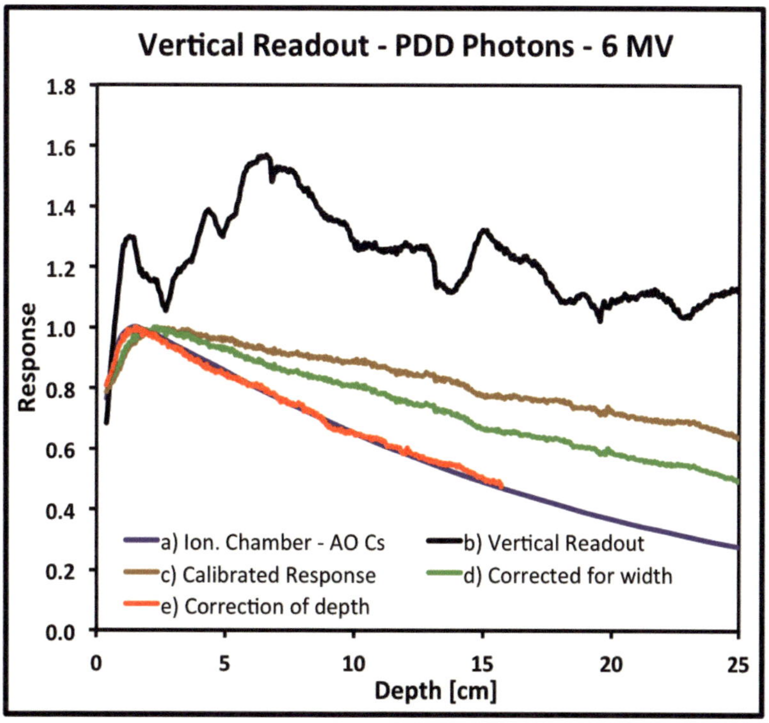


**Figure 8.** Percent dose depth (PDD) as a function of the depth. The blue curve (a) is the PDD measure through ionization chamber and water phantom at AO Cs; the black points (b) are the measured values in Volt; the brown points (c) are the (b) values, calibrated and normalized to the maximum value; the green points (d) are the (c) values corrected for width; the red points (e) are the (d) values after the correction of the depths.

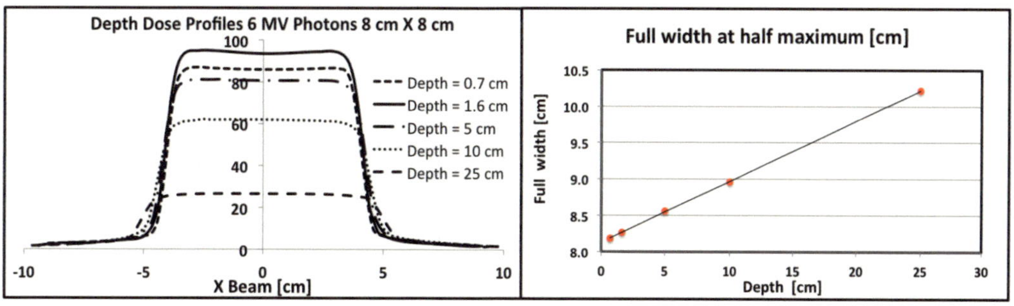


**Figure 9.** Left side: dose profiles at different depths measured through the ionization chamber and a water phantom in the center of the FOV. Right side: full width at half maximum of the previous profiles as a function of the depth.

The integrated doses along each fiber are measured through the scintillating fibers, and the region of the absorbed dose increases in width as a function of the depths. The behavior is shown in the left region of Fig. 9, where the profiles of the dose at different depth in water for a 6-MV photon beam are plotted. The full width at half maximum of the profiles increases linearly with the depth, as shown on the right of Fig. 9.

The correction factor $W_i$ guarantees the same integration length for all the fibers at different depths. The results are shown in curve d of Fig. 8. The slope is close to that of the curve obtained through the ionization chamber but they are not equal.

The second correction is reported in Equation (3):

$$\mathrm{Depth}_i^{H_2O} = IW_i \times \mathrm{Depth}_i^{C_8H_8} \text{ with } IW_i = \frac{\mathrm{ionization}(H_2O)_i}{\mathrm{ionization}(C_8H_8)_i} \tag{3}$$

In the assembled configuration to measure the PDD, shown in Fig. 7, both the layer of scintillating fibers and the passive sheets are mainly made up of transparent polystyrene ($C_8H_8$). The ionization is different from that in the water at the same depth. The correction factor $IW_i$ is needed to find the depth in water corresponding to the ionization measured in polystyrene.

The factor $IW$ has been measured by exposing two phantoms, a water and a transparent polystyrene phantom, in the same conditions. We used a beam of electrons of 9 MeV. The results are shown in Fig. 10. The two behaviors are different. The correction factor in depth IW has been evaluated to produce a full superposition.

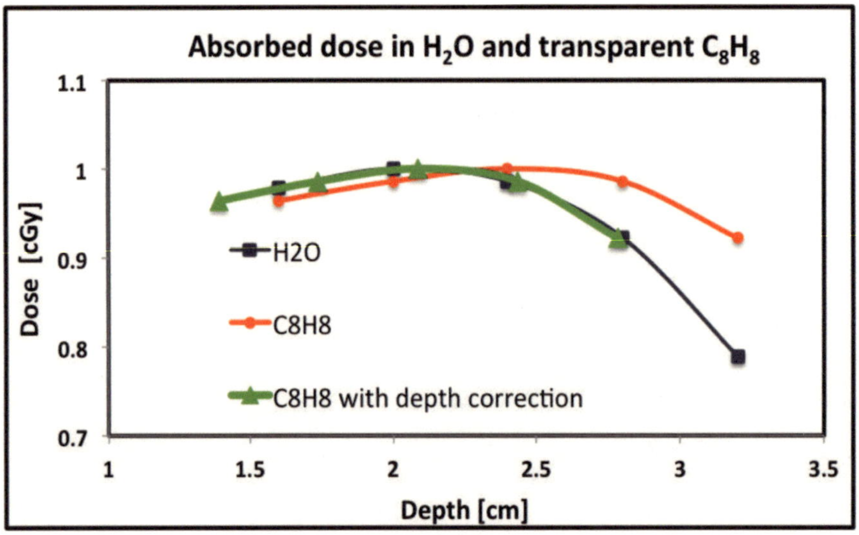


**Figure 10.** Dose measured through the ionization chamber, in water (black) and in a phantom of transparent polystyrene (red), at different depths. A correction factor for the depth has been evaluated to superimpose the two sets of data. The green points are the red points with the correction of the depths using Equation (3).

The layer of scintillating fibers can be used to measure the PDD in a vertical configuration of the exposure, taking into account that the integration of the dose along the fiber requires the corrections inserted in Equations (2) and (3).

http://dx.doi.org/10.5772/60421

# 8. Conclusions

The new techniques of delivering the therapeutic dose to the tumor tissues with an external beam require dosimeters that have better spatial and time resolutions than the standard detectors used in current clinical practice. The properties of scintillating fibers of dimensions less than 1 mm and their gluing in homogeneous layers allow these requirements to be obtained. The single-layer 1D can be used to reconstruct the dose profiles XY at a fixed depth, adopting the tomographic technique of filtered back-projection. The same configuration 1D may be used to measure the dose profiles in depth by applying the corrections inserted in Equations (2) and (3).

The final response is quite rapid. The data are acquired in few milliseconds and processed in 2–3 s, using a single layer. This characteristic together with the reduced thickness of two orthogonal layers in a 2D configuration (1 mm of water equivalent), allow the use of layers of fibers also online to measure the characteristics of an incident beam (a similar approach was published in [46]).

The positive results obtained and described in this chapter justify an investment of this approach and suggest some directions for future developments:

**a.** Acquisition can be speeded up by automating the process of rotation of the dosimeter.

**b.** The tomographic reconstruction can be improved by introducing iterative reconstruction methods available in the literature.

**c.** A 3D detector can be achieved by assembling layers 1D orthogonally to the beam axis in a solid phantom.

**d.** N.B. to realize point c: it is necessary to change the reading system by introducing MPPC or FPD to the arrays of photodiodes.

# Nomenclature

AAPM; American Association of Physicists in Medicine

AT ; acceptance testing

COM ; commissioning

EBR; external beam radiotherapy

EBTR; evidence-based timeline retrospective

FOV; field of view

FPD; flat panel detector

IMRT; intensity-modulated radiation therapy

MLC; multileaf collimator

MPPC; multipixel photon counters

PDD; percentage depth dose

QA; quality assurance

SRS; stereotactic radiosurgery

RT; radiotherapy

TBI; traumatic brain injury

UM; monitor units

VMAT; volumetric-modulated arc therapy

## Acknowledgements

We are grateful to the Italian INFN (National Institute of Nuclear Physics) for supporting the study of scintillating fiber dosimeters and the "Hospital of Cosenza" for allowing the use of the accelerator Varian to test the dosimeter.

## Author details

Ernesto Lamanna[1*], Bianco Cataldo[2], Giulia Marvaso[2], Marco D'Andrea[3] and Lidia Strigari[3]

*Address all correspondence to: lamanna@unicz.it

1 Department of Health Sciences, Magna Graecia University, Catanzaro, Italy

2 Department of Experimental and Clinical Medicine, Magna Graecia University, Catanzaro, Italy

3 Department of Medical Physics, Istituto Regina Elena, Roma, Italy

## References

[1] Raju MR. Particle radiotherapy: historical developments and current status. Radiat Res 1996; 145, 391–407.

[2] Slater JM. From X-rays to ion beams: a short history of radiation therapy. In: U. Linz. (Ed.), Ion Beam Therapy. Biological and Medical Physics, Biomedical Engineering. Berlin: Springer-Verlag, 2012. DOI: 10.1007/978-3-642-21414-1 1.

[3] Baskar R, Lee KA, Yeo R, Yeoh. Cancer and radiation therapy: current advances and future directions. Int J Med Sci 2012;9(3):193–9. DOI: 10.7150/ijms.3635. Available from http://www.medsci.org/v09p0193.htm

[4] Thariat J, Hannoun-Levi J-M, Sun Myint A, Vuong T, Gérard J-P. Past, present, and future of radiotherapy for the benefit of patients. Nat Rev Clin Oncol 2013;10, 52–60; published online 27 November 2012. Doi: 10.1038/nrclinonc.2012.203.

[5] Bernier J, Hall EJ, Giaccia A. Radiation oncology: a century of achievements. Nat Rev Cancer. 2004 Sep;4(9):737–47. Review.

[6] Bortfeld T, Jeraj R. The physical basis and future of radiation therapy. Br J Radiol. 2011 Jun;84(1002):485–98.

[7] Spezi E, Lewis G. An overview of Monte Carlo treatment planning for radiotherapy. Radiat Prot Dosimetry 2008;131(1):123–9.

[8] Brahme A. Development of radiation therapy optimization. Acta Oncol 2000;39(5): 579–95. Review.

[9] Goffman TE, Glatstein E. Intensity-modulated radiation therapy. Radiat Res 2002 Jul; 158(1):115–7. Review.

[10] Feng FY, Kim HM, Lyden TH, Haxer MJ, Feng M, Worden FP, Chepeha DB, Eisbruch A. Intensity-modulated radiotherapy of head and neck cancer aiming to reduce dysphagia: early dose–effect relationships for the swallowing structures. Int J Radiat Oncol Phys 2007;68: 1289–98.

[11] Wang-Chesebro A, Xia P, Coleman J, Akazawa C, Roach M 3rd. Intensity-modulated radiotherapy improves lymph node coverage and dose to critical structures compared to with three-dimensional conformal radiation therapy in clinically localized prostate cancer. Int J Radiat Oncol Phys 2006;66: 654–62.

[12] Mundt AJ, Lujan AE, Rotmensch J, Waggoner SE, Yamada SD, Fleming G, Roeske JC. Intensity-modulated whole pelvic radiotherapy in women with gynecologic malignancies. Int J Radiat Oncol Phys 2002;52:1330–7.

[13] Palma DA, Verbakel WF, Otto K, Senan S. New developments in arc radiation therapy: a review. Cancer Treat Rev 2010;36:393–9.

[14] Yu CX, Li XA, Ma L, Chen D, Naqvi S, Shepard D, et al. Clinical implementation of intensity-modulated arc therapy. Int J Radiat Oncol Biol Phys 2002;53:453–63.

[15] Leksell, L. The stereotaxic method and radiosurgery of the brain. Acta Chir Scand 1951;102:316–9.

[16] Salama JK, Kirkpatrick JP, Yin FF. Stereotactic body radiotherapy treatment of extracranial metastases. Nat Rev Clin Oncol 2012;9:654–5.

[17] Ezzell GA, Burmeister JW, Dogan N, et al. IMRT commissioning: multiple institution planning and dosimetry comparisons, a report from AAPM Task Group 119. Med Phys 2009;36(11):5359–73.

[18] Yu CX. Intensity-modulated arc therapy with dynamic multileaf collimation: an alternative to tomotherapy. Phys Med Biol 1995;40:1435–49.

[19] Phillips MH, Holdsworth C. When is better best? A multi-objective perspective. Med Phys 2011;38:1635.

[20] Ling C, et al. Commissioning and quality assurance of RapidArc radiotherapy delivery system. Int J Radiat Oncol Biol Phys 2008;72:575–81.

[21] Bedford JL, Warrington AP. Commissioning of volumetric modulated arc therapy. Int J Radiat Oncol Biol Phys 2009;73:537–45.

[22] Hussein M1, Rowshanfarzad P, Ebert MA, Nisbet A, Clark CH. A comparison of the gamma index analysis in various commercial IMRT/VMAT QA systems. Radiother Oncol 2013 Dec;109(3):370–6.

[23] Watanabe Y, Nakaguchi Y. 3D evaluation of 3DVH program using BANG3 polymer gel dosimeter. Med Phys 2013 Aug;40(8).

[24] Nakaguchi Y, Araki F, Ono T, Tomiyama Y, Maruyama M, Nagasue N, Shimohigashi Y, Kai Y. Validation of a quick three-dimensional dose verification system for pretreatment IMRT QA. Radiol Phys Technol. 2014 Sep 27.

[25] Fredh A, Scherman JB, Fog LS, Munck af Rosenschöld P. Patient QA systems for rotational radiation therapy: a comparative experimental study with intentional errors. Med Phys 2013 Mar;40(3).

[26] Cilla S, Azario L, Greco F, Fidanzio A, Porcelli A, Grusio M, Macchia G, Morganti AG, Meluccio D, Piermattei A. An in-vivo dosimetry procedure for Elekta step and shoot IMRT. Phys Med 2014 Jun;30(4):419–26.

[27] Marsolat F, Tromson D, Tranchant N, Pomorski M, Le Roy M, Donois M, Moignau F, Ostrowsky A, De Carlan L, Bassinet C, Huet C, Derreumaux S, Chea M, Cristina K, Boisserie G, Bergonzo P. A new single crystal diamond dosimeter for small beam: comparison with different commercial active detectors. Phys Med Biol 2013 Nov 7;58(21):7647–60.

[28] Di Venanzio C, Marinelli M, Milani E, Prestopino G, Verona C, Verona-Rinati G, Falco MD, Bagalà P, Santoni R, Pimpinella M. Characterization of a synthetic single crystal diamond Schottky diode for radiotherapy electron beam dosimetry. Med Phys 2013 Feb;40(2).

[29] Podesta M, Nijsten SM, Persoon LC, Scheib SG, Baltes C, Verhaegen F. Time dependent pre-treatment EPID dosimetry for standard and FFF VMAT. Phys Med Biol 2014 Aug 21;59(16):4749–68.

[30] Francescon P, Kilby W, Satariano N. Monte Carlo simulated correction factors for output factor measurement with the CyberKnife system-results for new detectors and correction factor dependence on measurement distance and detector orientation. Phys Med Biol 2014 Mar 21;59(6):N11–7.

[31] Archambault L, Arsenault J, Gingras L, Beddar AS, Roy R, Beaulieu L. Plastic scintillation dosimetry: optimal selection of scintillating fibers and scintillators. Med Phys 2005;32:2271–2278.

[32] Frelin AM, Fontbonne J-M, Ban G, Batalla A, Colin J, Isambert A, Labalme M, Leroux T, Vela A. A new scintillating fiber dosimeter using a single optical fiber and a CCD camera. IEEE Trans Nucl Sci 2006;53 (3):1113–1117.

[33] Lee B, Jang KW, Cho DH, Yoo WJ, Shin SH, Kim H, Yi JH, Kim S, Cho H, Park B, Moon JH, Kim S. Measurement of two-dimensional photon beam distributions using a fiber-optic radiation sensor for small field radiation therapy. IEEE Trans Nucl Sci 2008;55 (5):2632–2636.

[34] Staub D, Cailleret J, Guyonnet JL, Le TD, Wurtz J. Real-time radio-transparent dosimeter for X-ray imaging system. Nucl Instrum Methods Phys Res A 2004;525, 303–307.

[35] Fontbonne JM, Iltis G, Ban G, Battala A, Vernhes JC, Tillier J, Bellaize N, Brun CL, Tamain B, Mercier K, Motin JC. Scintillating fiber dosimeter for radiation therapy accelerator. IEEE Trans Nucl Sci 2002;49 (5):2223–2227.

[36] Guillot M, Gingras L, Archambault L, Beddar S, Beaulieu L. Toward 3D dosimetry of intensity modulated radiation therapy treatments with plastic scintillation detectors. J Phys Conf Ser 2010;250:1–5.

[37] Lacroix F, Beaulieu L, Gingras L, Guillot M. Clinical prototype of a plastic water-equivalent scintillating fiber dosimeter array for QA application. Med Phys 2008;35:3682–3690.

[38] Frelin A-M, Fontbonne J-M, Ban G, Batalla A, Colin J, Isambert A, Labalme M, Vela TLA. A new scintillating fiber dosimeter using a single optical fiber and a CCD camera. IEEE Trans Nucl Sci 2006;53 (3):1113–1117.

[39] Lee B, et al. Characterization of one-dimensional fiber-optic scintillating detectors for electron-beam therapy dosimetry. IEEE Trans Nucl Sci 2008;55 (5):2627–2631.

[40] Archambault L, Gingras L, Beaulieu L, Beddar AS. Measurement accuracy and Cerenkov removal for high performance, high spatial resolution scintillation dosimetry. Med Phys 2006;33,128–135.

[41] Lamanna E, Fiorillo AS, Bruno C, Santaniello A, Siaka YFT, Berdondini A, Bettuzzi M, Brancaccio R, Casali F, Morigi MP, Barca G, Castrovillari F. Dosimetry of high intensity electron beams produced by dedicated accelerators in intra-operative radiation therapy (IORT). IEEE Trans Nucl Sci 2009;56 (1):66–72.

[42] Lamanna E, Fiorillo AS, Gallo A, Trapasso A, Caroleo R, Brancaccio R, Barca G, Carnevale S, Castrovillari F, Tchuente Siaka YF. Dosimetric study of therapeutic beams using a homogeneous scintillating fiber layer. Nuclear Science Symposium and Medical Imaging Conference (NSS/MIC), 2011 IEEE, pp. 244–248, 23–29 Oct. 2011. ISBN: 978-1-4673-0119-0.

[43] Lamanna E, Fiorillo AS, Gallo A, Trapasso A, Caroleo R, Brancaccio R, Barca G, Carnevale S, Castrovillari F, Tchuente Siaka YF. Dosimetric study of therapeutic beams using a homogeneous scintillating fiber layer. IEEE Trans Nucl Sci 2013;60 (1):109–14.

[44] Quinto ET. Tomographic reconstructions from incomplete data-numerical inversion of the exterior Radon transform. Inverse Problems 1988;4 (3):867.

[45] Lyra M, Ploussi A. Filtering in SPECT image reconstruction. Int J Biomed Imaging 2011; ID 693795.

[46] Goulet M, Gingras L, Beaulieu L. Real-time verification of multileaf collimator-driven radiotherapy using a novel optical attenuation-based fluence monitor. Med Phys 2011;38:1459–1459.

Chapter 7

# Linac Twins in Radiotherapy

Marius Treutwein, Petra M. Härtl, Christian Gröger,
Zaira Katsilieri and Barbara Dobler

Additional information is available at the end of the chapter

http://dx.doi.org/10.5772/60427

**Abstract**

In a radiotherapy department having more than one linear accelerator, it is rather common to match the dose output of all machines. In particular, the recently developed flattening filter free mode requires new investigations regarding the feasibility of matching and the consequences for quality assurance and workload. This refers also to the beam model of the radiotherapy treatment planning system. Our results show that matching is possible not only for flat beams but also for flattening filter free mode. Therefore, the machines can substitute each other in the case of breakdown or service without new treatment planning even in the case of complex intensity-modulated radiotherapy or volumetric-modulated arc therapy. The quality assurance is reduced to only one data set for both the linear accelerators and the radiotherapy treatment planning system.

**Keywords:** linac twins, matched linacs, flatness filter free, FFF, quality assurance

## 1. Introduction

Electron linear accelerators (linacs) are the most common treatment machines in radiotherapy. Having two (or more) equal linacs (*linac twins* or also called *matched linacs*) enables a radiotherapy department to facilitate the workflow and to reduce the amount of quality assurance. The major part of the German standards (DIN) regarding quality assurance of medical linear accelerators has been reworked or has been published for the first time in the recent years due to technical developments. Similar updates have been published in other countries, e.g., by different task groups of the American Association of Physicists in Medicine (AAPM) [1-3]. However, these updates have again been overtaken by a technical novelty: the flattening filter free (FFF) mode. In modern linacs, it allows higher dose rates than used in the standard mode with flattening filter (up to three or four times), thus reducing treatment times at the cost of a nonflattened profile. This mode requires its own procedures for acceptance and quality assurance [4, 5].

The aims of this study are to setup a commissioning procedure and a quality assurance program for linac twins and to investigate if the time required for commissioning and quality assurance can be reduced as compared to two linacs of different types. This includes also the radiotherapy treatment planning system (RTPS). Although this investigation refers to the German standards and directives, the principles are valid for all countries. Although the concept of matched linacs has been mentioned earlier [6-10], the consequences for the quality assurance and standardization have not yet been regarded. This study will only investigate photon beam qualities. Characteristics of electron beams, portal imaging systems, and cone beam CT are not included. Although the concept and first results have been part of a congress proceeding [11], this chapter presents for the first time elaborated and generalized background, results, and discussion.

## 2. Materials and methods

### 2.1. Linacs

Tenders were invited to provide two linacs of the same type to replace the old Siemens Primus machines. We asked for linacs with two photon energies (6 and 15 MV) flattened beams (FB), additional FFF option for 6 MV, capability of intensity-modulated radiotherapy (IMRT) and volumetric-modulated arc therapy (VMAT), and five to six different electron energies between 4 and 22 MeV. Our requirement was that patients should be treatable at both machines with the same treatment plan. The first of the twin machines (Figure 1), an Elekta Synergy™ with Agility head, XVI cone beam CT, and Iview GT™ portal imaging, has been installed and commissioned according to earlier experiences [12] and has been running in the clinical routine for several months, but initially not FFF. The desktop software is Integrity 3.1. For the second linac, the installation has been completed in June 2014. The manufacturer specifies a routine for acceptance testing of linear accelerators [13], which refers to matched machines only pertaining to the beam quality.

### 2.2. Standards and guidelines for quality assurance

Although commissioning tests, the determination of basic performance characteristics, and consistency tests for linacs according the German standards [20] have to be accomplished for each machine, they can at least be set up identically without modifications for twin machines. This is also applicable for performance characteristics and consistency testing concerning special techniques as stereotactic radiotherapy [15, 16] and IMRT [17, 18], as well as electronic portal imaging devices (EPID) [14].

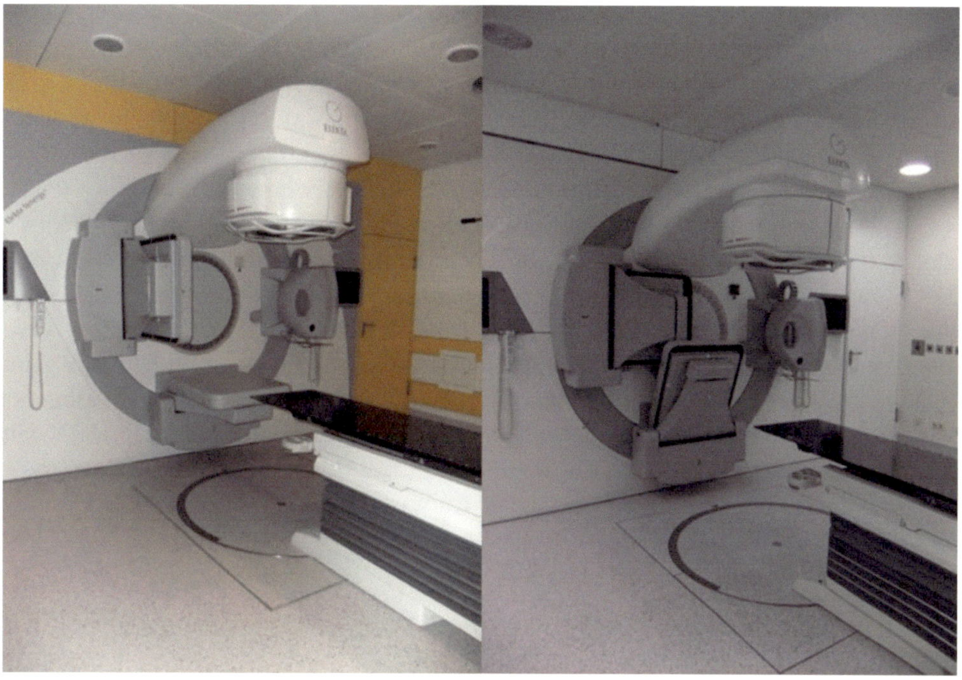

**Figure 1.** First and second of the linac twins: Elekta Synergy with Agility head.

The draft of the German standard for consistency tests of RTPS DIN 6873-5 [20] requires calculations for each treatment machine. Probably part 1 of DIN 6873 for commissioning of RTPS which is in development will demand this too. A technical report of the International Atomic Energy Agency (IAEA) [21] recommends checks for each photon and electron beam used in clinical planning and therefore each beam model and treatment machine. Having only one treatment machine model reduces time and effort for quality assurance and commissioning.

### 2.3. RTPS and water phantom

For commissioning of the linac model in the RTPS Oncentra® 4.3 (by Nucletron an Elekta Company), a set of geometrical data, absolute, and relative dose measurements have been measured [22] using a water phantom of the type Blue Phantom$^2$ of IBA company. It has been operated by the software OmniPro Accept 7.4, including a module for data export for Oncentra®. The data are processed by Elekta to create a model of the treatment unit, which takes several weeks according to our experience. Once the model is delivered by the company, it has to be validated by the customer. The RTPS comprises for this purpose the Beam Data Tool, which allows not only to compare measured and calculated dose distributions but also to adapt the size of the focus and the transmission of the collimator for final optimization. One aim of the study is to investigate if this procedure can be reduced to the validation process for the second linac.

The evaluation tools of the water phantom software were used to compare the measured dose distributions of both machines. For the graphical demonstrations, smoothing with least square algorithms was performed to get rid of some noise, and the curves were renormalized to the dose maximum on the central beam. The Beam Data Tool of the RTPS was applied for comparisons of measured and calculated dose distributions. For the calculations, the collapsed cone algorithm was used. The European Society for Radiotherapy and Oncology (ESTRO) gives tolerances as confidence limits for calculated doses [23]. The specifications of the manufacturer of the RTPS Oncentra® [24] are not always comparable, e.g., the shoulder region is not defined exactly, or dose deviations refer to different points, or different units are used (Table 1). We referred to the manufacturer's specifications in our evaluations as they have to be kept by the manufacturer's beam model and used the ESTRO's—modified to a gamma evaluation of 10% and 2 mm in the sharp gradient area—only for further investigations. Distance to agreement, dose deviation, and gamma evaluation [25] are the integrated evaluation options in the Beam Data Tool. The depth dose distributions and profiles in in-plane and cross-plane direction in depths of 5, 10, and 20 cm were evaluated for square field sizes of 2, 5, 10, 15, 20, 30, and 40 cm.

From the measured depth dose curves of fields of $10 \times 10$ cm$^2$ at a source surface distance (SSD) of 100 cm, the beam quality was derived as $Q$ = D200/D100, with D200 the dose in 20 cm depth and D100 in 10 cm.

| | Specifications Oncentra® | Recommendations ESTRO |
|---|---|---|
| Central 80% of field | ±3% | ±3% |
| Shoulder region | ±4% | ±2 mm |
| Regions outside the field | ±5% | ±30% local dose |
| Regions of sharp gradient | ±3 mm | ±2 mm or 10% |

**Table 1.** Accuracy of dose calculations in percent of calibration dose or mm distance deviation to correct dose value

### 2.4. Breakdown concept

The German directive "Strahlenschutz in der Medizin" [26], paragraph 2.3.4, requires a concept to ensure patient treatment even during machine down times (e.g., maintenance or breakdown). Linac twins allow shifting all patients from one machine to the other without calculating new treatment plans. Sjöström et al. [8] provided this as main argument for matching linacs. Depending on the tumor type and the patient state, such a transfer is also advised by the Board of Faculty of Clinical Oncology of the Royal College of Radiologists [27].

The record and verify system (Mosaiq®, version 2.50) can be configured in a manner that fields for one machine can be delivered at the other without warnings or password confirmation. Some VMAT and IMRT plans (FB and FFF) calculated with the beam model for the first machine were measured on the other to verify the exchangeability using the 2D array Matrixx Evolution phantom and software OmniPro-I'mRT, version 1.7 of IBA.

The procedure for these plan verifications has been described in detail in earlier publications [12, 28]. The patient plans were transferred to a cuboid phantom of solid water (RW3 of the PTW company) with the Matrixx phantom in the center in a horizontal plane. The dose distribution was recalculated without modification of any parameter. When the phantom was irradiated with the original plan, the software recorded the dose. The measured dose was compared to the calculated dose by gamma evaluation [25] with a dose tolerance of 3% of the maximum dose and a distance to agreement of 3 mm. The first clinical IMRT plans with 6MV and for testing of the feasibility some single IMRT plans with 6MV FFF and VMAT plans were evaluated.

# 3. Results

## 3.1. Beam quality and measured dose distributions

The results of the beam quality evaluation of the percentage depth dose curves are given in Table 2.

| | Linac 1 | Linac 2 |
|---|---|---|
| 6MV | 0.587 | 0.586 |
| 6MV FFF | 0.583 | 0.584 |
| 15MV | 0.648 | 0.648 |

**Table 2.** Beam quality $Q$ = D200/D100 for both machines at the three photon beam qualities.

Although these nearly identical values have been calculated only for field size 10 × 10 cm², the depth dose curves in Figures 2, 3, and 4 show that the beam quality is comparable with other field sizes too as it is derived from depth dose parameters. Most of the curves are congruent. The presented figures show only a selection of all measured photon depth dose distributions. Due to the high congruence, the colors are often not clearly defined as in the figure texts.

| Nominal field size in cm² | Linac 1 | Linac 2 |
|---|---|---|
| 2 × 2 | 2.06 | 2.05 |
| 5 × 5 | 5.08 | 5.05 |
| 10 × 10 | 10.08 | 10.01 |
| 15 × 15 | 15.08 | 14.96 |
| 20 × 20 | 20.10 | 19.96 |

**Table 3.** Evaluation of the field size of 6 MV photons in cross-plane direction as given by the profiles in Figure 5.

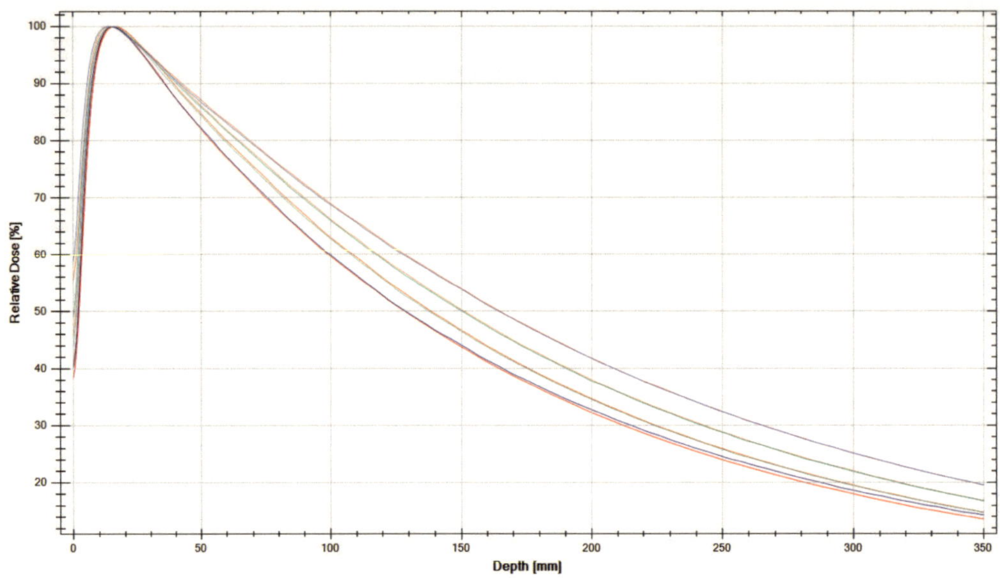


**Figure 2.** Depth dose curves for both machines of 6MV photons for square fields of 2, 10, 15, and 20 cm (left to right in the descending part). Red colors belong to the first, blue to the second linac.

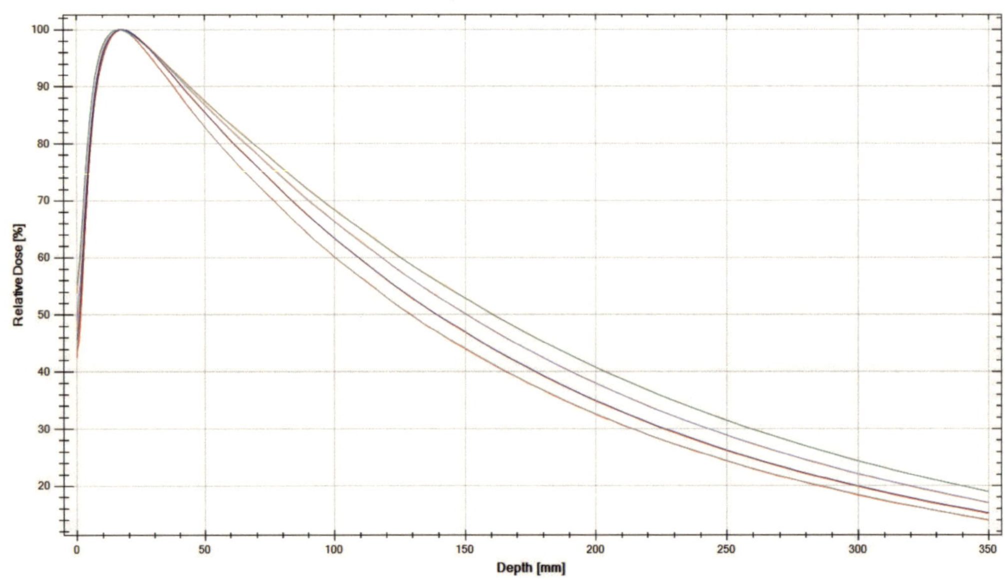


**Figure 3.** Depth dose curves for both machines of 6MV photons FFF for square fields of 2, 10, 15, and 20 cm (left to right in the descending part). Red colors belong to the first, blue to the second linac.

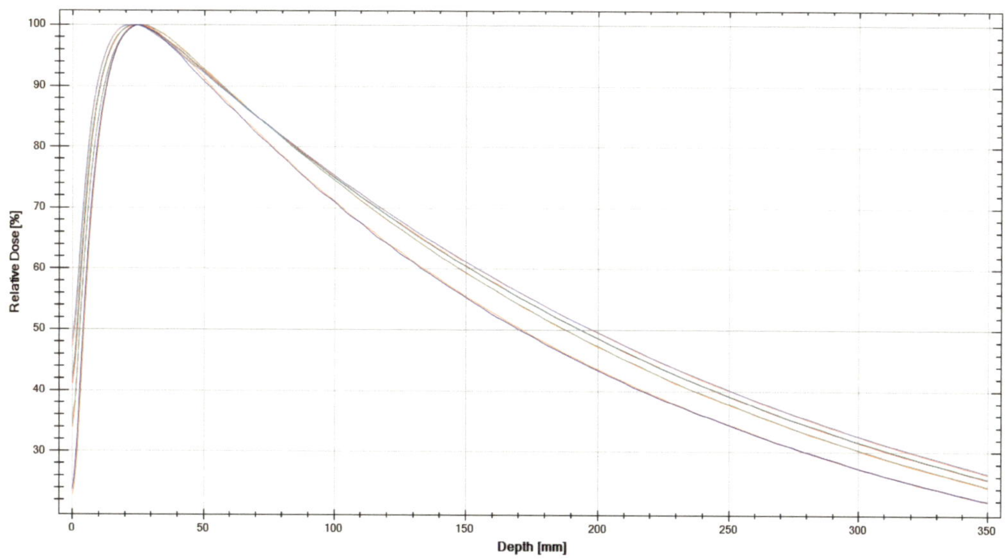


**Figure 4.** Depth dose curves for both machines of 15MV photons for square fields of 2, 10, 15, and 20 cm (left to right in the descending part). Red colors belong to the first, blue to the second linac.

The second group of figures shows profiles in the cross-plane direction in a depth of 10 cm. Others with further field sizes, in different depths, and in in-plane direction have been measured with similar results. The difference between the flattened and the flatness filter free mode is obvious; as the angular distribution of the bremsstrahlung is not compensated, the maximum dose is always on the central beam. The high congruence of the corresponding curves makes often only one of them visible, sometimes in a mixed color. Numerical values of the field size evaluation are given in Table 3.

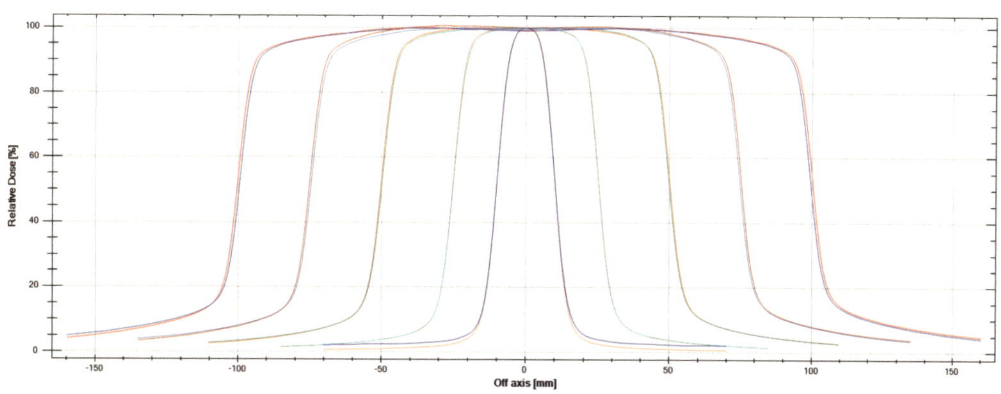


**Figure 5.** Profiles for both machines of 6MV photons for square fields of 2, 5, 10, 15, and 20 cm. Red colors belong to the first, blue to the second linac.

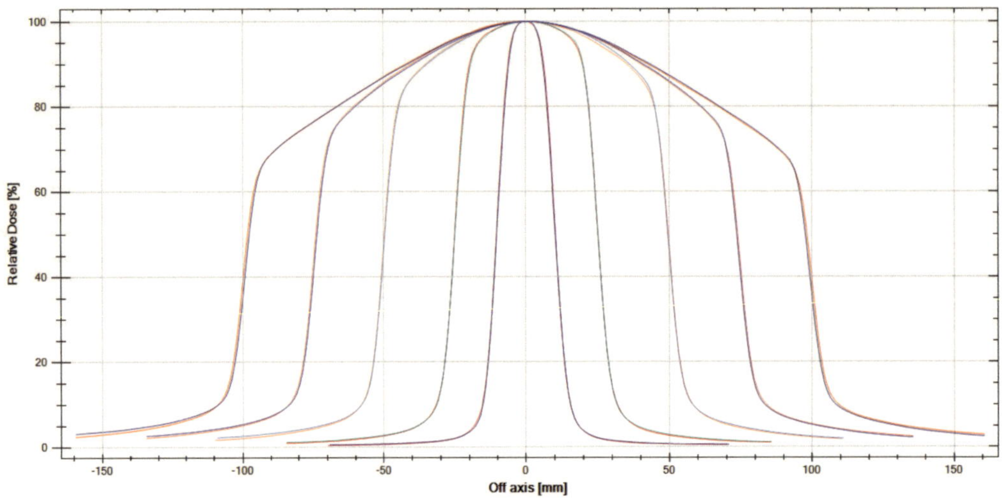


**Figure 6.** Profiles for both machines of 6MV FFF photons for square fields of 2, 5, 10, 15, and 20 cm. Red colors belong to the first, blue to the second linac.

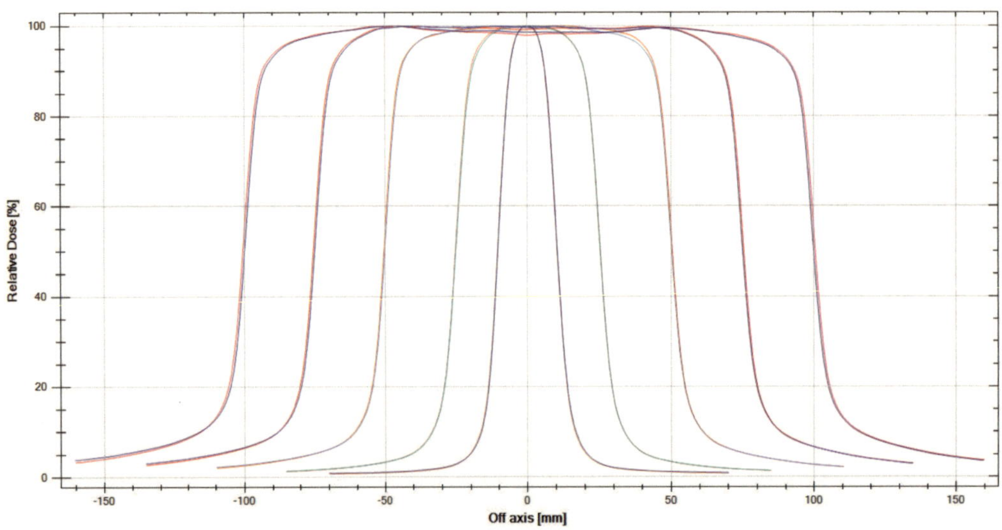


**Figure 7.** Profiles for both machines of 15MV photons for square fields of 2, 5, 10, 15, and 20 cm. Red colors belong to the first, blue to the second linac.

## 3.2. RTPS commissioning

As described above, no model for the second linac has been created in the RTPS, but all calculations were done with the first linac model and compared to the measurements of the second. Figure 8 shows the results of such a validation at an example of FFF profiles. The left ordinate refers to the measured and calculated profiles (green and orange), the right to the validation criterion. In the first step, both curves are compared using the dose deviation of the calculated dose from the measured dose. The horizontal orange lines show the limits of ±3%. They are not valid in the region of the field borders (high gradient). Therefore, the evaluation was repeated plotting the distance to the next point with the same dose with a distance to agreement of 3 mm in the second step. Here all points in this region are within the limits.

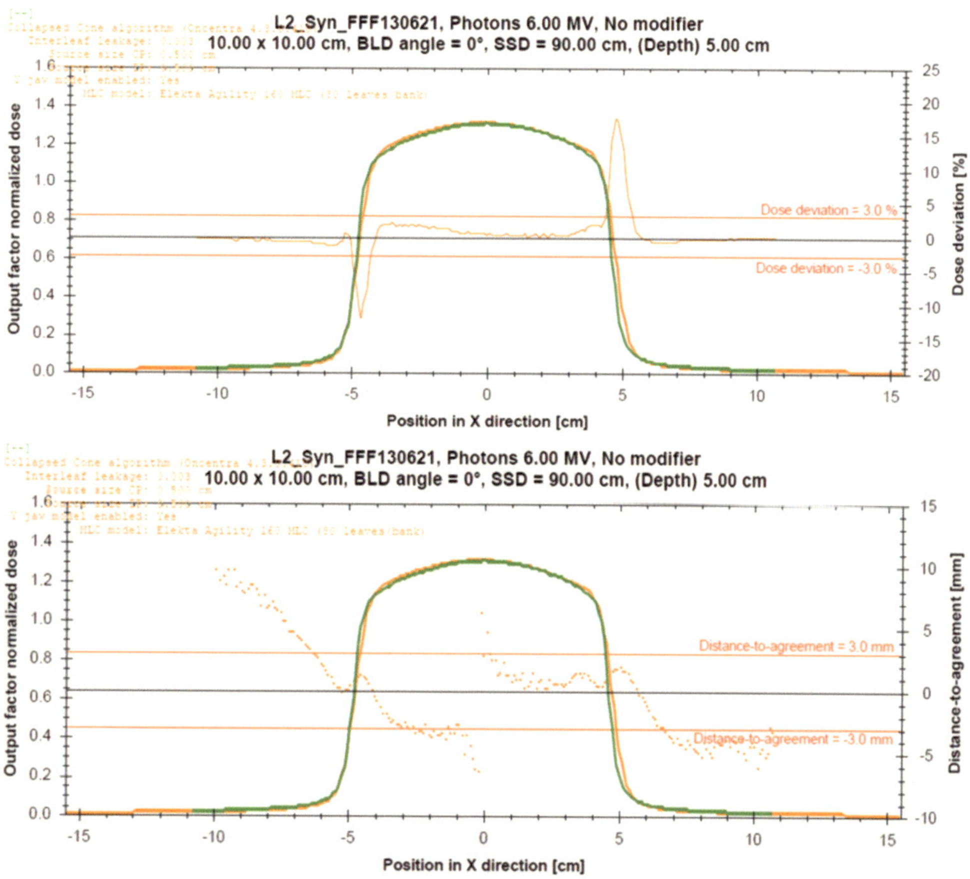


**Figure 8.** Profiles 6MV FFF as examples of the validation of the first linac model (calculated profiles in orange) in the RTPS for the second linac (measurements in green). In the upper part, a dose evaluation is shown (also in orange) referring to the right ordinate, in the lower part a distance to agreement evaluation.

All depth doses for 6MV, 15MV, and 6MV FFF were within the tolerances given by the specifications of the manufacturer. One example is shown in Figure 9. For a total amount of 42 analyzed profiles per energy, the specifications were met in the central region in every case. They were only excessed in single points in the gradient region of 6 MV for the smallest field size (2 cm square). The ESTRO recommendations were failed in single points of the two largest field sizes for the sharp gradient regions.

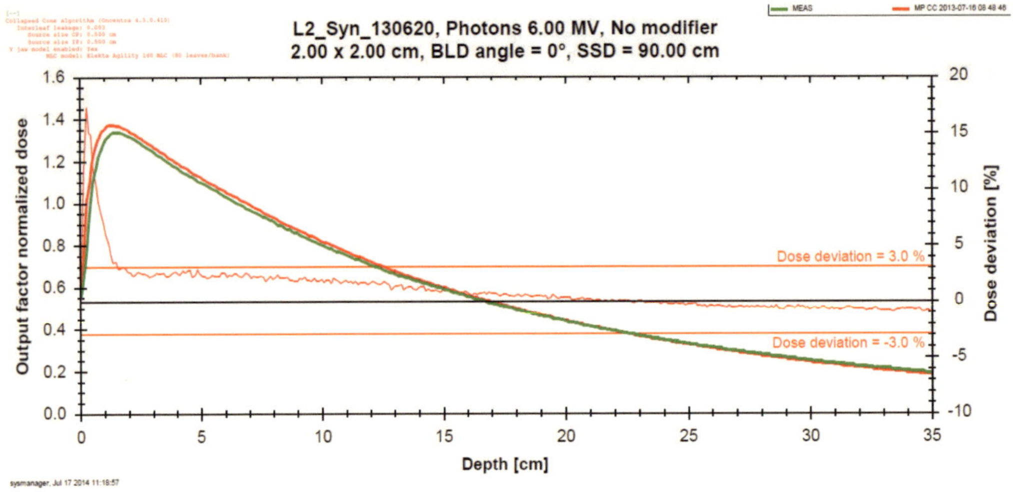


**Figure 9.** Depth dose curves of 6MV photons (calculation in red, measurement in green) for a field size of 2 cm square. The dose difference curve (also in red) refers to the right ordinate.

### 3.3. IMRT and VMAT plan verifications

Figures 10 and 11 show the evaluations of a VMAT plan with 6 MV and of an IMRT plan with 6MV FFF as it is presented on the screen of the OmniPro-I'mRT software. The upper left corner demonstrates the calculated dose distribution in the measurement plane, below the measured dose distribution can be seen. The upper-right corner presents profiles in both planes (calculated in red, measured in green). The position and the direction of these profiles are variable. The lower-right corner shows the gamma evaluation. Pixels in blue and white passed the evaluation. The number of pixels representing a specified value is given in the histogram below. Plans with a passing rate of 95% or more are accepted.

## 4. Discussion

The conformity of the depth dose distributions for the linac twins could be shown in the beam quality and the depth dose curves. The differences are in the order of repeated measurements at the same device. The quite unchanged beam quality of 6MV and 6MV FFF shows that not only the linacs are matched but also the FFF mode is matched to the flattened mode [29]. This is not self-evident; the beam hardening of the flattening filter must be compensated by the energy selection [30] that has been performed by the manufacturer.

The slight difference in the field size for the larger fields is within the specifications of the manufacturer [13]. However, this could be adjusted by the service engineer. From a practical point of view, the measured dose distributions are equivalent.

Therefore, the results allowed to continue with the validation of the model of the first machine in the RTPS for the employment at the second one. Deviations from the calculated dose above

http://dx.doi.org/10.5772/60427

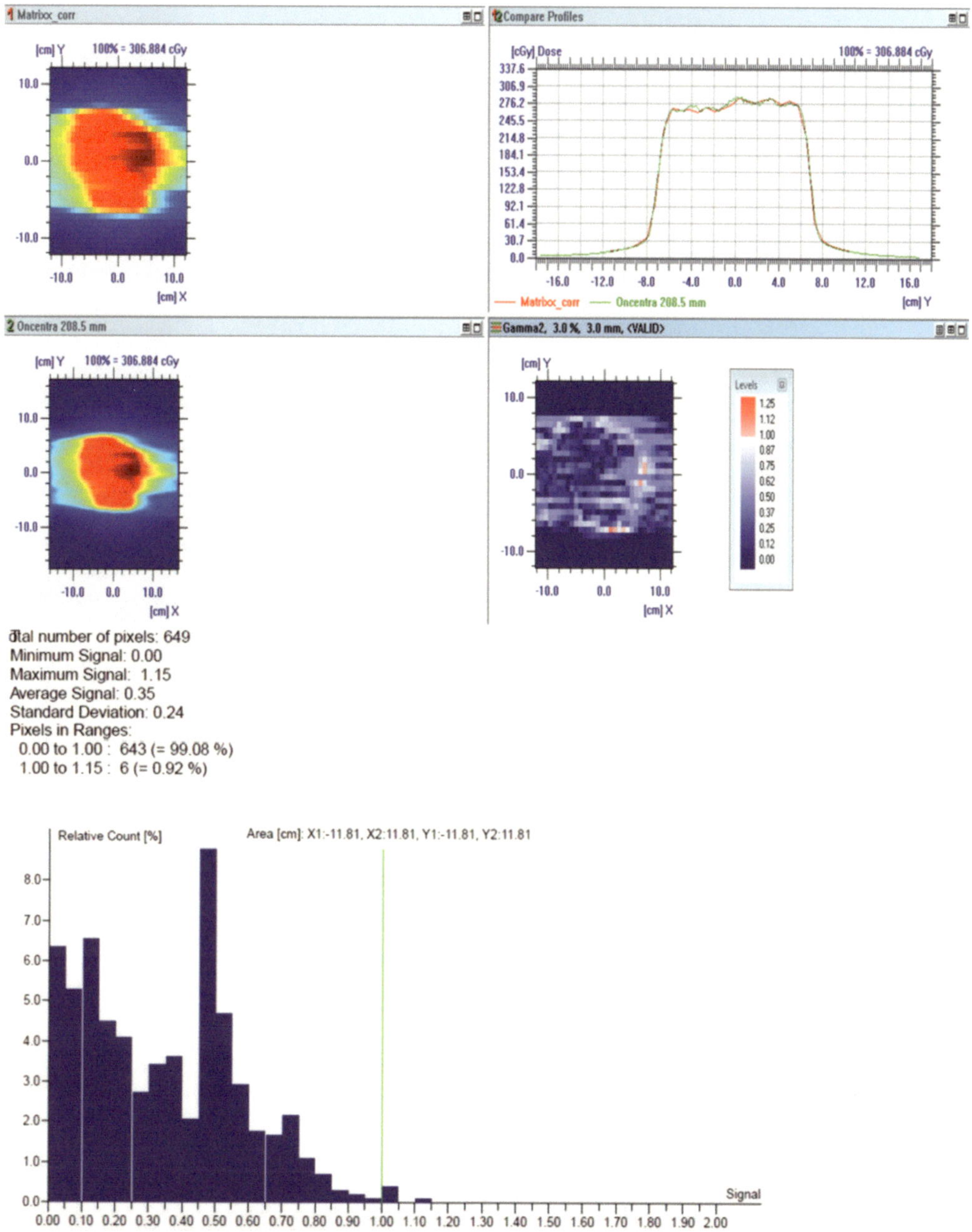


**Figure 10.** Plan verification of a VMAT plan with 6MV calculated for one machine and treated at the other. The gamma image was evaluated with 3% and 3 mm and showed a passing rate of 99%, which is indicated in the histogram below.

the specifications are exceptions in single points and have been assessed clinically to be acceptable. For example, exceeding the gamma criterion of 10% and 2 mm for the largest field size in the sharp gradient area can be traced back to the slightly different field size calibration

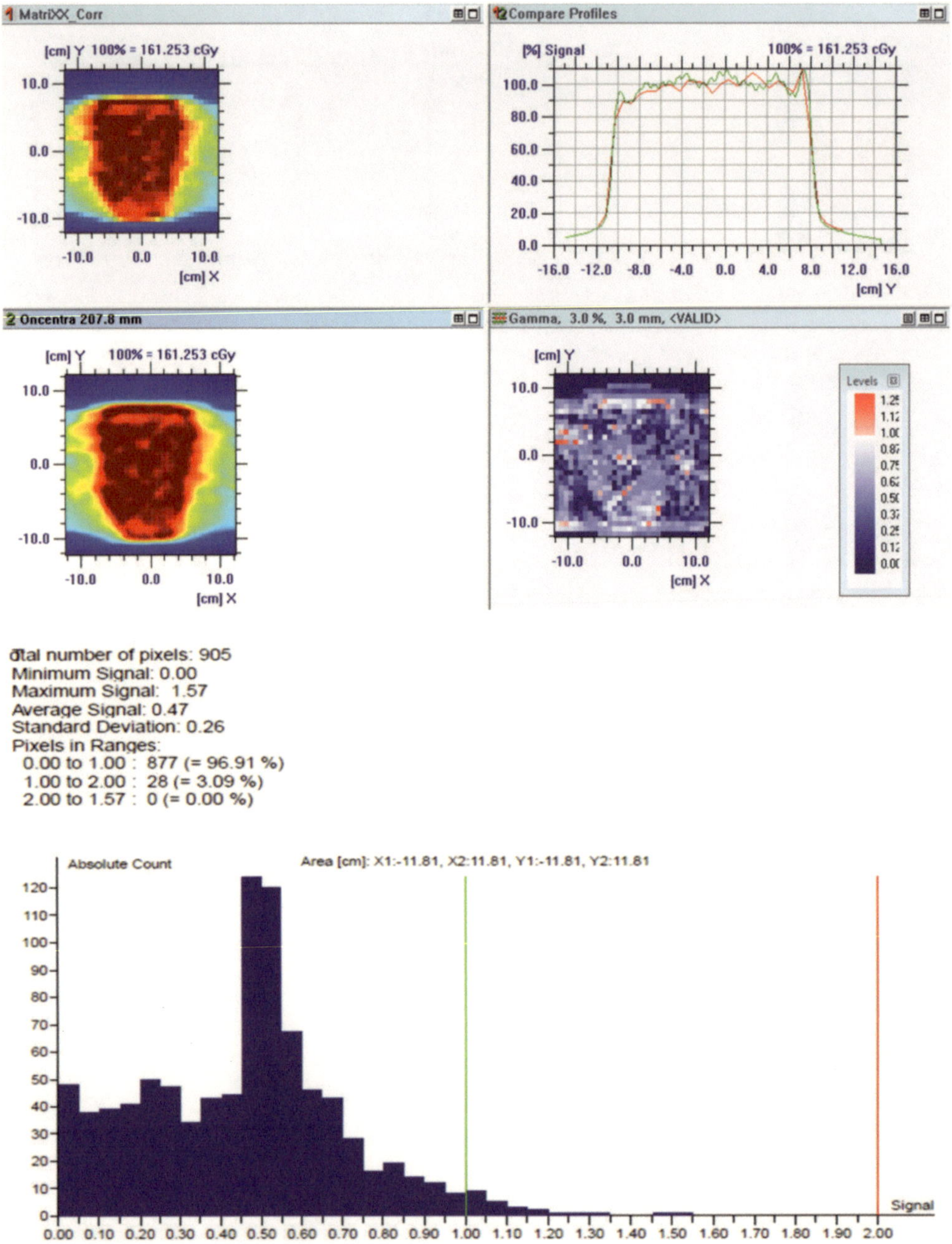


**Figure 11.** Plan verification of an IMRT plan with 6MV FFF, calculated for one machine and treated at the other. The gamma image was evaluated with 3% and 3 mm and showed a passing rate of 97% which is indicated in the histogram below.

of both machines, which has also be seen in the comparison of the measure data. Nevertheless, the specification of the manufacturer with a distance to agreement of 3 mm was met.

http://dx.doi.org/10.5772/60427

Thus, the compatibility of the linac twins has been proven for the depth dose distributions and profiles on both main axes in different depths and for the full range of collimator apertures. The acceptance test of the manufacturer for matched linacs, which only refers to the beam quality, is only a first step and runs too short as it has also been reported for the matching process of another manufacturer [8].

The evaluations of different IMRT and VMAT plans, which had been calculated for the first machine and irradiated at the second one, showed very good results. The passing rates were in the same range as they had been seen for verifications at the "original" machine. The beam model in the RTPS of the first machine has been demonstrated sufficient for the second machine even for plans of very high complexity as IMRT with 6MV FFF or VMAT. This means that the second linac can substitute the first one in cases of breakdown. Chang et al. [10] had similar satisfying results for three matched Varian linacs. However, they first measured the data of all three linacs and combined them by averaging to get composite beam data. Having a time interval of several months or more between the installation of different linacs, as it is given in the case of replacement of old machines, this procedure is not applicable.

## 5. Conclusion

It has been shown that the dose distributions for all photon energies and modes could be adjusted equivalent. The new FFF mode presents no exceptions. Plan verifications of complex IMRT and VMAT plans demonstrate the exchangeability of the linacs also for the FFF mode, allowing continued therapy during downtimes, e.g., service works. Our results confirm that the time and effort for commissioning and quality assurance can be reduced for linac twins:

- There will be only one set of quality checklists, including the tolerance values.
- One beam model for both machines is sufficient in the RTPS.

## Nomenclature

DIN; Deutsches Institut für Normung (German Institute for Standards)

ESTRO; European Society for Radiotherapy and Oncology

FFF; Flattening filter free

IMRT ; Intensity-modulated radiotherapy

Linac; Linear accelerator

RTPS; Radiotherapy planning system

VMAT; Volumetric-modulated arc therapy

## Acknowledgements

The authors thank Esther Illek for the planning and measurements of some VMAT and IMRT plans.

This work was supported by the German Research Foundation (DFG) within the funding program Open Access Publishing.


## Author details


Marius Treutwein*, Petra M. Härtl, Christian Gröger, Zaira Katsilieri and Barbara Dobler

*Address all correspondence to: marius.treutwein@ukr.de

Department of Radiotherapy, Regensburg University Medical Center, Regensburg, Germany


## References


[1] Klein EE, Hanley J, Bayouth J, Yin F, Simon W, Dresser S, Serago C, Aguirre F, Ma L, Arjomandy B, Liu C, Sandin C, Holmes T. Task Group 142 report: quality assurance of medical accelerators. Med Phys 2009;36(9):4197–212.

[2] Low DA, Moran JM, Dempsey JF, Dong L, Oldham M. Dosimetry tools and techniques for IMRT. Med Phys 2011;38(3):1313.

[3] Bissonnette J, Balter PA, Dong L, Langen KM, Lovelock DM, Miften M, Moseley DJ, Pouliot J, Sonke J, Yoo S. Quality assurance for image-guided radiation therapy utilizing CT-based technologies: a report of the AAPM TG-179. Med Phys 2012;39(4): 1946.

[4] Fogliata A, Garcia R, Knoos T, Nicolini G, Clivio A, Vanetti E, Khamphan C, Cozzi L. Definition of parameters for quality assurance of flattening filter free (FFF) photon beams in radiation therapy. Med Phys 2012;39(10):6455–64.

[5] Sahani G, Sharma SD, Sharma, P K Dash, Deshpande DD, Negi PS, Sathianarayanan VK, Rath GK. Acceptance criteria for flattening filter-free photon beam from standard medical electron linear accelerator: AERB task group recommendations. J Med Phys 2014;39(4):206–11.

[6] Marshall MG. Matching the 6-MV photon beam characteristics of two dissimilar linear accelerators. Med Phys 1993;20(6):1743–46.

[7] Hrbacek J, Depuydt T, Nulens A, Swinnen A, Van den Heuvel, Frank. Quantitative evaluation of a beam-matching procedure using one-dimensional gamma analysis. Med Phys 2007;34(7):2917.

[8] Sjöström D, Bjelkengren U, Ottosson W, Behrens CF. A beam-matching concept for medical linear accelerators. Acta Oncol 2009;48(2):192–200.

[9] Grattan MW, Hounsell AR. Analysis of output trends from Varian 2100C/D and 600C/D accelerators. Phys Med Biol 2011;56(1):N11–9.

[10] Chang Z, Wu Q, Adamson J, Ren L, Bowsher J, Yan H, Thomas A, Yin F. Commissioning and dosimetric characteristics of TrueBeam system: composite data of three TrueBeam machines. Med Phys 2012;39(11):6981–7018.

[11] Treutwein M, Härtl PM, Gröger C, Katsilieri Z, Dobler B. Linac twins with flatness filter free option in a radiotherapy department. In: Klöck S (ed.). Joint Conference of the SSRMP, DGMP, ÖGMP. Abstractbook. Zürich, Switzerland; 2014.

[12] Dobler B, Groeger C, Treutwein M, Alvarez-Moret J, Goetzfried T, Weidner K, Haertl P, Koelbl O. Commissioning of volumetric modulated arc therapy (VMAT) in a dual-vendor environment. Radiother Oncol 2011;99(1):86–89. doi:10.1016/j.radonc.2011.01.024.

[13] Elekta Limited. 2013. Elekta Digital Linear Accelerator—Customer Acceptance Tests. Elekta Limited, Crawley (1503568 02).

[14] Norm, DIN 6847-5, 2013-10. Medizinische Elektronenbeschleuniger-Anlagen—Teil 5: Konstanzprüfungen von Kennmerkmalen

[15] Norm, DIN 6875-1, 2004-01. Spezielle Bestrahlungseinrichtungen—Teil 1: Perkutane stereotaktische Bestrahlung, Kennmerkmale und besondere Prüfmethoden.

[16] Norm, DIN 6875-2, 2008-11. Spezielle Bestrahlungseinrichtungen—Teil 2: Perkutane stereotaktische Bestrahlung—Konstanzprüfungen.

[17] Norm, DIN 6875-3, 2008-03. Spezielle Bestrahlungseinrichtungen—Teil 3: Fluenzmodulierte Strahlentherapie—Kennmerkmale, Prüfmethoden und Regeln für den klinischen Einsatz.

[18] Norm, DIN 6875-4, 2011-4. Spezielle Bestrahlungseinrichtungen—Teil 4: Fluenzmodulierte Strahlentherapie—Konstanzprüfungen.

[19] Norm, DIN 6847-6, 2012-09. Medizinische Elektronenbeschleuniger-Anlagen—Teil 6: Elektronische Bildempfänger(EPID)—Konstanzprüfung.

[20] Norm-Entwurf, DIN 6873 Teil 5, 2013-04. Bestrahlungsplanungssysteme—Teil5: Konstanzprüfung von Kennmerkmalen.

[21] Commissioning and quality assurance of computerized planning systems for radiation treatment of cancer. 2004. Vienna: International Atomic Energy Agency (Technical reports series, no. 430).

[22] Nucletron. Radiation Commissioning and Quality Assurance. Nucletron, Veenendaal, NL (Oncentra External Beam v4.3 Oncentra Brachy v4.3, 192.740ENG-08).

[23] Mijnheer B, Olszewska A, Fiorino C, Hartmann G, Knöös T, Rosenwald J, Welleweerd H. Quality Assurance of Treatment Planning Systems—Practical Examples for Non-IMRT Photon Beams. 2004. Brussels, Belgium: ESTRO (ESTRO Booklet, 7).

[24] Nucletron. User Manual. Nucletron, Veenendaal, NL (Oncentra External Beam v4.3 Oncentra Brachy v4.3, 192.729ENG-08).

[25] Low DA, Harms WB, Mutic S, Purdy JA. A technique for the quantitative evaluation of dose distributions. Med Phys 1998;25(5):656–61.

[26] Bundesministerium für Umwelt, Naturschutz und Reaktorsicherheit30.11.2011. Strahlenschutz in der Medizin—Richtlinie zur Verordnung über den Schutz vor Schäden durch ionisierende Strahlen.

[27] Board of Faculty of Clinical Oncology. The timely delivery of radical radiotherapy: standards and guidelines for the management of unscheduled treatment interruptions. 3. Aufl. The Royal College of Radiologists; 2008. http://www.rcr.ac.uk/docs/oncology/pdf/BFCO%2808%296_Interruptions.pdf.

[28] Treutwein M, Hipp M, Koelbl O, Dobler B. Searching standard parameters for volumetric modulated arc therapy (VMAT) of prostate cancer. Radiat Oncol 2012;7108.

[29] Paynter D, Weston SJ, Cosgrove VP, Evans JA, Thwaites DI. Beam characteristics of energy-matched flattening filter free beams. Med Phys 2014;41(5):052103.

[30] Huang Y, Siochi RA, Bayouth JE. Dosimetric properties of a beam quality-matched 6 MV unflattened photon beam. J Appl Clin Med Phys 2012;13(4):3701.

Section 3

# Detection and Measurement

Chapter 8

# Ionizing Radiation Detectors

Marcia Dutra R. Silva

Additional information is available at the end of the chapter

http://dx.doi.org/10.5772/60914

**Abstract**

Ionizing radiation has always been present in the natural environment. However, this radiation is not easily detected and since it also possesses high ionizing power and penetration strength, it constitutes a risk to human health when it is found outside of its acceptable limits. The adverse effects of ionizing radiation on human health need to be systematically monitored in order to prevent damage, overexposure, or even death. The detection of the radiation depends on its particular interaction with a sensitive material, and different types of detectors, in different physical states (solid, liquid or gas), are used to measure selective types of ionizing radiation. New materials such as organic semiconductors, for instance, are being targeted for research and as potential candidates for new perspectives on ionizing radiation sensing.

**Keywords:** Radiation, high energy, detector

## 1. Introduction

Ionizing radiation has always been present in the natural environment. Sources of ionizing radiation are commonly found in water, air, soil, or manmade devices. However, ionizing radiation is situated in the electromagnetic spectrum outside the region of perception of the human eye - visible region - and it has no smell. Thus, it cannot be detected by the human senses. Since the ionizing radiation is not easily detected and it also possesses high ionizing power and penetration strength, it constitutes a risk to human health when it is found outside of its acceptable limits. The adverse effects of ionizing radiation on human health need to be systematically monitored in order to prevent damage, overexposure, or even death. The ability to identify sources of radiation, specific radioisotopes, and measure quantities of radiation is crucial to environmental monitoring, radiation protection, and development of security programs.

Ionizing radiation cannot be directly measured. The detection is done indirectly using an ionizing radiation sensitive material, which constitutes the basis when developing sensors or detectors of radiation. However, there is not a radiation detector that can measure all types of radiation efficiently. The interaction of radiation with matter depends on the nature of the radiation: the electromagnetic radiation, light charged particles, neutrons, or heavy charged particles. Therefore, a detector which efficiently measures a particular kind of radiation could be completely inappropriate for others. The nature of the sensitive material's response to the ionizing radiation and its energy range to be measured will determine the type of detector.

When the ionizing radiation interacts with a sensitive material constituting the detector device, it generates a signal, which can be a pulse, hole, light signal, and many others [1]. The detection of the radiation depends on the particular interactions with the sensitive material, and there are three main and well-established possibilities to relate and categorize the induced radiation with the generated signal in the detector, as shown below:

**i.** The generated signal from the incident radiation is created by the counting of the number of interactions occurring at the sensitive volume of the detector. In this case, the detector is called *counter.*

**ii.** The incident radiation generates a signal that measures the energy that has reached the detector. The detector is named *spectrometer*.

**iii.** The detector measures the average energy incident on a specific point of the sensitive volume, that is, the absorbed radiation dose. Such detectors are known as *dosimeters*.

*A priori* or *a posteriori* application of ionizing radiation detector will indicate which type is more suitable to use for a specific measurement. To measure the radiation in real time, as in the case of evaluating the average radiation of a given location, direct-read instruments such as gas detectors, scintillation detectors, or semiconductor detectors are used. In order to measure the radiation to which a person is exposed, detectors that can be further processed such as photographic emulsions and thermoluminescent dosimeters are used.

Radiation detectors have to two key principles: (i) ionization and (ii) excitation. In ionization-based detectors, electron-ion pairs are generated by enough energy when ionizing radiation reaches atoms of a sensitive material and removes orbital electrons (Figure 1).

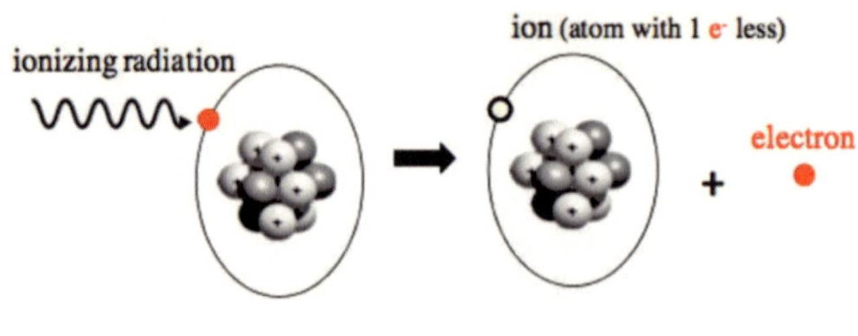


**Figure 1.** Ionization Process.

http://dx.doi.org/10.5772/60914

In excitation-based detectors, bounded electrons are raised to an excited state in the atom or molecule when part of the radiation energy is transferred to them (Figure 2). The electron excited to these states returns to its ground state emitting light in the UV-Visible region.

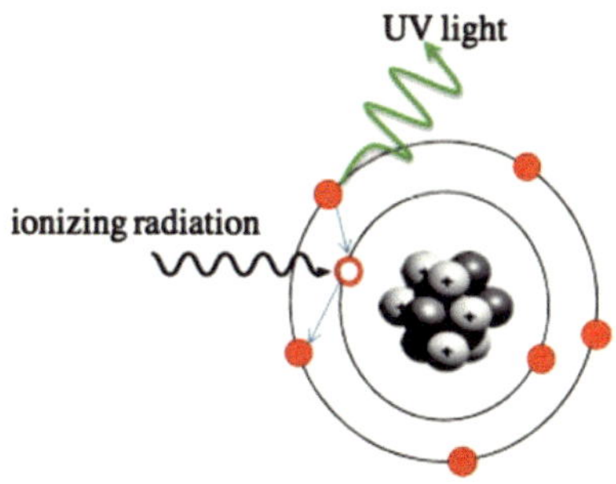


**Figure 2.** Excitation process.

## 2. Detectors

### 2.1. Gas-filled detectors

When a high-energy radiation passes through a medium, it undergoes ionization and releases charges that depend on the excitation radiation energy. In gas detectors, the ionization appears as electron-ion pairs and these charge carriers can be attracted and collected by electrodes [2-4].

In gases, ionized particles can travel more freely than in a liquid or a solid. Therefore, in gas counters the space between the electrodes is filled with a gas and when a voltage is applied an electric field is created by the potential difference between the electrodes. Electrons and positively charged gas atom of each ion pair accelerate to anode and cathode, respectively, resulting in an electric signal (current) in the circuit that can be correlated to radiation exposure and displayed as a value (Figure 3).

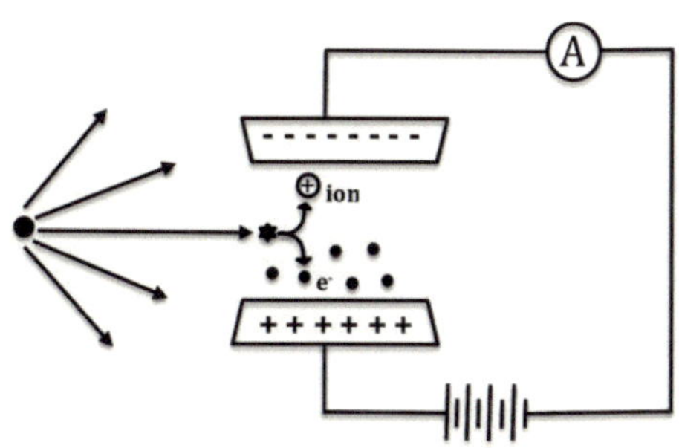


**Figure 3.** Current mode.

Another detection possibility is to acquire the incident radiation signal through pulses (pulse counting mode). In this case, the number of ion-electron pairs generated corresponds to the intensity of the detected pulse (Figure 4). The ionization chamber, proportional counters, and Geiger-Muller counters are examples of gas detectors. Typically, ionization chambers are used in the current mode while proportional counters and Geiger-Muller use the pulse mode to measure the radiation.

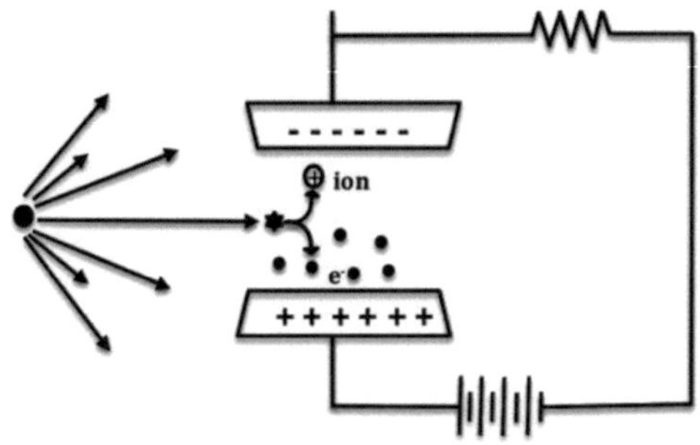


**Figure 4.** Pulse mode counting.

The average energy W required to produce an electron-ion pair varies (20-45eV) depending on the gas used. The average energy W can be expressed as [5]:

$$W = \frac{E_i}{\langle N \rangle} \quad (1)$$

where $E_i$ is the energy deposited by the incident radiation and N the average number of electron-ion pairs formed.

The number of ion pairs generated varies according to the applied voltage for constant incident radiation. The voltages can vary widely depending upon the detector geometry and the gas type and pressure. The different voltage regions are indicated schematically in Figure 5. There are six main practical operating regions, where three are useful to detect ionizing radiation.

**Region (1):** At low voltage, the electric field is not large enough to accelerate electrons and ions. Then, many electrons and ions produced in gas recombine before they reach the electrodes and they are not collected. In this area, the size pulse increases as applied voltage increases, and the recombination rate decreases to the point where it becomes zero. This first region is called recombination and is not useful for counting radiation.

**Region (2):** In the ionization region, the voltage is high enough and each ion pair generated reaches the electrodes. However, the number of the ion pairs does not change when voltage is increased and the curve is flat. Then, the number of ion pairs produced by the radiation is nearly equal to the number of ion pairs collected by the electrodes because there is no recom-

http://dx.doi.org/10.5772/60914

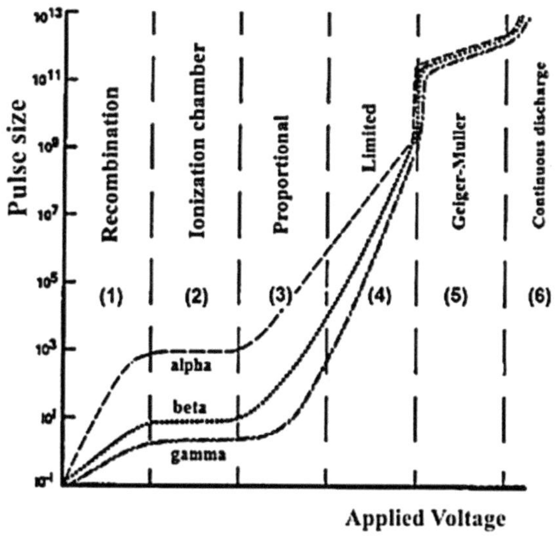


**Figure 5.** Six-region curve for gas-filled detectors.

bination and the voltage is not high enough to produce gas amplification. The ionization chambers work in this region.

**Region (3):** In the proportional region, there is a gas amplification that causes more ion pairs to reach the electrodes than ion pairs are initially produced by radiation. The electrons from the primary ionization acquire enough energy between collisions to produce additional ionizations due to strong electric field. These secondary ions formed are also accelerated causing an effect known as Townsend avalanches, which creates a single large electrical pulse.

Primary ionization

$$X + p = X^{+} + p + e^{-}$$

Secondary ionization

$$X + e^{-} = X^{+} + e^{-} + e^{-}$$

where X is the gas atom, p is the charge particle traversing the gas, and e is the electron.

The number of ion pairs collected divided by the number of ion pairs produced by the primary ionization provides the gas amplification factor. For example, if 50, 000 ion pairs are collected and 10, 000 ion pairs were initially produced, the gas amplification factor is 5. The gas amplification factor varies according to the applied voltage across the electrodes and it also varies with the geometry of the detector. However, it is constant at a specific voltage and for any kind of radiation or energy of radiation. Then, if a voltage increases the gas amplification factor increases proportionally, but if a voltage remains constant the gas amplification factor also does not change. Because of this amplification process, proportional counters are extremely sensitive (<10KeV) while ionization chambers are limited by discriminate particles of >10 keV energy.

The pulse height depends on the detected particle energy. Therefore, different energies of radiation can be distinguished by analyzing the pulse height. For instance, the size pulse from an alpha particle, for a fixed applied voltage, will be larger than the signal from a beta particle. Thus, particle identification and energy measurement are possible by using proportional counters.

**Region (4):** In the limited proportional region, the gas amplification factor is not constant for a given voltage setting and there is no proportionality of the output signal to the deposited energy at a given applied voltage. Additional avalanches occur, leading to additional ionizations and nonlinear effects take place. The nonlinearities observed are due to the high positive ion concentration, which leads to distortion in the electric field. Free electrons due to their high mobility are quickly collected by the electrodes while positive ions are slowly moving. Then, clouds of positive ions are created near the electrode, leading to distortions in gas multiplication. This region is usually avoided as a detection region.

**Region (5):** At high voltages, the electric field is so strong that avalanches of electron-ion pairs are produced and reach the electrodes. A strong signal is produced by these avalanches with shape and height independently of the primary ionization and the energy of the detected photon. This region is called the Geiger-Muller region, and the large output pulse is the same for all photons. Therefore, the energy or even incident radiation particle cannot not be distinguished by GM detectors.

**Region (6):** At still higher voltages (above GM region), the electric field strength is so intense that it itself produces ionization in the gas and completes avalanching. Continuous electric discharges occur between the electrodes and the detectors that operate in this region can be damaged. Therefore, no practical detection of radiation is possible.

### 2.2. Scintillation detectors

Scintillators are materials that exhibit luminescence when excited with ionizing radiation. The scintillation mechanism can be explained by means of the energy-band theory. In this model, a band gap separates the valence band (filled band) of conduction band (usually empty). Thus, via the ionization process, an electron can be excited from the valence band to the conduction band or to the energy states located close to the mid-gap (impurities). An exciton is formed when the electron removed remains electrostatically bonded with the hole left in the valence band. The electron excited to these states decays to the ground state emitting light in the visible range of the electromagnetic spectrum [6]. This visible light interacts with the photocathode and electrons are emitted by photoelectric effect and/or Compton scattering, producing a current in the circuit. However, scintillation detectors produce currents of low intensity and only after the advent of photomultiplier tubes has its use become feasible. In this way, the electrons emitted by photocathode are multiplied by the dynodes in the photomultiplier tube and collected in the anode. As a result, a measurable electrical current is acquired. The output pulse of electrons of a scintillation detector is proportional to the energy of the original radiation.

A good scintillator material is highly efficient in converting incident radiation energy into light. The scintillator must also be transparent to its own light emissions and it must have a short decay time because the transparence is important to a good light transmission to reach the electrode, and the short decay time allows fast response.

Decay time is the time required for scintillation emission to decrease to $e^{-1}$ of its maximum and it can be described as the sum of two exponential components [7, 8]:

$$i(t) = \frac{\omega}{\tau_0} e^{-t/\tau_0} + \frac{1-\omega}{\tau_1} e^{-t/\tau_1} \tag{2}$$

where $\tau_0$ and $\tau_1$ are the decay time constants of the fast and slow components of a scintillator, respectively, and $\omega$ is the weight of the fast component.

The fast component is related to the fluorescence and the slow component is related to phosphorescence or afterglow. These two types of radiative processes (photon emission processes) are well-established in the literature and they are illustrated by the Perrin-Jablonski diagram in Figure 6. The fluorescence occurs in the de-excitation process between singlet electronic states (same spin multiplicity), and it is responsible for the majority of emitting radiative processes due to short decay time ($10^{-9}$s). The phosphorescence occurs in de-excitation process between different multiplicity states (triplet-singlet), in times the order of $10^{-3}$s. The singlet states are represented by $S_n$ and triplet states by $T_n$, where n = 0, 1, 2, 3..., and n = 0 corresponds to the ground state [8, 9]. Other type of delayed emission is the delayed fluorescence (DF), which is a reverse intersystem crossing T1->S1, it is induced thermally or by collisions. Afterglow competes with the scintillation process leading to a decrease of total efficiency of conversion of ionizing radiation into light, and it should be avoided in scintillation detectors [10].

Scintillation detectors are composed of two basic types of detector materials: organic and inorganic. Inorganic scintillators have scintillation properties due to their crystalline structure or due to activators (impurities), which enable scintillation process. Organic scintillators do not need crystal structure or activators because each molecule can act as a scintillation center. The difference in their behavior comes to the different ranges of energy levels excited by the incident radiation. Inorganic scintillators usually respond more slowly than organic scintillators, but they are more efficient than organic materials for detecting ionizing radiation because of their greater density and higher average atomic number. However, organic materials are more flexible and cheaper than inorganic material, leading to numerous scientific efforts to increase their performance in recent decades.

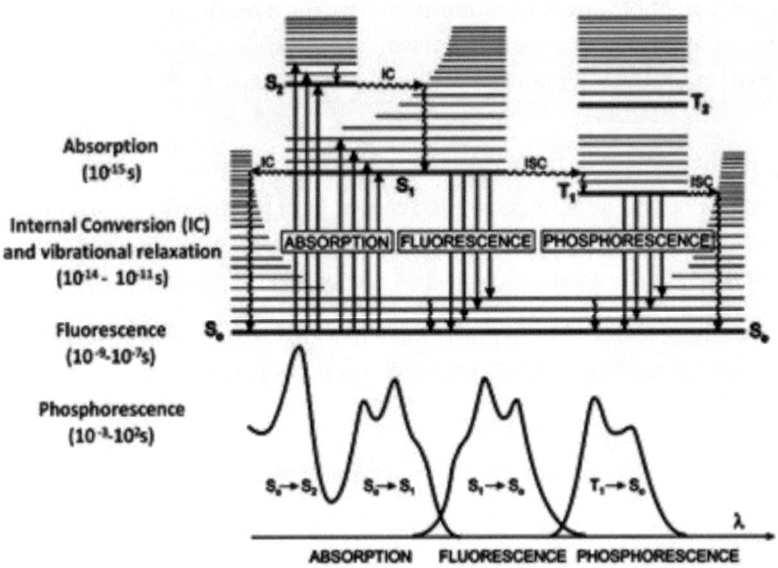


**Figure 6.** Perrin-Jablonski diagram.

Currently, the scintillation detectors have excellent sensitivity to excitation energy and fast response time. Different types of scintillators, in different physical states (solid, liquid, or gas), are used to measure selective types of ionizing radiation. They are widely used in medical applications for image generation (X-rays and tomography), as well as high-energy physics experiments, plant laboratories, airports security (X-rays machines), and radiation sensing for nuclear installations.

## 2.3. Semiconductor detectors

Semiconductors are materials, inorganic or organic, which have the ability to control their electronic conduction depending on chemical structure, temperature, illumination, and presence of dopants. The name semiconductor comes from the fact that these materials usually present an intermediate conductivity between conductors and insulators. Consequently, they have an energy gap less than 4eV [11]. In solid-state physics, energy gap or band gap is an energy range between valence band and conduction band where electron states are forbidden. The valence band is the region where electrons are connected to the lattice atoms. The conduction band is the region that contains the energy levels where free electrons can move through the crystal structure [12, 13]. The width of the forbidden energy band is what categorizes the material as conductor, semiconductor, or insulator (Figure 7).

There are many semiconductors in nature and others synthesized in laboratories; however, the best known are silicon (Si) and germanium (Ge). Silicon has been considered precursor to the revolution that has occurred in recent decades in the electronic area. However, germanium is more used than silicon for radiation detection because the average energy necessary to create an electron-hole pair is 3, 6eV for silicon and 2, 6eV for germanium, which provides the latter a better resolution in energy. In addition, in gamma spectroscopy, germanium is preferred due to its atomic number being much higher than silicon and which increases the probability of γ-ray interaction.

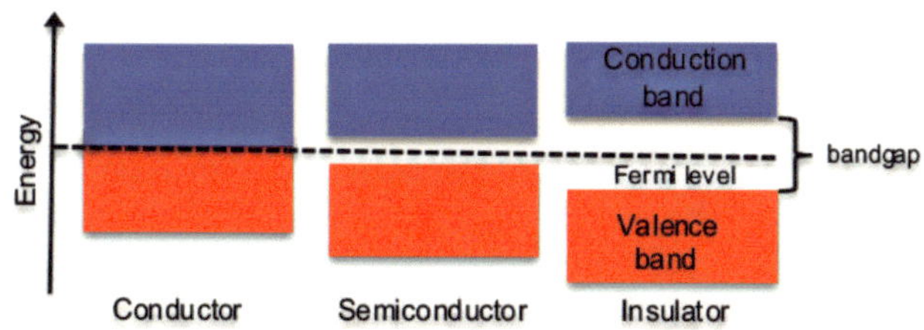


**Figure 7.** Band structure for electron energies in solids.

In semiconductor detectors, also called solid-state detectors, charge carriers are produced and collected by electrodes as in ionization chambers. However, the charge carriers are electrons and holes and not electrons and ions as in ionization chambers. When semiconductor detectors are subjected to high-energy radiation, electron-hole pairs are produced and converted into electric current.

The electron mobility in a gas counter is thousands of times greater than that of the ions. In fact, the electron mobility in semiconductors is roughly equal that of the holes and both types of carriers contribute to conductivity.

Conductivity is the inverse of resistivity and it is defined by

$$J = \sigma E \tag{3}$$

where $J$ is the current density (A/m2), $\sigma$ is the conductivity [A/(V.m)], and $E$ is the electric field (V/m).

Another expression for the current density is:

$$J = eNv \tag{4}$$

where $N$ is the number of charge carriers, $e$ is the elementary charge, and $v$ is the speed of carriers.

The following equation is obtained by using Equations 3 and 4:

$$\sigma = eN\frac{v}{E} \tag{5}$$

The ratio $\boldsymbol{v}/\boldsymbol{E}$ is called carrier mobility μ:

$$\mu = \frac{v}{E} \tag{6}$$

The expression for the conductivity becomes:

$$\sigma = e\left(N_e\ \mu_e + N_h \mu_h\right) \tag{7}$$

where $N_e$ and $N_h$ are carrier concentrations and $\mu_e$ and $\mu_h$ are the mobilities of electrons and holes, respectively, and according to this equation, the conductivity changes if the mobility of charge carriers and/or their concentrations change. Thus, both terms in the right-hand side of Equation 7 contribute to the conduction in semiconductor detectors.

A small energy is required to create an electron-hole pair in semiconductor materials (~3 eV for germanium) compared to the energy needed to create an electron-ion pair in gases (~30 eV for typical gas detectors) or to create an electron-hole pair in scintillators (~100eV) [14]. As a consequence, a great number of electron-hole pairs are produced and reach the electrodes, increasing the number of pairs per pulse and, then, decreasing both statistical fluctuation and signal/noise in the preamplifier. This generates a big advantage over other detectors and the output pulse provides much better energy resolution. Moreover, the small sensitive area used to detect radiation (few millimeters) and the high speed of charge carriers provide an excellent charge collection time (~$10^{-7}$ s).

The energy resolution, R, determines the ability of the system to distinguish two energies that are very close to each other, and that constitute an important parameter in the spectral detection of ionizing radiation (Figure 8). It is commonly defined as:

$$R = \frac{FWHM}{H_0} \tag{8}$$

where FWHM is the full-width-at-half-maximum and $H_0$ is the peak centroid.

In order for a semiconductor to act as a radiation detector, the active area to radiation must be free of excess electrical charges (depleted). The depletion region can be formed through the use of very high purity materials like High Purity germanium (HPGe) or PN junctions. PN junctions are obtained when an n-type semiconductor (excess of electrons) is placed in contact with a p-type semiconductor (excess of holes). Then, electrons and holes diffuse from n-region to p-region and from p-region to n-region, respectively, and they recombine around the interface. The ions, which are left behind by electrons and holes that were recombined, create an electric field that will attract more electrons and holes until there is no more charge carriers to recombine (Figure 9).

http://dx.doi.org/10.5772/60914

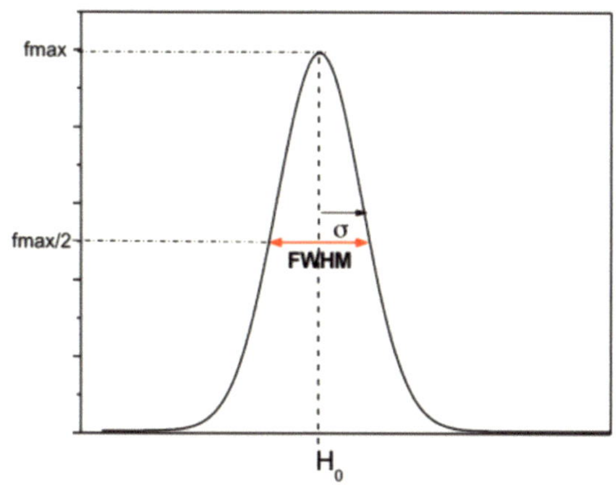


**Figure 8.** FWHM for a Gaussian distribution. In this case, the FWHM results related to the $\sigma$ as FWHM = *2.35* $\sigma$.

At this moment, if the ionizing radiation interacts with the semiconductor in this depleted region, electrons are raised to the conduction band leaving behind holes in the valence band and producing a large number of electron-hole pairs. If a voltage is applied across the semiconductor, these carriers are readily attracted to the electrodes and current flows into circuit resulting in a pulse. The size of the pulse is directly proportional to the number of carriers collected, which is proportional to the energy deposited in the material by the incident radiation.

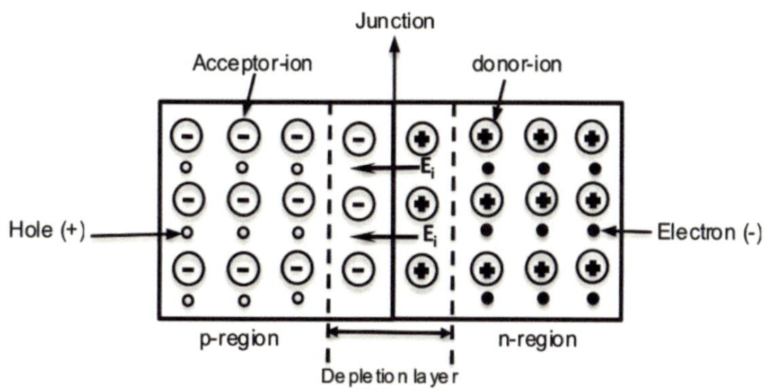


**Figure 9.** PN junction.

In semiconductors, if the temperature increases, electrons can be thermally excited from the valence band to the conduction band. Consequently, some semiconductor detectors must be cooled so as to reduce the number of electron-hole pairs in the crystal in the absence of radiation. Although solid-state detectors can be manufactured much smaller size than those of equivalent gas-filled detectors and they have short response time, seconds compared to the hours of TLD detectors, they are still expensive because they need to be cooled. Thus, they are used when higher resolution is required; if higher efficiency is necessary, scintillation detectors are used.

Different semiconductor materials and device arrangements are used, depending on the type of radiation to be measured and the aim of application. The types of radiation that can be measured with semiconductor detectors comprise a large range of the electromagnetic spectrum: <1 eV to ~10 MeV for photons and energies above keV for charged particles. Commonly, semiconductor detectors are employed for beta particles or gamma radiation because the heavy charged particles cause more radiation damage. They are widely used in nuclear power station electronic dosimeters and portable survey instruments in gamma spectroscopy systems.

### 2.4. Thermoluminescent dosimeters

The amount of radiation absorbed by the human body can be determined through radiation dosimetry. A dosimeter has to correlate the absorbed radiation with biological effects induced in humans. The physical quantity that quantifies this relationship is called *absorbed dose*. The absorbed dose, D, of any ionizing radiation can be considered as the amount of energy given to the medium by ionizing particles or photons per unit of mass *dm* [2, 14, 15]:

$$D = \frac{d\bar{\epsilon}}{dm} \tag{9}$$

where $d\bar{\epsilon}$ is the average energy deposited by the radiation on a point P.

The SI unit of absorbed dose is the Gray (Gy), defined as: 1Gy = 1 J/kg. The obsolete units for dose are the rad (radiation absorbed dose) and the centigray (cGy): 1rad = $10^{-2}$ J/kg = 1cGy.

Thermoluminescent dosimeters (TLDs) are the foremost used devices for personal dosimetry. They are composed of crystal devices that emit light when are heated. The TLD reading device is able to calculate the amount of light released during heating, which can then be correlated with the absorbed dose received and stored by the TLD dosimeter.

A useful model of the thermoluminescence mechanism is provided in terms of the band model for solids. Thus, when a thermoluminescent crystal is exposed to ionizing radiation, electrons are quickly promoted to their conduction band through direct excitation process. However, some electrons are trapped by metastable states and when the material is subjected to thermal stimulation, they have enough energy to leave the trap states and recombine with holes that were left in the valence band. The excess of energy in this process is conserved by radiative deactivations emitting light, which is proportional to the absorbed ionized dose [16].

Figure 10 shows a model of energy bands with electronic transitions in thermoluminescence process.

http://dx.doi.org/10.5772/60914

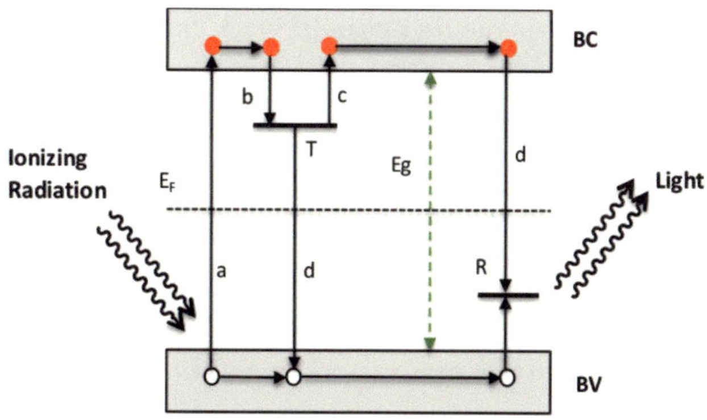


**Figure 10.** Model of energy bands in thermoluminescence process. (a) excitation and generation of electron-hole pair, (b) trapping, (c) de-trapping by thermal stimulation, (d) recombination. T is the center of the trap, R is the recombination center, $E_F$ is the Fermi level, and $E_g$ is the bandgap.

The heating of the TLD dosimeter to assess the accumulated radiation dose is done in temperature ramps and each temperature value is associated with a value of the light intensity (Figure 11). Thus, through thermoluminescence photons emission it is feasible to establish a curve of thermoluminescence intensity versus temperature that is called TL glow curve. The area under the TL glow curve is directly proportional to the number of emitted photons and, thereby, proportional to radiation dose received.

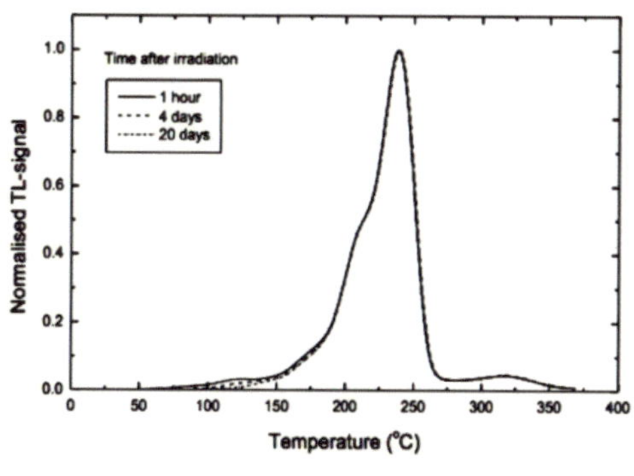


**Figure 11.** TL glow curve of LiF:Mg, Ti measured with a TLD reader at a low heating rate.

Thermoluminescent crystals possess good levels of deeper traps that require greater thermal energy to release the carrier, thus they can accumulate energy for a longer period of time. Many materials are purposely doped to create impurity levels; others such as LiF (lithium fluoride) already have natural impurities and intrinsic defects. Other substances are used as materials for thermoluminescent dosimetry, for example, CaSO4:Dy (calcium sulfate doped with dysprosium); the CaSO4:Mn (calcium sulfate doped with manganese); and CaF2 (fluorite).

A thermoluminescent crystal can be used as dosimeter only if it presents high emission efficiency, good stability on temperature ranges of work, high resistance to environmental variations and linear radiation dose-response.

The principal advantages of TLD dosimeters are:

- High sensitivity over a wide dosage range.
- Small and varied forms.
- Can be used several times.
- They are equivalent to human tissues.
- They have high degree of accuracy and precision in the measurements.

Among the disadvantages are:

- Infeasibility of rereading.
- The necessary instrumentation for the measurement has high cost.
- Sensitivity varies with the time after irradiation.
- Readings and the results are not immediate.
- They present fading for sensitivity to light and moisture.

### 2.5. Chemical detectors

In chemical dosimetry, the ionizing radiation produces chemical changes in the medium that can be measured by using a suitable measuring system. Oxidation, reduction, and chemical dissociation are the principal mechanisms of chemical detectors.

The intensity of these changes is characterized by radiation chemical yield (G), which is defined as a number of molecules, ions, atoms, or free radicals of product or dissolved reaction components for 100 eV of absorbed energy, or even defined as the mean number of moles produced/destroyed by mean energy transmitted to the matter [2]:

$$G = \frac{\overline{n}}{\overline{E}} \tag{10}$$

where $\bar{n}$ is the mean moles number and $\bar{E}$ is the mean transmitted energy. $G$ SI unit is $mol.J^{-1}$.

The most widely used chemical dosimetry standard is the Fricke dosimeter. The Fricke dosimetry system provides a reliable means for measurement of absorbed dose to water by ferrous ions oxidation. The dosimeter consists of a solution with 1 mmol/l ferrous sulfate (or ferrous ammonium sulfate), 1 mol/l NaCl, and 0.4 mol/l sulfuric acid. When the Frick solution is irradiated, the ferrous ions, Fe2+, are oxidized by radiation to ferric ions, Fe3+ [17]. The ferric ion concentration is determined by spectrophotometry, which measures absorption peaks at wavelengths of 224 nm and 304 nm. In this case, G-value is defined as the number of moles of ferric ions produced per joule of the energy absorbed in the solution. The usual range of the Fricke dosimeter is from 30Gy to 400 Gy.

### 2.6. Calorimetric detectors

Calorimetric methods measure the dose of radiation by measuring the temperature increase in a medium. Although the basic principles of calorimeters are very simple, they have technical problems to ionizing radiation sensing and they have been viewed as complex to make and operate [18]. Small temperature response to low dose of radiation and necessity of extremely thermic isolation are some problems of this type of detector. Therefore, few laboratories use these detectors; however, efforts have been made in order to increase their performance.

### 2.7. New materials to ionizing radiation sensing

Despite the well-established known techniques and detectors for ionizing radiation, the field still has a lack of new materials and sensor devices. The use of ionizing radiation in industrial processes, in clinics, hospitals, universities, and research centers has increased considerably and consistently in the past few years. In addition, the inspection and monitoring of aircrews is a current concern and should be mandatory to all flights in the near future. Thus, the development of new materials sensitive to ionizing radiation and robust devices, faster and more accurate, is of crucial importance to this research field and its direct applications.

In the last two decades, there was an effort to combine the energy sensitivity found in semiconductor devices with the low cost and flexibility of organic semiconductor-based conjugated polymers. In this fashion, oligomers and polymers such as PPV (poly p-phenylene vinylene) [19], MEHPPV (poly (2-methoxy, 5- (2 -ethyl-hexoxy) -p-phenylene vinylene) [20, 21], P3HT (poly 3-hexylthiophene) [22, 23], and pentacene [24] have become the target of research and are potential candidates for new perspectives to ionizing radiation sensing. Use of these materials, which have known properties and have been studied, have played an important role in the study of ionizing radiation effects on polymers (Figure 12).

In the interaction of high-energy radiation with semiconductors, primarily there occur excitations and ionizations that generate ions and electrons. The electrons generated (primary electrons) will interact again with the environment and generate secondary excitations that will produce electron-hole pairs. Therefore, the efficiency of the material with the highly energetic radiation will depend on its stopping power or absorption efficiency, the limited capacity of producing electron traps and its ability to grow large areas. Semiconductor polymers generally have high efficiency luminescence and absorption in the UV-Vis region; they can also form films producing large areas, and, hence, they constitute a new alternative in the area of radiation detectors.

In the field of electromagnetic radiation, there are several possible interactions of the most energetic radiation with matter: mainly, the photoelectric effect, Rayleigh scattering, Compton effect, and production of electron-positron pairs. Eventually, these interactions can lead to temporary or permanent modifications. These changes are called effects of degradation. They

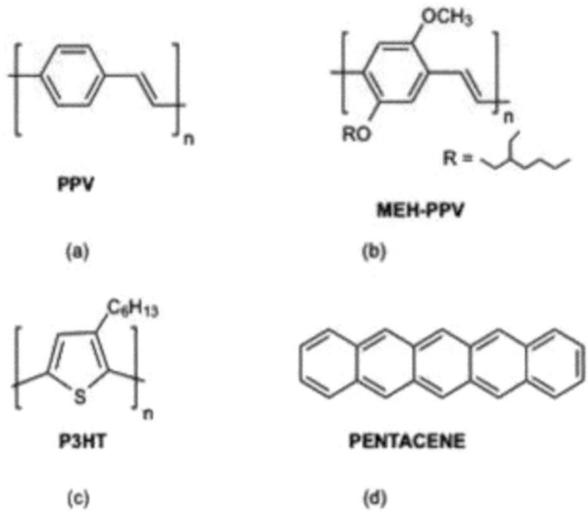


**Figure 12.** Chemical structure of polymers: (a) PPV (b) MEHPPV (c) P3HT and (d) pentacene.

may be superficial when there is change only in the physical appearance (color, transparency, etc.) or they may be structural.

Polymer degradation effects have been reported such as scission [25], cross-linking [26, 27], and photobleaching [28]. In scission, there occurs break of the main chain into smaller molecules, reducing its molecular weight. In cross-linking, due to link between two polymer chains or between two big radicals, there is a formation of an insoluble portion with increasing molecular weight. Decrease in viscosity and increase of ductility are effects of scission. Increase of hardness, viscosity, and brittleness are some of the macroscopic effects of cross-linking. Photobleaching occurs when the fluorescent signal of a fluorophore disappears permanently due to photon-induced chemical damage and covalent modification.

Degradation effects are often considered problems such as the oxidation effects of medical implants based on polyethylene after irradiation sterilization, for example[29]. However, the ionizing radiation degradation effects are not sometimes bad results. Many times they are desirable, as in the creation of integrated circuits, decreasing the molecular weight to make a material compatible with the other, in polysaccharides, for use in health care products, cosmetics, textile and food industry, or even to increase viscosity or resistance materials, for instance [30-32].

Polymer interaction with gamma radiation has been studied since the 1970s and different effects have been observed depending on the chemical structure of the polymer and the energy range used for irradiation process. The mechanisms involved in the interaction of gamma radiation with polymers have not been fully elucidated, but changes in conductivity and optical properties have been reported mainly in polyaniline [33] and on PPV and its derivatives. The results indicate the feasibility of using semiconductor polymers as gamma radiation detectors.

http://dx.doi.org/10.5772/60914

The interest in the use of conductive polymers in this area is due to the adjustability of its luminescence properties and conductivity, and they also have a lower cost than inorganic semiconductors. However, the use of polymers as radiation sensors is recent and few studies are reported in the literature. Among them, the highlighting results are P3HT as the active layer of OLEDs and OFETs for sensing radiation [20], and the MEH-PPV in halogenated solutions for detection of low doses of gamma radiation.

Studies using MEH-PPV have demonstrated that the use of solutions is effectively more sensitive to gamma radiation than solid state. Current knowledge shows that polymeric materials are more sensitive to gamma radiation when solubilized in halogenated solvents [34]. The halogens are well-known to have large cross section for interaction with gamma radiation.

The main results obtained on irradiated P3HT devices were a significant improvement in conductivity with increasing gamma irradiation dose. Polythiophenes irradiated with gamma radiation go to a polaronic state and then stabilized for a bipolaronic and neutral state of the chain, where they remain in the oxidized state. Undoubtedly, the result enables the P3HT as radiation sensor and it leads to a great leap regarding the use of OLEDs and OFETs devices in the space area. However, the order of radiation dose used on P3HT (kGy) is very high for using in personal dosimetry (order of ten grays), for example.

In contrast, studies of MEH-PPV with gamma radiation at this order of dose have shown significant results compatible with personal dosimetry area. However, the results were limited to the use of the polymer in solution, due to the effect being dependent on the solvent. In other words, the effect is indirect: the radiation breaks the solvent chain and the radicals derived from solvent attack break the polymer chain. The attack occurs at the vinylene, breaking the double bond and leading to the conjugation break displayed as blue shift in optical measurements. This experimental result has been corroborated by theoretical studies and the attack mechanism on vinylene is well-established [35]. Figure 13 shows MEHPPV after ionizing radiation interaction.

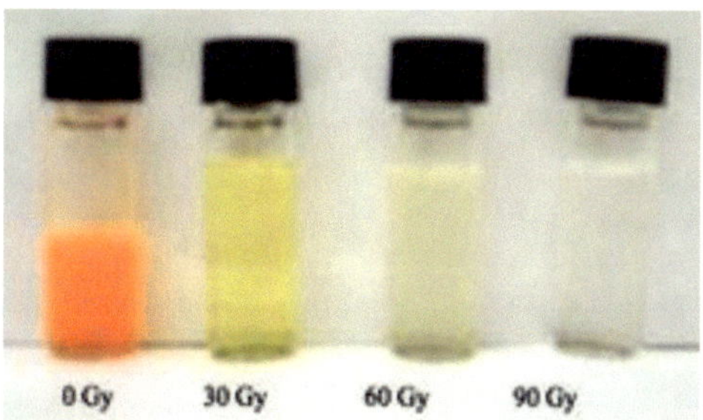


**Figure 13.** MEHPPV nonirradiated (0Gy) and irradiated at 30, 60, and 90Gy doses of gamma radiation.

The principal disadvantage of MEH-PPV in the interaction with the gamma radiation is its limitation of use in optoelectronic devices due to the effect of this range of dose not included the utilization in film. Moreover, with the breaking of the chain and the conjugation length, the MEH-PPV has not sufficient extension of conjugation for a good conduction of electrons and neither has any chain doping as P3HT, which is also a limiting factor for their use in optoelectronic devices. Thus, its use is limited as optical sensor radiation in solution and can not be reused. Recalling that, despite limited use, this type of sensor provided an important advance due to measurement method be cheaper and affordable than other types of detectors. Semiconductor detectors based in conjugated polymers do not need be cooled, the instrumentation used for reading is simple, and polymers are cheaper and easier to process.

## 3. Applications

In short, a radiation detector is a device used to track, detect, or identify high-energy particles or radiation from natural or artificial sources such as cosmic radiation, nuclear decay, particle accelerators, and X-rays.

Since it is not possible for a single detector to measure all types of radiation efficiently, various types of detectors made of different materials are used in the sensing of specific types of radiation. The main types of radiation detectors and applications are summarized in Table 1.

| Instrument Types | Detection Principle | Applications |
|---|---|---|
| **Gas-filled detectors** | Ionization of air or other gases (ionization chambers). | Direct measurement of exposure or exposure rates. |
| | Ionization of gas with multiplication of electrons in detector (proportional counters and Geiger-Muller). | Detection of individual events, i.e., alpha or beta particles and secondary electrons, for measuring activity (in samples or on surfaces); detecting low intensities of X-rays or gamma radiation. |
| **Scintillators** | Light emission (solids). | Photons; energy spectrometry; e.g., NaI (Tl). |
| | | Alpha particles; detection only (ZnS (Ag)). |
| | Light emission (liquid). | Detection of low-energy gamma and beta emitters, for measuring activity (in low-activity sources). |
| **Semiconductor detectors** | Ionization, excitation. | Detection of individual events (alpha and beta particles); e.g., diodes and silicon barrier detectors. |
| | | Detection and energy measurement of photons or particles; primarily for laboratory use; gamma spectroscopy; X-rays; dosimeters; e.g., Germanium detectors. |
| **Thermoluminescent detector (TLD)** | Excitation of crystal; light release by heating. | Personal and environmental exposure monitoring. |
| **Photographic film** | Ionization of Ag Br. | Personal exposure monitoring. |

**Table 1.** Instrument types and its applications.

http://dx.doi.org/10.5772/60914

Despite the variety of sensor devices, the demand for new materials that can detect ionizing radiation efficiently and at as low a cost as possible is essential to the development of this area. In this context, many other materials like polymeric semiconductors, for instance, have been target of research in the last years (section 2). They constitute promising materials for radiation sensing, although no polymeric device for radiation detection is still available for practical use at a large scale.

Unfortunately, the field lacks new sensor devices that are more practical, fast, and accurate for the maintenance and safety of human life in the natural environment as well as in the complex areas of modern civilization.

## Author details

Marcia Dutra R. Silva

Address all correspondence to: marciadrs@yahoo.com.br

Materials Spectroscopy Group - GEM, Physics Institute - INFIS, Federal University of Uberlandia - UFU, Brazil

## References

[1] Tsoulfanidis N. Measurement and Detection of radiation. 2nd ed. Francis T&, editor. 1995. p. 614.

[2] Khan FM. The Physics of Radiation Therapy. Lippincott Williams & Wilkins; 2012. p. 592.

[3] Dendy PP, Heaton B. Physics for Diagnostic Radiology. CRC Press; 2011. p. 695.

[4] Attix FH, Rogers DWO. Introduction to Radiological Physics and Radiation Dosimetry. Med. Phys. 1987. p. 692.

[5] Simon Wedlund C, Gronoff G, Lilensten J, Ménager H, Barthélemy M. Comprehensive calculation of the energy per ion pair or W values for five major planetary upper atmospheres. Ann. Geophys. 2011;29:187-195.

[6] Price WJ. Nuclear Radiation Detection. Mac Gray-Hill, editor. 1958. p. 382.

[7] Zhong W-L, Li Z-H, Yang C-G, Cao J. Measurement of decay time of liquid scintillator. Nucl. Instruments Methods Phys. Res. Sect. A Accel. Spectrometers, Detect. Assoc. Equip. 2008;587:300-303.

[8] Valeur B. Molecular Fluorescence Principles and Applications. Weinheim: Wiley-VCH; 2001. p. 381.

[9] Condon EU. On the Theory of Intensity Distribution in Band Systems. Berkeley: University of California, Berkeley; 1926. p. 78.

[10] Rodnyi PA. Physical Processes in Inorganic Scintillators. CRC Press 1997, editor. Taylor & Francis; 1997. p. 240.

[11] Yu PY, Cardona M. Fundamentals of Semiconductors. 1999. p. 617.

[12] Ashcroft NW, Mermin ND. Solid state physics. Saunders College; 1976. p. 826.

[13] Kittel C. Introduction to solid state physics. 5 ed. Wiley; 1976. p. 599.

[14] Knoll GF. Radiation Detection and Measurement. 3rd ed. Wiley, editor. 2010. p. 860.

[15] (2007) I. The 2007 Recommendations of ICRP. Ann. ICRP 37(2-4), ICRP Publ. 103.

[16] Horowitz YS. Thermoluminescence and thermoluminescent dosimetry. Horowitz YS, editor. Florida: CRC Press; 1984.

[17] Frick H, Hart EJ. Chemical Dosimetry. In: Attix FH, Roesch WC, editors. Radiat. Dosim. II. New York: Academic Press; 1966. p. 167-239.

[18] Kase KR, Bjärngard BE, Attix FH. The Dosimetry of Ionizing Radiation. Academic Press; 1985. p. 384.

[19] Silva MDR, Gançalves AA, Silva R a., Marletta A. Gamma radiation effects on absorbance and emission properties of layer-by-layer PPV/DBS films. J. Non. Cryst. Solids. Elsevier B.V.; 2010;356:2429-2432.

[20] Raval HN, Tiwari SP, Navan RR, Rao VR. Determining ionizing radiation using sensors based on organic semiconducting material. Appl. Phys. Lett. 2009;94:123304.

[21] Silva E a. B, Borin JF, Nicolucci P, Graeff CFO, Netto TG, Bianchi RF. Low dose ionizing radiation detection using conjugated polymers. Appl. Phys. Lett. 2005;86:131902.

[22] Zotti G, Zecchin S. Decay of electrochemically injected polarons in polythiophenes. Bipolaron stabilization by structural factors. Synth. Met. 1997;87:115-118.

[23] Wei Z, Xu J, Pu S, Du Y. Electrosyntheses of high-quality freestanding poly(fluorene-co-3-methylthiophene) films with tunable fluorescence properties. J. Polym. Sci. Part A Polym. Chem. 2006;44:4904-4915.

[24] Raval HN, Sutar DS, Nair PR, Rao VR. Investigation of effects of ionizing radiation exposure on material properties of organic semiconducting oligomer - Pentacene. Org. Electron. Elsevier B.V.; 2013;14:1467-1476.

[25] Thompson LF, Willson CG, Bowden MJ. Introduction to microlithography. American Chemical Society; 1994. p. 527.

[26] Mitsui, H., Hosoi, F. & Kagiya T. y-radiation-Induced Cross-Linking of Polyethylene. Polym. J. 4:79-86.

[27] Mitsui, H., Hosoi, F. & Kagiya T. Accelerating Effect of Acetylene on the y-Radiation-Induced Cross-Linking of Polyethylene. Polym. J. 1974;6:20-26.

[28] Rabek JF. Photodegradation of polymers: physical characteristics and applications. Springer; 1996. p. 212.

[29] Saum; Kenneth Ashley, Sanford; William Michael, Dimaio, Jr.; William Gerald, Howard J. EG. United States Patent US6562540. 2003.

[30] Delides CG, Panagiotalidis CZ, Lega-Panagiotalidis OC. The Degradation of Cotton by Ionizing Radiation. Text. Res. J. 1981;51:311-317.

[31] Yoshii F, Zhao L, Wach RA, Nagasawa N, Mitomo H, Kume T. Hydrogels of polysaccharide derivatives crosslinked with irradiation at paste-like condition. Nucl. Instruments Methods Phys. Res. Sect. B Beam Interact. with Mater. Atoms. 2003;208:320-324.

[32] Ershov BG. Radiation-chemical degradation of cellulose and other polysaccharides. Russ. Chem. Rev. IOP Publishing; 1998;67:315-334.

[33] Laranjeira JMG, Khoury HJ, de Azevedo WM, de Vasconcelos EA, da Silva EF. Polyaniline nanofilms as a monitoring label and dosimetric device for gamma radiation. Mater. Charact. 2003;50:127-130.

[34] Bronze-Uhle ES, Borin JF, Batagin-Neto A, Alves MCO, Graeff CFO. MEH-PPV hypsochromic shifts in halogenated solvents induced by $\gamma$-rays. Mater. Chem. Phys. 2012;132:846-851.

[35] Lavarda FC, Graeff CFO. Ionizing radiation induced degradation of poly (2-methoxy-5- (2 ′ -ethyl- hexyloxy) -1, 4-phenylene vinylene) in solution Ionizing radiation induced degradation of poly. J. Appl. Phys. 2011;110:073510.

Section 4

# Industrial Application

Chapter 9

# Ionizing Radiations in Entomology

Valter Arthur, Andre Machi and
Thiago Mastrangelo

Additional information is available at the end of the chapter

http://dx.doi.org/10.5772/60409

**Abstract**

Radiation in the form of particles ($\alpha$ or $\beta$ particles and neutrons) or electromagnetic waves (gamma or X-rays) can induce biological effects in insect cells like in other living cells. Ionization and chemical damages to organic molecules can be caused directly (mostly by particulate types of radiation) or indirectly by free radicals. Radioinduced ions and radicals, most of them coming from water radiolysis, may react with neighboring molecules to produce secondary DNA radicals or even chain reactions, particularly in lipids, and most of the significant biological effects results from damage to DNA. Currently, more than 300 species of arthropods, mostly of economic importance, have already been subjected to irradiation studies for basic research, pest control applications, and disinfestation of commodities (quarantine and phytosanitary purposes). This chapter focused on insect sterilization and disinfestation by ionizing radiations in view of the socioeconomic impacts. The release of insects that are sterile after exposure to radiation aiming to control or eradicate pest populations revealed to be a revolutionary tactic in the area-wide management of pests, and many successful cases with the application of the sterile insect technique can be found around the globe. The use of ionizing radiations to inhibit the spread of quarantine insects represents an important alternative postharvest control, and the development of generic radiation treatments has resulted in a significant increase in the international use of phytosanitary irradiation for trade in horticultural products and other commodities

**Keywords:** Radiation, sterile insects, phytosanitary irradiation

## 1. Introduction

The radioentomology can be defined as a branch of science that deals with the effects of ionizing radiations over insects and the study of insects using nuclear techniques. Radioentomological studies have been extremely useful in elucidating many entomological problems that were previously considered hard to solve or even insoluble due to limitations posed by conventional methods available.

The first radiobiological experiments performed with insects were initiated at the end of the 19th century. One of the first bioassays was performed by Professor Axenfelt in 1897 with house flies, but due to the methodology used, the results were not conclusive [1]. In 1911, Hunter made a series of experiments exposing several arthropods to X-rays, like *Sitophilus oryzae* L., *Culex pipiens* L., and some species of ticks, but no effects upon fertility or the tested life stages were observed. In the fall of 1912, Morgan and Runner performed experiments at Florida with the cigarette beetle *Lasioderma serricorne* F. with an X-ray machine aiming to sterilize cigar boxes in commercial scale. Their results, however, were also negative, as the beetle presented normal development.

According to Runner [2], the negative results from previous tests were caused by the fact that the equipments used were too rudimental. Most X-ray tubes that were tested were unable to operate continuously without neither fluctuation of intensity nor alteration of penetration power, being impossible to establish precisely the radiation dosage. Runner then executed new experiments with *L. serricorne*, using a device improved by W.D. Coolidge, whose X-ray tubes received a pure electron discharge, intensity and penetration power did not vary, and start and running voltages were the same. All these characteristics resulted in a homogeneous irradiation, and sterilization could be reached with high doses.

More detailed investigations on the genetic effects caused by ionizing radiations began with Muller's demonstration that genetic damage and a larger number of dominant lethal mutations could be induced in *Drosophila melanogaster* Meigen by X-rays [3]. He demonstrated, for instance, that an X-ray dose around 49 Gy applied on spermatic cells of *D. melanogaster* increased 100-fold the mutation frequency per generation.

However, entomologists became really aware of the extension of Muller's discovery only after 1950, when Muller made a great effort to publicize the biological effects of radiation. That moment of the 20th century could be considered as the rising of radioentomology.

Currently, there are almost 3000 references in literature, published continuously for the past seven decades. One of the most complete sources of information about radiation effects on the major groups of insects is the International Database on Insect Disinfestation and Sterilization (IDIDAS; http://www-ididas.iaea.org/ididas/). This website was developed with the aim to collect data of radiation doses for sterilization and disinfestations of arthropods, also performing a comparative analysis and quality assurance check on existing data [4]. IDIDAS have provided scientists a basis for literature searches to better plan experiments and became a comprehensive entry to the scientific literature for regulatory authorities to evaluate sterilization or disinfestation methods.

Over 300 species of arthropods, mostly of economic importance, have already been subjected to irradiation studies for basic research, pest control applications, and disinfestation of commodities (quarantine and phytosanitary purposes) [4]. In addition, insects may be labeled with stable or radioactive isotopes for radioecology or feeding studies. Nevertheless, this chapter will focus on insect sterilization and disinfestation by ionizing radiations in view of the socioeconomic impacts.

## 2. Effects of ionizing radiations in insects and radiation sources

Ionizing radiations can be emitted in the decay process of unstable nuclei or by de-excitation of atoms in nuclear reactors, X-ray devices, cyclotrons, and other equipments. Radiation in the form of particles ($\alpha$ or $\beta$ particles and neutrons) or electromagnetic waves (gamma or X-rays) can induce random biological effects in cells of insects likewise to other living cells [5, 6].

The chemical damage to organic molecules from the absorbing medium through which the radiation pass can be caused directly (mostly by particulate types of radiation) or indirectly by free radicals (i.e., atoms or molecules carrying at least one unpaired orbital electron in the outer shell), secondary electrons, or other charged particles [7]. The radioinduced ions and radicals, most of them coming from the water radiolysis, may react with neighboring molecules to produce secondary DNA radicals or even chain reactions, particularly in lipids. Most significant biological effects result from damage to DNA, which is the critical target in living organisms. Some radioinduced lesions in DNA are single-strand breaks in the phosphodiester linkage, double-strand breaks, base damage, protein–DNA cross-links, and protein–protein cross-links. The double-strand breaks in DNA double helix are believed to be the most important type of lesion produced in chromosome by ionizing radiation, cracking the chromatin into different pieces that may result in cell killing or mutation. Examples of lethal aberrations to the cell are the dicentric and ring (which are chromosome aberrations) and the anaphase bridge (a chromatid aberration). Two relevant aberrations that are usually not lethal to the cell are symmetrical translocation and small deletions. These changes and mutations left in the genetic code will influence base pairing, coding, transcription, and gene expression [5, 7].

According to the law of Bergonie and Tribondeau, cells that are dividing are more radiosensitive. Thus, cells that have a high mitotic rate and a long mitotic future, such as the reproductive cells, stand among the most radiosensitive cells [8]. Radioinduced changes in DNA of germ cells of insects can result in physiologically compromised gametes, aspermia, infertility, and even inability to mate. Sterilization can also be a result of fragmentation in germ cell chromosomes that generated random dominant lethal mutations, translocations, deletions, and other aberrations, which will lead to the production of imbalanced gametes and early zygotic death. The later type of sterilization is explored by the sterile insect technique (SIT), a genetic control method that relies essentially on the transfer of competitive sperm from released irradiated males to wild females [9, 10].

Somatic cells are more radioresistant than germ cells since they are usually differentiated cells, which explains why lethal radiation doses must be higher than sterilizing doses [11]. In general, insects are less resistant to radiation than bacteria, protozoa, and viruses, although more radioresistant than higher vertebrates [12, 13, 14]. Dyar's rule serve to explain this difference in sensitivity to radiation, as insects have a discontinuous growth and most of the cells divide only during the molting process [15].

The radiosensitivity varies widely among and within insect orders (Figure 1) [11]. Bakri et al. [4] highlights that the comparison of radiosensitivity between insect species must clearly take into account the end result measured, like sterilization, death, or inability to reach the next life stage. Lepidopterans exhibit more resistance to be sterilized by ionizing radiation (mean sterilization doses ranging between 40 and 400 Gy) [11] because some species may present a more complex sperm transfer, spermatophore formation, lower ability for mating after irradiation, production of eupyrene and apyrene sperm, and resistance to the induction of dominant lethal mutations due to the presence of holokinetic chromosomes (diffuse centromere) [16].

Besides the inherent differences in radioresistance between species and insect orders, many other factors can influence the sensitivity to radiation. These factors can be physical or biological conditions.

The other biological conditions that can influence insect radiosensitivity are as follows:

**a.** Age/developmental stage: in general, adults are more radioresistant than pupae, which in turn are more resistant than larvae and eggs [11].

**b.** Sex: female insects are usually more radiosensitive than males [17].

**c.** Size and weight: large long-lived adults of some species, with higher moisture content, may be more radiosensitive than small short-lived adults [18].

**d.** Nutritional stage: starvation may increase the radiosensitivity [19].

**e.** Diapause: diapausing larvae of some species could be more radiosensitive [20].

**f.** Genetic differences: strains of some species adapted to diverse environments could develop different radioresistances [21].

The main physical factors that can modify insect radiosensitivity are as follows:

**a.** Atmosphere: radioinduced damages are fewer with hypoxia [22].

**b.** Temperature: radioresistance may increase at lower temperatures [23].

**c.** Irradiation dose rate: as the dose rate is lowered and the exposure time extended, more sublethal damage can be repaired [7].

**d.** Dose fractionation: when splitting a radiation dose in time, cells are allowed to repair sublethal damage during the intervals between doses [24].

**e.** Radiation type: radiations with a higher linear energy transfer (LET), like $\alpha$ particles and neutrons, are more effective in inducing biological effects [7].

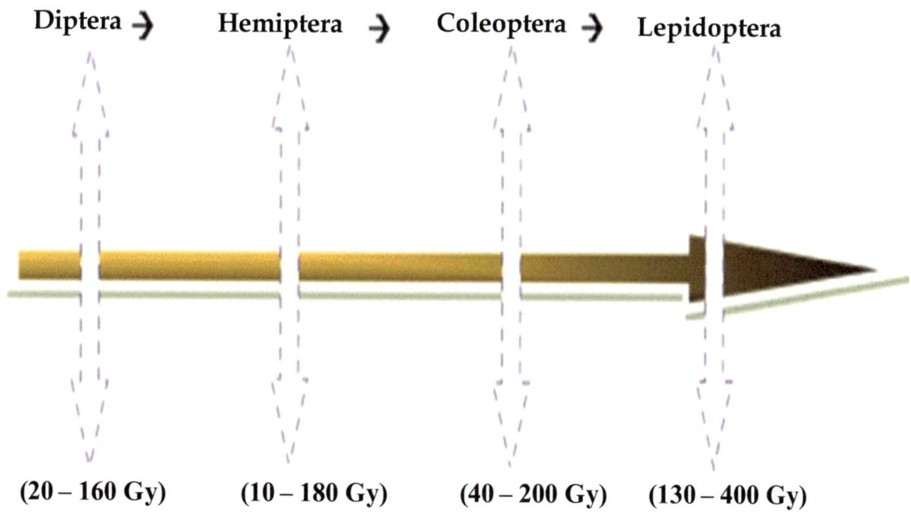


**Figure 1.** Decrease in radiosensitivity based on estimated sterilization doses for different insect orders (IDIDAS, 2015).

As aforementioned, radiations with a high LET are more effective in inducing biological effects, but their penetration can be limited. A typical alpha particle, for example, has high LET, but its penetration range is of only about 3 cm in air or 0.04 mm in tissue [7]. Neutrons also produce dense ionized tracks, but they can travel great distances in air as they carry no charge, requiring thick hydrogen-containing materials, such as concrete or water, to block them. Nevertheless, the application of neutron in radioentomological projects and pest control is constrained due to the easy induction of radioactivity in irradiated materials and the availability of neutron sources, which are usually restricted to nuclear reactors.

Researchers have preferably applied gamma or X-rays and high-energy electrons in studies involving pest control and disinfestation of commodities. As these radiations have similar relative biological effectiveness (RBE), most studies have indicated not significant differences in the biological damage induced by them for most doses and insect life stages [25, 26]. The insects are not rendered radioactive when irradiated with these sources by ensuring that the incident radiation is below 10 million electron volts (MeV) for high-energy electrons and less than 5 MeV for photons (gamma or X-rays) [27].

High-energy electrons are generated by electron accelerators, not involving any type of radioisotope. Likewise, most X-ray machines do not use radioisotopes, and X-rays are generated basically by the rapid deceleration of a beam of electrons before a material of high atomic number (e.g., tungsten or gold). The major advantages of these radiation sources are that no radioactive waste is produced, no radiation is produced when switched off, and the dose rate from electron accelerators can be hundred times greater than from gamma irradiators [11].

Despite these advantages, the types of irradiator used most frequently by radioentomologists for the past four decades have been those equipped with the radioisotopes $^{60}Co$ or $^{137}Cs$ as source of gamma rays. $^{60}Co$ has a half-life of 5.3 years and emits two gamma photons of 1.17 and 1.33 MeV, while $^{137}Cs$ has a half-life of 30.1 years and emits a monoenergetic photon of 0.66 MeV. The gamma irradiators used in pest control programs or for disinfestation of commodities are commonly of two types: large-scale panoramic irradiators or self-contained dry storage irradiators (Figure 2). The choice of radiation source is based considering basically costs, penetration, and irradiated material throughput [11]. Panoramic irradiators allow the irradiation of entire rooms and large number of samples or products can be irradiated at the same time. In self-contained irradiators, such as the most common irradiator used for insect sterilization, the Gammacell-220 (MDS Nordion International Inc., Ottawa, Canada), the canister containing the samples is lowered from the loading position to the shielded chamber with the radiation sources. The production of the Gammacell-220 was discontinued since 2008. On its place, appeared new models whose irradiation chamber contains a single source, lowering the overall costs, and the sample rotates through its own axis in front of the radiation source.

**Figure 2.** Types of gamma irradiators used in pest control trials or for disinfestation of commodities at the Center for Nuclear Energy in Agriculture (CENA, São Paulo, Brazil): (left) large-scale panoramic Gammabeam-650 irradiator; (right) self-contained Gammacell-220 irradiator.

## 3. Sterile insect technique

One of the main applications of ionizing radiations in Entomology is the production of sterile insects by the sterile insect technique (SIT). The SIT can be defined as a control tactic that uses area-wide inundative releases of sterile insects to reduce the fertility of a field population of the same species [28]. This technique is usually used as one of the components of area-wide integrated pest management programs, where the density of the target insect pest population is initially reduced by other control methods, like cultural or chemical control [29, 30].

The idea of releasing insects of the same species to introduce sterility into wild populations was independently conceived on the 1930s by three researchers: A.S. Serebrovskii at the USSR, F.L. Vanderplank at Tanzania, and E.F. Knipling from the United States [31]. Serebrovskii used chromosomal translocations to induce inherited partial sterility in *Musca domestica* L. and *Calandra granaria* L., but his research was not continued in the USSR during World War II [32]. Vanderplank tried to use hybrid sterility to combat tsetse flies, after obtaining low fertility from cross-matings between *Glossina morsitans* Westwood and *Glossina swynnertoni* Austen, but the detailed results were not published until his death [33]. At the United States Department of Agriculture (USDA), Knipling and colleagues [31, 34, 35, 36] exploited Muller's discovery that ionizing radiation could induce dominant lethal mutations, and their studies continued despite the tribulations during the World War II, resulting in an approach that was applied to eradicate the New World Screwworm, *Cochliomyia hominivorax* Coquerel, from the United States and Central America.

The SIT does not apply to all insects species. Innumerous factors must be considered before the adoption of the technique: (a) the species must reproduce sexually (even low levels of parthenogenesis can derail the technique); (b) the technique can be impractical for species that are vector of serious diseases, nuisance pests, or those which are highly destructive in the adult stage; (c) mass rearing procedures must be available; (d) the released sterile insects must present adequate dispersion; (e) the sterilization must not compromise the competitiveness of the males; (f) females must preferably mate only once or irradiated sperm must be very competitive; and (g) the population density of the target pest must be low, making economically feasible the release of a dominant population of sterile males over an extended period of time [34, 37].

Knipling et al. [38] realized that the degree of sterility introduced into the wild population by the sterile males must be sufficiently high to overcome the rate of increase of the wild females in order to provoke an overall reduction in the target population. As the ratio of sterile to fertile insects increases asymptotically as the density of the wild population declines to low levels, Knipling advocated that the sterile insects should be released when the wild population was at a seasonal low or after its decimation by weather events or other control methods. Most of the successful programs that released sterile insects were applied when field populations were at low densities [29].

Basically, the SIT involves the mass rearing of the target species, exposing the insects to ionizing radiation to induce sexual sterility, and then releasing the irradiated insects into the target population. The released sterile males mate with wild females, preventing the generation of a fertile offspring [10, 39].

The production of high quality insects in sufficient numbers using mass-rearing techniques is one of the main steps of the technique [40]. Methods to rear insects on artificial diets have been developed for more than 1000 species so far [41–45]. The production must be timely and cost effective, taking advantage of economies of scale whenever possible [46–49], and maximum attention must be paid to the factors that can affect quality of the insects produced [50].

Since the 1950s, most of the insect pest control programs that integrate the SIT have applied radioisotope irradiators loaded either with $^{60}Co$ or $^{137}Cs$, sterilizing the insects, therefore, with gamma rays [11, 51, 52]. Sterilization doses for hundreds of insect species can be found at IDIDAS database [53]. As absorbed dose is a key parameter for the success of the technique, the facilities that sterilize insects must have an accurate dosimetry system [11]. Due the growing complexities of the transboundary shipment of radioisotopes and the fear of "dirty bombs" after the September 11 attacks, some studies have supported the adoption of other practical alternatives for the sterilization of insects, such as X-ray irradiators [26, 54–58].

Studies aiming to develop procedures for handling and chilling adult insects or to provide food and water prior to release are continually performed. After sterilization, the insects can be released via static-release receptacles, ground-release methods, or most commonly from the air [59]. One of the most efficient methods of release is the aerial release of chilled irradiated insects or bags containing the adults, especially when aircraft flight paths are guided by a global positioning system (GPS) linked to a computer-controlled release mechanism [59, 60].

The SIT has been used mostly against species that are highly harmful to agriculture or public health or which elimination would have significant economic benefits. Currently, about 38 facilities are making research on SIT or sterilizing millions of insects per week for national area-wide integrated pest control programs [53]. Effective programs integrating the SIT have been performed against screwworm flies, tropical fruit flies, some species of tsetse flies, the pink bollworm *Pectinophora gossypiella* Saunders, and the codling moth *Cydia pomonella* L.

One of the best examples of application of the SIT was the phenomenal successful eradication campaigns conducted against the New World Screwworm, *C. hominivorax*, in the American continent. This fly can be sterilized as pupae 24 h before adult emergence with 40 Gy [61, 62]. The economic losses to livestock caused by *C. hominivorax* in the United States during the 1930s were significant [63]. After the field pilot tests at the Sanibel Island (1951–1953) and the Curaçao Island (1954) [64], eradication campaigns using suppression techniques and sterile insects were implemented in the Southeastern (1957–1959) and Southwestern (1962–1966) United States. As fertile flies continued infesting the United States coming from Central America, the eradication campaigns advanced through Mexico. Using sterile flies reared in the mass-rearing facility from COMEXA (*Comisión México-Americana para la Erradicación del Gusano Barrenador del Ganado*) at Tuxtla Gutiérrez, Mexico, the eradication of *C. hominivorax* was achieved until the Isthmus of Tehuantepec in 1984. With the interest of Central American countries and as fewer sterile flies would be required to maintain a buffer zone at Panama (150 million sterile flies/week were needed in the Isthmus of Tehuantepec, while only 40 million/week would be required in Panama), national eradication campaigns continued with the aerial release of more than 20 million sterile flies/week [65] during more than two decades (Figure 3). Panama was finally declared free from *C. hominivorax* in 2006 and a biological barrier of 30,000 km$^2$, maintained by the weekly release of 50 million sterile flies, was set at the Darien Gap [65, 66]. With this eradication effort, all warm-blooded animals became free of this deadly parasite in the United States, Mexico, Belize, Guatemala, Honduras, El Salvador, Nicaragua, Costa Rica, Panama, some Caribbean Islands, and additionally Libya, North Africa, after an outbreak [36, 67]. The economic benefits of these campaigns trespassed US$1 billion per year [68].

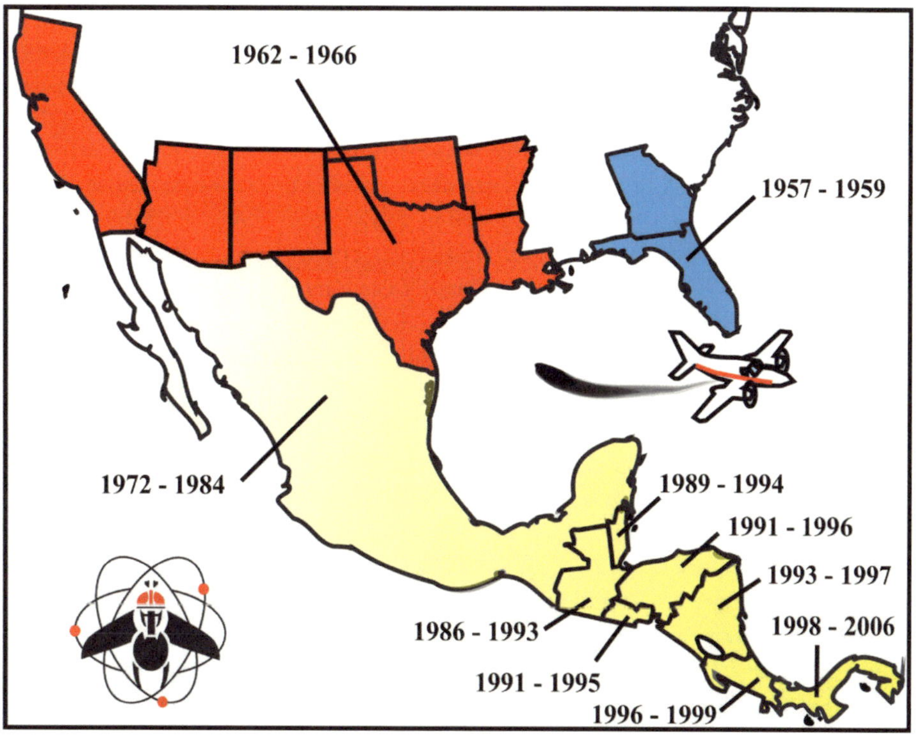


**Figure 3.** Expansion of the eradication campaigns that used aerial releases of irradiated flies against the New World Screwworm in North and Central America.

Many species of fruit flies are major economic pests due to the direct and indirect damages caused to fruit growers and difficulties imposed to international trade of fruits and vegetables [69]. Because of that, some species, especially tephritid fruit flies, have been target of programs that integrate the SIT. Fruit flies from the Tephritidae family can be generally sterilized at 90–150 Gy, and *Bactrocera* spp. are usually sterilized at 30–90 Gy [11, 53]. The first large-scale program stopped the invasion of the Mediterranean fruit fly (medfly) *Ceratitis capitata* Wiedemann from Central America into southern Mexico in the 1970s [70, 71]. After the invasion of Costa Rica by the medfly in 1955 and its expansion up to southern Mexico in 1976, the Government of Mexico started working with Guatemala and the United States to establish a large area-wide program using the SIT against this pest [71]. Using 500 million sterile flies/week from the rearing facility at Metapa, Mexico, and, currently, almost 2 billion sterile males/week [69, 72] from the biofactory located at El Piño, Guatemala, the MOSCAMED program has kept the United States, Mexico, and half of Guatemala free of the medfly for over 35 years. To prevent the establishment of the medfly in the continental United States through infested imported fruits, sterile males are regularly released in the Los Angeles Basin and Florida [31, 73]. During the 1980s and 1990s, the SIT was employed to eradicate the melon fly *Bactrocera cucurbitae* Coquillett in all of Japan's southwestern islands [74]. Significant SIT programs against the medfly and *Anastrepha* species have also been developed in several provinces of Argentina, some of which have become pest-free areas [75, 76].

Sterilization doses for flies from the Glossinidae family range from 50 to 120 Gy [53], and some SIT trials have been conducted on tsetse flies, which are vectors of trypanosomosis ("sleeping sickness"), in African countries during the 1970s and 1980s. However, as most programs had not been conducted area-wide, the pest-free status of most of the areas could not be maintained [31]. For example, three tsetse species (*G. morsitans submorsitans* Newstead, *Glossina palpalis gambiensis* Vanderplank, *G. palpalis palpalis* Robineau-Desvoidy) were eradicated at the same time in 3,000 km$^2$ from Burkina Faso through insecticide application and trapping suppression, followed by ground release of irradiated adults [77]. *G. palpalis palpalis* was eradicated in 1,500 km$^2$ of Nigeria with traps and insecticide-impregnated targets followed by ground releases of sterile adults [78]. In 1994–1997, *Glossina austeni* Newstead was eradicated from Unguja Island of Zanzibar (1,650 km$^2$) by using attractive devices, treating livestock with insecticide and aerial releases of irradiated adults, ceasing the transmission of trypanosomosis [79, 80]. The government of Ethiopia started the Southern Tsetse Eradication Project (STEP) in 2009, aiming to eradicate two species of tsetse flies over a 25,000 km$^2$ area in the Southern Rift Valley [81, 82], and after area-wide suppression activities, the mass-rearing facility in the Kality suburb of Addis Ababa had supplied in 2012 up to 60,000 sterile males/week to be released over the Deme Basin region. Since 2012, very good progress is also being made in the eradication of *G. palpalis gambiensis* on the Niayes area in Senegal with aerial releases of sterile males [60], and the annual increases of cattle sales after eradication were estimated in more than € 2,800/km$^2$ for the farming communities.

Despite some difficulties when applying the SIT against moths [83], like high mean sterilization doses (usually higher than 100 Gy) and appropriate air-handling and filtering in the mass-rearing facilities, radiobiological studies have been conducted for more than 30 lepidopteran species [84] and two SIT programs are still operational.

Since 1968, the pink bollworm, *Pectinophora gossypiella* Saunders, has been excluded from the San Joaquin Valley, USA, by a containment program [85] (http://www.cotton.org/tech/pest/bollworm/index.cfm), which releases adults that emerge from pupae irradiated with 100–150 Gy at the rearing facility in Phoenix, Arizona. The cost of this program has been around US $12.5/ha/season for each cotton grower (but control costs would increase by US$200/ha per grower if the program was not in place, besides an additional 2.2 million kg of pesticide that would have to be used every year) [83].

Populations of the codling moth, *Cydia pomonella* L., from British Columbia are being kept at insignificant levels since 1997 and individuals of this pest have not been detected in 37% of the orchards since 2009 due to the Okanagan-Kootenay suppression program that integrates the SIT (newly emerged males are partially sterilized with 100–250 Gy and chilled moths are released). Growers used to pay a tax of US$169/ha/year, and the application of insecticides in the province was reduced 82% since then [83, 86, 87].

## 4. Radiation as quarantine treatment against insect pests

One major concern in exporting agricultural commodities is to prevent the introduction or spread of exotic quarantine pests. Phytosanitary measures are used to disinfest commodities of pests, providing quarantine security [88]. The fumigant gas methyl bromide used to be the most common treatment for agricultural commodities [89] due the low cost, effectiveness against a wide range of insects, rapid dispersion, and minimal impact on commodity quality [90]. However, with the imminent phasing out of methyl bromide as mandated by the Montreal Protocol [91], the interest in alternative phytosanitary treatments has raised [92, 93]. The use of ionizing radiations as a way to inhibit the spread of quarantine insects represents an important alternative postharvest control, reducing the need for chemical fumigants and other toxic products [94].

Hallman [95] stated that the objective of using ionizing radiations as a phytosanitary treatment is not to obtain acute mortality of the insects but to prevent development or reproduction, as most commodities do not tolerate the usual dose ranges required to achieve immediate mortality (usually ≥1 kGy). Actually, the U.S. Food and Drug Administration (FDA) has approved radiation up to 1 kGy to control insects in foods and to extend the shelf life of fresh fruits and vegetables [96]. Thus, a phytosanitary irradiation treatment must be effective against the most tolerant insect stage that could be present on the commodity [97], and the inhibition of further development should be considered as a measure of efficacy of phytosanitary irradiation [98].

Some regulators may consider this a disadvantage since other commercially applied quarantine treatments, which are generally based on heat, cold or methyl bromide fumigation, do reach acute mortality. When inspectors find live quarantine pests from these treatments, the entire consignment can be rejected or retreated regardless of certification of treatment because the inspectors may assume that the treatment was not properly done, the shipment was contaminated with infested commodity or the cargo was reinfested after treatment. Furthermore, live adults found in survey traps could trigger restrictive and costly regulatory responses in importing countries [99].

Nevertheless, phytosanitary irradiation can be a viable commercial insect control technique. The advantages of radiation include the fact that pest insects cannot develop resistance, the absence of residual radioactivity, and few significant changes in the physicochemical properties of the treated products for most doses applied [100].

Another advantage of phytosanitary irradiation compared with other treatments is the possibility of using generic doses (i.e., one dose serves for a group of insects and commodities, although not all have been tested for efficacy), which facilitate the development and application of the treatment [94].

Radioentomologists are constantly looking for a generic radiation dose to serve as quarantine treatment, i.e. a dose that could control a broad group of pests without adversely affecting the quality of a wide range of commodities [101]. This dose would necessarily be set at the minimum absorbed dose required for the most tolerant organism within the insect group considered [102]. Due the high radiotolerance of the Angoumois grain moth (*Sitotroga cerealella* Olivier), Hallman and Phillips [102] suggested that a generic dose of 600 Gy for all insects in ambient atmospheres would be efficacious to attend quarantine purposes. Currently, some of the generic phytosanitary irradiation treatments are 150 Gy for all hosts of Tephritidae, 150 Gy also for mangoes and citrus fruits exported from Mexico to the United States, 250 Gy for all arthropods on mango and papaya shipped from Australia to New Zealand [103], 300 Gy for all arthropods on mango shipped from Australia to Malaysia, 350 Gy for all arthropods on lychee shipped from Australia to New Zealand, and 400 Gy is applied for Mexican guavas, Indian mangoes, and dragon fruit (*Hylocereus undatus* Britton and Rose) from Vietnam exported to the United States [94, 99]. Hallman [88] also presented a number of cases indicating the usefulness of generic doses for important pest groups such as mealybugs, scales, and weevils.

In 2006, the USDA approved irradiation at a generic dose of 150 Gy for any tephritid fruit fly and 400 Gy for all insects except pupae and adult of Lepidoptera [88, 104, 105]. Subsequent studies lead the USDA-APHIS to approve minimum doses for 23 insect pests [106], including 10 tephritid fruit fly species, 6 lepidopteran species, 4 curculionid species, and 1 mite species. These approved specific doses for fruit flies range between 60 and 150 Gy, between 100 and 250 Gy for lepidopterans, between 92 and 300 Gy for Coleoptera, and 300 Gy for the spider mite [106].

The International Plant Protection Convention (IPPC) also accepted the 150 Gy minimum absorbed dose for Tephritids as an international standard for phytosanitary treatment of these quarantine pests, including it in the International Standards of Phytosanitary Measures (ISPM #28) together with 13 species-specific treatment procedures [107]. The IPPC, however, did not accept the generic dose of 400 Gy for all insects (except pupae and adult of lepidopterans). The IPPC does not approve at first some irradiation treatments due to perceived problems with the study or the presence of live adults after irradiation (an issue that must be carefully addressed).

The development of methods to determine whether quarantine pests have been irradiated could help to resolve the issue of presence of live adults after exposure to radiation. Biomarkers based on the molecular processes of irradiation-induced DNA damage and repair would have internationally broad application to confirm the irradiation status of pests found on commodities and for the detection of sterile insects. Siddiqui et al. [108] discovered a protein in the Queensland fruit fly, *Bactrocera tryoni* Froggatt, that was modified due to radiation, with a higher amount of modified protein at higher radiation doses. The authors also tested the doses approved for disinfestation and SIT. Leifert et al. [109] reported highly specific antibodies that allowed the sensitive detection of proteins from irradiated *B. tryoni* using even standard commercial technologies, such as western blot or ELISA assays.

According to Follett [110], current research on phytosanitary irradiation is focused on development of specific doses for quarantine lepidopterans not covered by the generic treatments, shortening treatment time through the reduction of dose levels for specific pests and commodities, the development of generic doses below 400 Gy for economically important groups of quarantine insects other than fruit flies, and deep investigations on commodity tolerance and novel methods to reduce damages and extend shelf life. The author also discussed that future research should be dedicated to reduce the present barriers to the wider use of phytosanitary irradiation, like the 1 kGy limit, restrictions on the use of modified atmosphere and the small number of countries that approve the use of phytosanitary irradiation. For example, the development of small-scale X-ray machines could provide farmers and packinghouses with in-house treatment capability, accelerating the adoption of phytosanitary irradiation. A recent change in U.S. import regulations has permitted the irradiation upon entry, allowing exporting countries to explore new markets without investing in expensive irradiation facilities [111].

## 5. Conclusion

The use of ionizing radiations allowed the rise of a new branch of the study of insects in the middle of the 20th century, the radioentomology. The release of insects that are sterile after exposure to radiation aiming to control or eradicate pest populations revealed to be a revolutionary tactic in the area-wide management of pests, and many successful cases of the application of the sterile insect technique can be found around the globe. Furthermore, the development of generic radiation treatments has resulted in a significant increase in the international use of phytosanitary irradiation for trade in horticultural products and other commodities.

## Author details

Valter Arthur, Andre Machi and Thiago Mastrangelo*

*Address all correspondence to: thiagomastrangelo@gmail.com

Center for Nuclear Energy in Agriculture (CENA/USP), Brazil

## References

[1] Hunter WD. Results of experiments to determine the effect of Roentgen rays upon insects. Journal of Economic Entomology 1912;5(1):188–192.

[2] Runner GA. Effect of Roentgen rays on the tobacco, or cigarette, beetle, and the results of experiments with a new form of Roentgen tube. Journal of Agricultural Research 1916;6:383–388.

[3] Muller HJ. Artificial transmutation of the gene. Science 1927;66:84–87.

[4] Bakri A, Heather N, Hendrichs J, Ferris I. Fifty Years of radiation biology in entomology: lessons learned from IDIDAS. Annals of Entomological Society of America 2005;98:1–12.

[5] O'Brien RD, Wolfe LS. Radiation, Radioactivity and Insects. New York: Academic Press; 1964.

[6] L'Annunziata MF. Radioactivity: Introduction and History. Elsevier: The Netherlands; 2007.

[7] Hall EJ, Giaccia E (eds.). Radiobiology for the Radiologist, 6th edition. Philadelphia: J.B. Lippincott Company; 2006.

[8] Casarett, AP. Radiation biology. Englewood Cliffs, NJ: Prentice-Hall; 1968.

[9] Robinson AS. Mutations and their use in insect control. Mutation Research 2002;511:113–132.

[10] Vreysen MJB, Robinson AS. Ionising radiation and area-wide management of insect pests to promote sustainable agriculture. Agronomy for Sustainable Agriculture 2011;31:233–250.

[11] Bakri A, Mehta K, Lance DR 2005. Sterilizing insects with ionizing radiation. Sterile insect technique: principles and practice in area-wide integrated pest management. In: Dyck VA, Hendrichs J, Robinson AS (eds.). Springer; 2005. pp. 233–269.

[12] Rice TR, Baptist JP. Ecologic effects of radioactive emissions from nuclear power plants. In: Thomas CC (ed.). Human and Ecological Effects of Nuclear Power Plants. Springfield, IL, USA: L. A. Saga; 1974. pp. 373–439.

[13] Whicker FW, Schultz, V. Radioecology: Nuclear Energy and the Environment. Boca Raton: CRC Press; 1982.

[14] Harrison FL, Anderson SL. Taxonomic and development aspects of radiosensitivity. In: Amiro B, Avadhanula R, Johansson G, Larsson CM, Luning M (eds.). Proceedings of the Symposium: Ionizing Radiation, pp. 65–88, the Swedish Radiation Protection Institute (SSI) and the Atomic Energy Control Board (AECB) of Canada, Stockholm, Sweden, 20–24 May 1996.

[15] Hutchinson JMC, McNamara JM, Houston AI, Vollrath F. Dyar's rule and the investment principle: optimal moulting strategies if feeding rate is size-dependent and growth is discontinuous. Philosophical Transactions of the Royal Society B: Biological Sciences 1997;352:113–138.

[16] Robinson AS. Genetic basis of the sterile insect technique. In: Dyck VA, Hendrichs J, Robinson AS (eds.). Sterile insect technique: principles and practice in area-wide integrated pest management. Berlin, Germany: Springer; 2005. pp. 95–114.

[17] Hooper GHS. The effect of ionizing radiation on reproduction. In: Robinson AS, Hooper G (eds.). World Crop Pests: Fruit Flies, Their Biology, Natural Enemies and Control. New York: Elsevier Publishers; 1989. pp. 153–164.

[18] Willard WK, Cherry DS. Comparative radiosensitivity in the class Insecta. Journal of Theoretical Biology 1975;52:149–158

[19] Drummond RO, Medley JG, Graham OH. Engorgement and reproduction of lone star ticks (*Amblyomma americanum* L.) treated with gamma radiation. International Journal of Radiation Biology 1966;10:183–188.

[20] Mansour M. Gamma irradiation as a quarantine treatment for apples infested by codling moth (Lepidoptera: Tortricidae). Journal of Applied Entomology 2003;127:137–141.

[21] Azizyan, A. Study the comparative resistance of codling moth populations at different geographical regions to gamma irradiation. Progress Report. Research Contract ARM 11024/R1. Co-ordinated Research Program (CRP). Vienna, Austria: IAEA; 2003.

[22] Fisher K. Irradiation effects in air and in nitrogen on Mediterranean fruit ßy (Diptera: Teprhitidae) pupae in Western Australia. Journal of Economic Entomology 1997;90:1609–1614.

[23] Barbosa AP. Interaction of gamma radiation and temperature on the determination of the sterilizing dose of some stored products pests. MSc thesis, Sao Paulo University, Escola Superior de Agricultura Luiz de Queiroz, Piracicaba, Brazil; 1976.

[24] Tamhankar AJ, Shantharam K. Sterile insect technique (SIT) for control of spotted bollworm of cotton: field experiments confirm dose fractionation improves mating competitiveness of sterile males. In: Ramamoorthy N, Ananthakrishnan M, Nandakumar AN (eds.). Proceedings, Conference: Applications of Radioisotopes and Radiation Technology in the 21st Century, NAARRI International Conference, 12–14 December 2001, Mumbai, India. 2001.

[25] Dohino T, Tanabe K, Hayashi T. Comparison of lethal effects of electron beams and gamma rays on eggs of two spotted spider mite, *Tetranychus urticae* Koch (Acari: Tetranychidae). Research Bulletin of the Plant Protection Service, Japan 1994;30:69–73.

[26] Mastrangelo T, Parker AG, Jessup A, Pereira R, Orozco-Dávilla D, Islam A, Dammalag T, Walder JMM. A new generation of X ray irradiators for insect sterilization. Journal of Economic Entomology 2010;103:85–94.

[27] Codex. Recommended international code of practice for the radiation processing of foods. CAC/RCP 19-1979, Rev. 1-2003. Roma, Italy: FAO. 2003.

[28] FAO. Food and Agricultural Organization. Glossary of phytosanitary terms. Provisional additions. Rome: FAO/IPPC. 2005.

[29] Mangan RL. Population suppression in support of the Setrile Insect Technique. In: Dyck VA, Hendrichs J, Robinson AS (eds.). Sterile Insect Technique. Principles and

Practice in Area-wide Integrated Pest Management. Dordrecht, The Netherlands: Springer; 2005. pp. 407–426.

[30] Vreysen MJB, Robinson AS, Hendrichs J. Area-wide Control of Insect Pests, from Research to Field Implementation. Dordrecht, The Netherlands: Springer; 2007.

[31] Klassen W, Curtis CF. History of the sterile insect technique. In: Dyck VA, Hendrichsc J, Robinson AS (eds.). Sterile Insect Technique—Principles and Practice in Area-wide Integrated Pest Management. Netherlands: Springer; 2005.pp. 3–38.

[32] Serebrovskii AS. Theoretical foundations of the translocation method of pest control. Nauka: Moscow, Russian Federation; 1971.

[33] Vanderplank FL. Experiments in the hybridization of tsetse flies ("*Glossina* Diptera") and the possibility of a new method of control. Transactions of the Royal Entomological Society 1947;98:1–18.

[34] Knipling EF. Possibilities of insect control or eradication through the use of sexually sterile males. Journal of Economic Entomology 1955;48:459–462.

[35] Knipling EF. Sterile-male method of population control. Science 1959;130:902–904.

[36] Lindquist DA, Abusowa M, Klassen W. Eradication of the New World screwworm from the Libyan Arab Jamahiriya. In: Proceedings of the Management of Insect Pests: Nuclear and Related Molecular and Genetic Techniques. FAO/IAEA International Symposium, 19–23 October 1992, Vienna, Austria. STI/PUB/909. Vienna, Austria: IAEA; 1993. pp. 319–330.

[37] Lance DR, McInnis DO. Biological basis of the sterile insect technique. In: Dyck VA, Hendrichs J, Robinson AS (eds.). Sterile Insect Technique. Principles and Practice in Area-wide Integrated Pest Management. Dordrecht, The Netherlands: Springer; 2005. pp. 69–94.

[38] Knipling EF, Laven H, Craig GB, Pal R, Kitzmiller JB, Smith CN, Brown AW. Genetic control of insects of public health importance. Bulletin of the World Health Organization 1968;38:421–438.

[39] Dyck VA, Hendrichs J, Robinson, AS. Sterile insect technique: principles and practice in area-wide integrated pest management. Berlin: Springer; 2005. p. 787.

[40] Parker AG. Mass-rearing for sterile insect release. In: Dyck VA, Hendrichs J, Robinson AS (eds.). Sterile Insect Technique. Principles and Practice in Area-wide Integrated Pest Management. Dordrecht, The Netherlands: Springer; 2005. pp. 209–232.

[41] Smith CN. Insect Colonization and Mass Production. New York: Academic Press; 1966.

[42] Singh P, Moore RF. Handbook of Insect Rearing, Volumes I and II. Amsterdam, The Netherlands: Elsevier Science Publishers BV; 1985.

[43] Ochieng'-Odero JPR. Techniques of insect rearing for the development of integrated pest and vector management strategies. Volumes 1 and 2. Proceedings of the International Group Training Course on Techniques of Insect Rearing for the Development of Integrated Pest and Vector Management Strategies, 16 March–3 April 1992, Nairobi, Kenya. ICIPE Science Press; 1994.

[44] Cohen AC. Insect Diets: Science and Technology. Boca Raton: CRC Press; 2003.

[45] Sørensen JG, Matthew F. Addison, John S. Mass-rearing of insects for pest management: challenges, synergies and advances from evolutionary physiology. Crop Protection 2012;38:87–94.

[46] Hendrichs J, Robinson AS, Cayol JP, Enkerlin W. Medfly areawide sterile insect technique programmes for prevention, suppression or eradication: the importance of mating behavior studies. Florida Entomologist 2002, 85: 1–13. http://www.fcla.edu/FlaEnt/fe85p1.pdf

[47] Enkerlin, W. Economics of area-wide SIT control programs. In: Recent Trends on Sterile Insect Technique and Area-wide Integrated Pest Management—Economic Feasibility, Control Projects, Farmer Organization and *Bactrocera dorsalis* Complex Control Study. Okinawa, Japan: Research Institute for Subtropics; 2003;1–10.

[48] Caceres C, Cayol JP, Enkerlin W, Franz G, Hendrichs J, Robinson AS. Comparison of Mediterranean fruit fly (*Ceratitis capitata*: Tephritidae) bisexual and genetic sexing strains: development, evaluation and economics. In: Barnes BN (eds.). Proceedings, Symposium: 6th International Symposium on Fruit Flies of Economic Importance, 6–10 May 2002, Stellenbosch, South Africa, Irene: Isteg Scientific Publications; 2004. pp. 367–381.

[49] Barnes BN, Rosenberg S, Arnolds L, Johnson J. Production and quality assurance in the SIT Africa Mediterranean fruit fly (Diptera: Tephritidae) rearing facility in South Africa. Florida Entomologist 2007;90:41–52.

[50] Calkins CO, Parker AG. Sterile insect quality. In: Dyck VA, Hendrichs J, Robinson AS (eds.). Sterile Insect Technique. Principles and Practice in Area-wide Integrated Pest Management. Dordrecht, The Netherlands: Springer; 2005. pp. 269–296.

[51] Bushland RC, Hopkins DE. Sterilization of screw-worm flies with X-rays and gamma rays. Journal of Economic Entomology 1953;46:648–656.

[52] Lindquist AW. The use of gamma radiation for the control or eradication of the screwworm. Journal of Economic Entomology 1955;48:467–469.

[53] International Database on Insect Disinfestation and Sterilization. IDIDAS. 2015. International Atomic Energy Agency, Vienna, Austria. http://www-infocris.iaea.org/IDIDAS/start.htm. (accessed 5 Jan 2015).

[54] Helinski MEH, Parker AG, Knols BGJ. Radiation biology of mosquitoes. Malaria Journal, 2009;8(2): S6. doi:10.1186/1475-2875-8-S2-S6

[55] Viscarret MM, Conte CA, Paladino LZC, López SN, Segura DF, Muntaabski I, Lanzavecchia SB, Cladera JL. Rearing of the fruit fly parasitoid *Diachasmimorpha longicaudata* (Hymenoptera: Braconidae) on X-ray irradiated larvae of *Ceratitis capitata* (Diptera: Tephritidae). Biocontrol Science and Technology 2012;22:1429–1441.

[56] Cagnotti CL, Viscarret MM, Riquelme MB, Botto EN, Carabajal LZ, Segura DF, López SN. Effects of X-rays on *Tuta absoluta* for use in inherited sterility programmes. Journal of Pest Science 2012;85:413–421.

[57] Yamada H, Parker AG, Oliva CF, Balestrino F, Gilles JRL. X-ray-induced sterility in *Aedes albopictus* (Diptera: Culicidae) and male longevity following irradiation. Journal of Medical Entomology 2014;51:811–816.

[58] Ndo C, Yamada H, Damiens DD, N'do S, Seballos G, Gilles JRL. X-ray sterilization of the *An. arabiensis* genetic sexing strain 'ANO IPCL1' at pupal and adult stages. Acta Tropica 2014;131:124–128.

[59] Dowell RV, Worley J, Gomes PJ. Sterile insect supply, emergence, and release. In: Dyck VA, Hendrichs J, Robinson AS (eds.). Sterile Insect Technique. Principles and Practice in Area-wide Integrated Pest Management. Dordrecht, The Netherlands: Springer; 2005. pp. 297–324.

[60] Mubarqui RL, Perez RC, Kladt RA, Lopez JLZ, Parker A, et al. The smart aerial release machine, a universal system for applying the sterile insect technique. PLoS ONE, 2014;9:e103077. doi:10.1371/journal.pone.0103077

[61] Bushland RC, Hopkins DE. Experience with screwworm flies sterilized by X-rays. Journal of Economic Entomology 1951;44:725–731.

[62] Crystal MM. Sterilization of screwworm flies (Diptera: Calliphoridae) with gamma rays: restudy after two decades. Journal of Medical Entomology 1979;15:103–108.

[63] Baumhover AH. Eradication of the screwworm fly, an agent of myiasis. Journal of the American Medical Association 1966;196:240–248.

[64] Baumhover AH, Graham AJ, Bitter BA, Hopkins DE, New WD, Dudley FH, Bushland RC. Screw-worm control through release of sterilized flies. Journal of Economic Entomology 1955;48:462–466.

[65] Wyss JH. Screw-worm eradication in the Americas—overview. In: Tan KH (ed.). Areawide Control of Fruit Flies and Other Insect Pests. Penerbit Universiti Sains Malaysia, Pulau Pinang, Malaysia 2000. pp. 79–86.

[66] Mastrangelo T, Welch J. An overview of the components of AW-IPM campaigns against the New World screwworm. Insects 2012. pp. 930–955.

[67] Lindquist DA, Abusowa M, Classen W. The New World screwworm fly in Libya: a review of its introduction and eradication. Medical and Veterinary Entomology 1992;6:2–8.

[68] Vargas-Terán M, Hofmann HC, Tweddle NE. Impact of screwworm eradication programmes using the sterile insect technique. In: Dick VA, Hendrichs J, Robinson AS (eds.). Sterile Insect Technique, Principles and Practice in Area-wide Integrated Pest Management. The Netherlands: Springer; 2005. pp. 629–650.

[69] Enkerlin, WR. Impact of fruit fly control programmes using the sterile insect technique. In: Dyck VA, Hendrichs J, Robinson AS (eds.). Sterile Insect Technique. Principles and Practice in Area-wide Integrated Pest Management. Dordrecht, The Netherlands: Springer; 2005. pp. 651–676.

[70] Rhode RH, Dimon J, Perdomo A, Gutierrez J, Dowling Jr CF, Lindquist DA. Application of the sterile-insect technique in Mediterranean fruit fly suppression. Journal of Economic Entomology 1971;64:708–713.

[71] Hendrichs J, Ortiz G, Liedo P, Schwarz A. Six years of successful medfly program in Mexico and Guatemala. In: Cavalloro R (eds.). Proceedings, Symposium: Fruit Flies of Economic Importance. CEC/IOBC International Symposium, 16–19 November 1982, Athens, Greece. A. A. Balkema, Rotterdam, The Netherlands; 1983. pp. 353–365.

[72] Rendón P, McInnis D, Lance D, Stewart J. Medfly (Diptera: Tephritidae) genetic sexing: large-scale field comparison of males-only and bisexual sterile fly releases in Guatemala. Journal of Economic Entomology 2004;97:1547–1553.

[73] IAEA. International Atomic Energy AgencyModel business plan for a sterile insect production facility. Vienna. (Project INT/5/145 Insect Pest Control Using the Sterile Insect Technique) 2008.

[74] Kuba H, Kohama T, Kakinohana H, Yamagishi M, Kinjo K, Sokei Y, Nakasone T, Nakamoto Y. The successful eradication programs of the melon fly in Okinawa. In: McPheron BA, Steck G (eds.). Fruit Fly Pests. A World Assessment of Their Biology and Management. Delray Beach, FL, USA: St. Lucie Press; 1996. pp. 543–550.

[75] Guillen GD, Sanchez R. Expansion of the National Fruit Fly Control Programme in Argentina. In: Vreysen MJB, Robinson AS, Hendrichs J (eds.). Area-wide Control of Insect Pests, from Research to Field Implementation. Vienna: FAO/IAEA Programme of Nuclear Techniques in Food and Agriculture; 2007. pp. 653–660.

[76] Cladera JL, Vilardi JC, Juri M, Paulin LE, Giardini MC, Gomez Cendra PV, Segura DF, Lanzavecchia SB. Genetics and biology of *Anastrepha fraterculus*: research supporting the use of the sterile insect technique (SIT) to control this pest in Argentina. BMC Genetics, 15(Suppl 2):S12;2014.

[77] Politzar H, Cuisance D. An integrated campaign against riverine tsetse, *Glossina palpalis gambiensis* and *Glossina tachinoides* by trapping and the release of sterile males. Insect Science and its Application 1984;5:439–442.

[78] Takken W, Oladunmade MA, Dengwat L, Feldmann HU, Onah JA, Tenabe SO, Hamann HJ. The eradication of *Glossina palpalis palpalis* (Robineau-Desvoidy) (Diptera:

Glossinidae) using traps, insecticide-impregnated targets and the sterile insect technique in central Nigeria. Bulletin of Entomological Research 1986;76:275–286.

[79] Feldmann U, Mramba F, Parker AG, Dyck VA, Vreysen MJB, et al. Application of the sterile insect technique in Zanzibar to eradicate tsetse flies, the vectors of trypanosomosis. In: Ruane J, Dargie JD, Mba C, Boettcher P, Makkar HPS, Bartley DM, Sonnino A (eds.). Biotechnologies at Work for Smallholders: Case Studies from Developing Countries in Crops, Livestock and Fish. FAO, Rome, Italy; 2013. pp 125–132.

[80] Vreysen MJB, Saleh K, Mramba F, Parker A, Feldmann U, et al. Sterile insects to enhance agricultural development: the case of sustainable tsetse eradication on Unguja Island, Zanzibar, using an area-wide integrated pest management approach. PLoS Neglected Tropical Diseases 2014;8:e2857. doi:10.1371/journal.pntd.0002857

[81] Alemu TB, Kapitano S, Mekonen G, Aboset M, Kiflom B, Bancha G, Woldeyes K, Bekele, Feldmann U. Area-wide control of tsetse and trypanosomiasis; Ethiopian experience in the southern rift valley. In: Vreysen MJB, Robinson AS, Hendrichs J (eds.). Area wide control of insect pests. Dordrecht, the Netherlands: Springer; 2007. pp. 325–335.

[82] Abd-Alla AMM, Parker AG, Vreysen MJB, Bergoin M. Tsetse salivary gland hypertrophy virus: hope or hindrance for tsetse control? PLoS Neglected Tropical Diseases 2011;5:e1220. doi:10.1371/journal.pntd.0001220.

[83] Bloem KA, Bloem S, Carpenter JE. Impact of moth suppression/eradication programmes using the sterile insect technique or inherited sterility. In: Dyck VA, Hendrichs J, Robinson AS (eds.). Sterile Insect Technique. Principles and Practice in Area-wide Integrated Pest Management. Dordrecht, The Netherlands: Springer; 2005. pp. 677–700.

[84] Carpenter JE, Bloem S, Marec F. Inherited sterility in insects. In: Dyck VA, Hendrichs J, Robinson AS (eds.). Sterile Insect Technique. Principles and Practice in Area-wide Integrated Pest Management. Dordrecht, The Netherlands: Springer; 2005. pp. 115–146.

[85] Walters M, Morrison NI, Claus J, Tang G, Phillips CE, et al. Field Longevity of a Fluorescent Protein Marker in an Engineered Strain of the Pink Bollworm, *Pectinophora gossypiella* (Saunders). PLoS ONE 2012;7:(6)e38547. doi:10.1371/journal.pone.0038547.

[86] Bloem S, McCluskey A, Fugger R, Arthur S, Wood S, Carpenter JE. Suppression of the codling moth Cydia pomonella in British Columbia, Canada using an area-wide integrated approach with an SIT component. In: Vreysen MJB, Robinson AS, Hendrichs J (eds.). Area-wide control of insect pests. From research to field implementation. Dordrecht, The Netherlands: Springer; 2007. pp. 591–601.

[87] Judd G, Thompson D. Taking a Flexible Approach to Mating Disruption in British Columbia: Abstracts from the Orchard Pest Management Conference, Portland, Jan 2012, Washington; 2012.

[88] Hallman GJ. Phytosanitary applications of irradiation. Comprehensive Reviews in Food Science and Food Safety 2011;10:143–51.

[89] MBTOC. Assessment report of the methyl bromide technical options committee. Nairobi, Kenya: UNEP 2007. 2006;485 p.

[90] Aegerter AF, Folwell RJ. Economic aspect of alternatives to methyl bromide in the postharvest and quarantine treatment of selected fresh fruit. Crop Protection 2000;19:161–8.

[91] Environmental Protection Agency. EPA, 1999: Protection of stratospheric ozone: incorporation of Montreal Protocol Adjustment for a 1999 Interim Reduction in class I, group VI controlled substances. http://www.epa.gov/fedrgstr/EPA-AIR/1999/June/Day-01/a13803.htm (accessed Jan 15).

[92] Ferrier P. Irradiation as a quarantine treatment. Food Policy 2010;35:548–55.

[93] Hallman GJ, Thomas DB. Evaluation of the efficacy of the methyl bromide fumigation schedule against Mexican fruit fly (Diptera: Tephritidae) in citrus fruit. Journal of Economic Entomology 2011;104:63–68.

[94] Hallman GJ. Generic phytosanitary irradiation treatments. Radiation Physics and Chemistry 2012;81:861–866.

[95] Hallman GJ. Ionizing irradiation quarantine treatment against oriental fruit moth (Lepidoptera: Tortricidae) in ambient and hypoxic atmospheres. Journal of Economic Entomology 2004;97:824–827.

[96] U.S. FDA. Irradiation in the production, processing and handling of food: final rule. United States Food and Drug Administration, Fed. Reg. 2004;69:76844–76847.

[97] Hallman GJ. Expanding radiation quarantine treatments beyond fruit flies. Agricultural and Forest Entomology 2000;2:85–95.

[98] Food and Agricultural Organization, FAO. Guidelines for the use of irradiation as a phytosanitary measure. ISPM #18. Food and Agricultural Organization, Rome, Italy; 2003.

[99] Hallman GJ, Levang-Brilz NM, Zettler JL, Winborne IC. Factors affecting ionizing radiation phytosanitary treatments, and implications for research and generic treatments. Journal of Economic Entomology 2010;103:1950–1963.

[100] Lapidot M, Saveanu S, Padova R, Ross I. Insect disinfestation by irradiation. In: IAEA (ed.). Insect Disinfestation of Food and Agricultural Products by Irradiation, Vienna; 1991. pp. 93–103.

[101] Follett PA, Neven LG. Current trends in quarantine entomology. Annual Review of Entomology 2006;51:359–385.

[102] Hallman GJ, Phillips TW. Ionizing irradiation of adults of Angoumois grain moth (Lepidoptera: Gelechiidae) and Indianmeal moth (Lepidoptera: Pyralidae) to prevent

reproduction and implications for a generic irradiation treatment for insects. Journal of Economic Entomology 2008;101:1051–1056.

[103] MAF. Ministry of Agriculture and Fisheries, New Zealand 2009. Import health standard commodity sub-class: fresh fruit/vegetables mango, *Mangifera indica* from Australia. http://www.biosecurity.govt.nz/flies/ihs/mango-au.pdf.

[104] Hallman GJ, Arthur V, Blackburn CM, Parker AG. The case for a generic phytosanitary irradiation dose of 250 Gy for Lepidoptera eggs and larvae. Radiation Physics and Chemistry 2013;89:70–75.

[105] Hallman GJ, Parker AG, Blackburn CM. The case for a generic phytosanitary irradiation dose of 400 Gy for Lepidoptera that infest shipped commodities as pupae. Radiation Physics and Chemistry 2013;89:70–75.

[106] USDA–APHIS. Animal and Plant Health Inspection Service. Treatment Manual. T-105-2-7. 2013;5-2-71.

[107] IPPC. International Plant Protection Convention. ISPM#28: Phytosanitary Treatments for Regulated Pests. Food and Agriculture Organization, Rome, Italy; 2011.

[108] Siddiqui MS, Filomeni E, François M, Collins SR, Cooper T, Glatz, RV, Taylor PW, Fenech M, Leifert WR. Exposure of insect cells to ionising radiation *in vivo* induces persistent phosphorylation of a H2AX homolog (H2AvB). Mutagenesis 2013;1–11. doi:10.1093/mutage/get030

[109] Leifert WR, Glatz RV, Siddiqui MS, Collins SR, Taylor PW, Fenech M. Development of a test to detect and quantify irradiation damage in fruit flies. Final Report for Horticulture Australia Ltd. Project VG09160. 2013.

[110] Follett P. Phytosanitary irradiation for fresh horticultural commodities: generic treatments, current issues, and next steps. Stewart Postharvest Review 2014;10:1–7.

[111] Bustos-Griffin E, Hallman GJ, Griffin RL. Phytosanitary irradiation in ports of entry: a practical solution for developing countries. International Journal of Food Science and Technology 2015;50: 249–255.

Chapter 10

# Ionizing Radiation Disinfestation Treatments against Pest Insects

Abdurrahman Ayvaz and Semih Yilmaz

Additional information is available at the end of the chapter

http://dx.doi.org/10.5772/60923

**Abstract**

Pesticides are often considered a suitable solution for controlling pests. However, the use of chemicals is very costly, and their residues have always the potential to pollute soil, air, and ground water and also pose significant risks to the natural ecosystems and nontarget organisms. Considering all these, irradiation could offer substantial and charming option for eliminating the export commodity fumigation uses for the undesirable effects of chemicals. Gamma rays, high-energy electrons, and X-rays are among the ionizing radiation sources utilized practically in sterile insect releasing programs using "self-contained" and "non-self-contained or panoramic" irradiators. When applying radiation sources, dosimetry should be adjusted to ensure quarantine security for large groups of insect pests. Because of growing concerns related to health problems and environmental pollutions, chemical sanitizing treatments are faced with a lot of regulatory restrictions, so irradiation reveals best choice for this purpose. The sterile insect technique (SIT) may have indispensable consideration for integrated pest management (IPM) of many important insect pests, including agricultural, veterinary, and medicinal importance. On the other hand, to overcome the obstacles of SIT treatments, genetic engineering techniques were supposed to ease the development of transgenic insects for sustainable tactics to control pest populations. Thus, genetic means should be an integral part of SIT treatments in controlling important pest populations.

**Keywords:** Ionizing radiation, dosimetry, sterile insect release, genetic sexing strain, $F_1$ sterility,

## 1. Introduction

Chemical pesticides have been the most widely used insect control methods, especially after the Second World War together with the invention of synthetic chemical pesticides. Pesticides are often considered a suitable solution for controlling pests. However, the use of chemicals is very costly and has polluted almost every part of our environment. Pesticide residues are found in soil, air, and ground water and they poses significant risks to the natural ecosystems and nontarget organisms.

There are two overriding problems facing insect controlling specialists. These concerns are the rapid development of resistance and environmental pollutions resulting from pesticide use. It was reported that irradiation could offer substantial and charming option for eliminating the export commodity fumigation uses for the undesirable effects of chemicals [1, 2]. In many countries, the direct control of stored product insects in wheat and wheat flour through radiation treatments is regarded as an approved method and would soon be approved for all grain products and other dry foods [3]. To this end, research is needed to continue for improving the methods. Although irradiation quarantine disinfestation treatment has been in progress for decades, it is not so common to use these tactics because radiation cannot kill the insects abruptly, and there are great concerns with regard to radiation applications among peoples. Due to the relationship with radioactivity and nuclear technology, consumers and industrial organizations have significant concerns about the radiation applications in food preservations, whereas even at the highest doses, radioactivity cannot be induced using these sources in food or insects exposed [4]. Accordingly, the development of radiation methods in controlling the agricultural products is so slow, and the adoption of these practices by the public and commercial organizations takes time. Informing the public awareness on the issue of the reliability of this method will enable more widespread use of these applications and will provide more acceptances by people. If safer and more secure products are obtained as a result of food irradiation and consumers are satisfied with the nutritional adequacy, their attitudes can be positive and they will buy the products without hesitation.

The superiority of irradiation in protecting agricultural products can be summarized as follows: it reduces product loss after harvesting. In terms of treating the products uniformly, it is more advantages over the fumigation treatments. It leaves no residues on the products and a best alternative to chemicals ensuring product quality standards in international trade [5]. It is also an important strategy for improving the hygienic quality of the agricultural products. The future inclination in quarantine measures against insect pests will mainly focus on the following issues: (1) determining specific doses for the insects resistant to radiation such as lepidopteran pests, (2) reducing radiation doses and abate treatment periods to maintain the product quality, (3) developing generic treatments lower than 400 Gy for important quarantine insects, and (4) developing information on value-added irradiated fresh products [6]. The standardized radiation treatments will facilitate safer trade between countries. The measures taken with radiation aim to prevent adult insect emergence. In this way, the risk of introducing exotic plant pest into new ecological areas during trade between countries can efficiently be prevented [6]. If there are eggs, larvae, and pupae in agricultural products, they are intended to be sterile. By examining numerous studies, these goals can be achieved with relatively lower doses for the pests belong to Diptera, Homoptera, and Coleoptera. Lepidopteran pests require higher doses than other groups. Radiation resistance of insects increase with advanced developmental stages. The tolerance of male insects is higher than that of females.

This review aims to provide information in presenting advances related to irradiation quarantine treatments against pest insects, assessing the worries in this field, discuss apprehensions with the applications, stressing the future trends, and explaining the mode of action of radiation on pest insects.

## 2. Ionizing radiation sources

Ionizing radiation has been classified into X-rays, $\gamma$-rays, $\alpha$-rays, $\beta$-rays, and neutron radiation [7]. Nature and background are the main sources of ionizing radiation. Of these, cosmic radiation can be classified into various forms according to its origin, energy and type, and flux density of the particles. Tree main sources of cosmic rays are galactic cosmic radiation, solar cosmic radiation, and radiation from the earth's radiation belts (Van Allen belts) [8]. Gamma radiations have the possibility to ionize the atoms but not affect the nucleus, so they cannot induce radioactivity on irradiated materials [9]. Among the ionizing radiation sources, gamma rays, high-energy electrons, and X-rays are the types used practically in sterile insect releasing programs [10-12]. However, $\alpha$ particles are not suitable for insect sterilization due to their high linear energy transfer and weak penetrability. On the other hand, neutrons are more effective in insect sterilizing, but their radioactivity induction in irradiated materials makes them impractical for sterile insect technique (SIT) programs [13-15]. It should be taken into consideration that for the fitness of insects, the acceptable level of energy for SIT applications is less than 5 MeV for Gamma rays (from 60Co and 137Cs) or X-rays and less than 10 MeV for electrons [4, 16-18]. Due to their similar relative biological effectiveness values, a different type of radiation source does not exert significant difference in their lethal effects on particular insects [4]. Cobalt-60 and cesium-137 radioisotopes are the most commonly used gamma radiation sources for SIT programs [8].

## 3. Comparison of irradiators

Irradiators with several hundreds to thousands of Curies of a high-energy gamma or beta emitter are large self-shielded devices. The basic components of an irradiation unit (gamma-ray or electron) are composed of the following: (1) the control systems related to radiation source are referred as "irradiators," (2) a product transport system, and (3) a shielding for protecting human health and environment from radiation [4].

Two major types of irradiator are "self-contained" and "non-self-contained or panoramic." In the former, primary beam is entirely shielded during use and storage conditions. In the latter, primary beam is not contained [19]. Irradiator design varies from small, which is suitable for radiation studies, to very large, which is convenient for hundreds of tons of product throughput daily. The activity level of the radiation source and the methods for the translocation of the products in the radiation field are the main differences between various irradiators [4].

Both cobalt and cesium are widely used as source rods in gamma irradiators [4]. Sterilization of insects is usually carried out with gamma rays from self-contained irradiators. In most self-contained irradiators, the position of irradiation is in the center of an annular array of long parallel pencils that include the encapsulated radiation source. Within this irradiation compartment, the doses are provided uniformly. Although self-contained irradiators provide a high-dose rate with a small irradiation volume (1–4 L), this design is suitable for small-scale programs that apply the SIT [4].

Panoramic irradiators are used efficiently for large-volume irradiation. In this design, the radiation source includes either several Co-60 rods lined up in a plane or a single rod that can be moved up and down into a wide chamber. Because gamma rays are emitted in all directions from isotopic sources, the high-energy utilization efficiency can be achieved through surrounding by insects, and irradiation can be applied to several containers at the same time [4]. Large-scale commercial irradiators are mostly not practical for practicing dosage-determining research due to differences between maximum and minimum absorbed doses received. Therefore, the determination of minimum absorbed dose required for an irradiation quarantine application is best performed using small irradiators [20].

Electrons and X-ray are the two modes of accelerator-generated radiation in electron and X-ray irradiators. The two characteristics of the principal electron-beam are the energy of particles (MeV), which affects the penetration of electrons, and the average current (in mA), which affects the rate of absorbed dose. Contrary to gamma rays, electron beams from accelerator-generated radiation are quite focused, and insects are continuously moved in conveyors through the beam. Due to the deeper penetration of X-rays compared to electrons, it allows the use of larger containers of insects for treatments [4].

Although sterilizing effects of electron and gamma rays are similar, the factors determining the source selection for SIT programs mostly depend on penetration, cost, product throughput, presence of experts, and safety factors [21]. Besides, gamma irradiators are normally cheaper and easier to run when compared with accelerators. However, due to their safety when switched off, the reliability and the public acceptance of electron accelerators are higher than gamma source [22-24]. The emission power of 100 kCi of Co-60 gamma-ray source is more or less equivalent to that of a 1.5-kW electron accelerator. The commercial accelerators have usually higher power capacities (5–10 MeV electrons), thus rendering them unsuitable for SIT applications. Although X-ray irradiators have the advantages over the gamma irradiators and accelerators, the effectiveness of transforming electrons into X-rays is nearly 7% for 5 MeV electrons. Thus, a great majority of the power is wasted while heating the converter [25]. When these conditions are all considered, gamma irradiators can be thought as mostly used in nearly all SIT programs [4].

## 4. Dosimetry

Dosimetry is the radiation absorbed dose for sterilizing and is of major importance for programs that comprise the release of sterile insects [26]. Dosimeters are frequently used in producing sterile insects for such tasks as absorbed-dose mapping, process control, and qualification of the irradiator [4]. Some of them are convenient for routine use at SIT programs [27]. Insects receiving very low doses may not adequately be sterile, and those that absorb very high may be uncompetitive. In such cases, the effectiveness of the program that requires a greater number of sterile insects to be released may essentially be decreased [28]. In executing the analysis, variation in both dose-dependent sterility and competitiveness data are required at the same time. For the competitiveness data to be realistic, the tests should be performed in field cages or open plots [26]. Given the importance of dosimetry in SIT programs, selecting a convenient dosimetry system has a critical importance [26]. Methods for calibrating regular dosimetry systems and for determining radiation fields for insect sterilization are described in periodically updated ISO/ASTM standards [27, 29-31] and in IAEA technical reports [9]. Gray (Gy) is used as the absorbed dose unit, which is equivalent to a joule of absorbed energy per kilogram of sample [9]. Therefore, in newly planned programs, dosimetry system needs to be established for adequately measuring the absorbed dose and estimating the associated confidence interval [27].

## 5. Doses achieve quarantine security

Irradiation is a quarantine treatment with the potential to disinfest a variety of fresh commodities of great number of quarantined pests. Many insect groups from the orders Diptera, Coleoptera, and Homoptera can be controlled with relatively low doses without damaging host plants of economic importance [20]. Other insects in Lepidoptera are controlled by moderate doses (0.2–0.3 kGy), which are tolerated by some major commodities, such as apples, cherries, and blueberries [20]. These doses need further evaluation using adequate numbers of insects to accomplish the degree of confidence required in quarantine treatments [20].

Moreover, because effective irradiation doses against most insects and mites do not affect the characteristics of commodities, this technology is ideal in developing "generic" treatments [32]. A generic quarantine treatment should provide quarantine security for large insect groups. For example, it can be applied to all pests belonging to Diptera, or to tephritid fruit flies in the genus *Bactrocera*. Before recommending generic treatments, effective irradiation doses should be evaluated in controlling the wide range of species belonging to a taxon [32] (Table 1).

| Pest group | Objective | Dose (kGy) |
|---|---|---|
| Stored product moths | Adult sterilization | 0.1–1 |
| Stored product beetles | Adult sterilization | 0.05–0.4 |
| Pyralidae and Tortricidae | Late-pupa sterilization | 0.2–0.3 |
| Noctuidae and Tortricidae | Prevent adult emergence | 0.1–0.3 |
| Scarab beetles | Adult sterilization | 0.05–0.15 |

**Table 1.** Doses to achieve quarantine security for various pest insects [20]

## 6. Advantages of irradiation over other postharvest treatments

The advantages of irradiation in controlling agricultural products can be outlined as follows: It is an effective and important tactic in controlling postharvest food losses. It is more advantageous compared to fumigation treatments due to its uniform penetration in the products and also time saving. It does not leave residues in commodities and a best alternative over chemical pesticides ensuring product quality standards in international trade. It is also an important strategy for improving the hygienic quality of the agricultural products [5]. The penetration power and the dose uniformity of the radiation treatment to treat products of different sizes and shapes and also to prevent the formation of resistance make the radiation treatments superior to chemicals [33]. Besides, radiation can reach pathogen organisms in areas of fruits not accessible to chemicals [34].

Because of growing concerns related to health problems and environmental pollutions, chemical sanitizing treatments are faced with a lot of regulatory restrictions. Thus, irradiation offers the most viable alternative for eliminating these concerns [5]. It was also reported that the minimum dose (150 Gy) required for disinfestation of fruit fly to satisfy quarantine regulations (0.15 kGy) does not adversely affect the physicochemical and nutritional value of most fruits and vegetables [35]. If the application is done properly, the efficacy of the irradiation process is guaranteed. It does not cause a significant amount of temperature increase during application; radiation does not leave residues. It is safe and removes concerns that may arise in terms of human health and environment. It is possible to apply for packaged products. However, some other disinfestation methods such as heat, cold, and fumigation treatments can be used in controlling pest insects in the commodities. For controlling pest species, irradiation treatments should be developed irrespective of commodity. Most products can have tolerance to irradiation at doses killing the pest; however, other methods cannot guarantee quality of the host commodities [4].

## 7. A generic quarantine treatment

Introducing exotic pest insects through the improvement of international world trade in agricultural commodities becomes increasingly important problem day by day. This new problem will cause extra costs for control programs and quarantine restrictions [36]. A generic quarantine treatment is one that provides quarantine security for a broad group of pests [37]. The International Consultative Group on Food Irradiation (ICGFI) was the first group to formalize a recommendation for a generic treatment. In 1986, based on irradiation data for many tephritid fruit fly species and a limited number of other insect pests, ICGFI proposed a dose of 150 Gy for fruit flies and 300 Gy for other insects [38]. Before generic treatments can be recommended, information is needed on effective irradiation doses for a wide range of insects within a taxon [36]. Data from all available insects are used in developing generic treatments because they serve as representatives for their respective groups [39]. According to a rule published in the United States in 2006, a dose of 150 Gy generic radiation was determined for all tephritid fruit flies and 400 Gy for all other insects, except for pupae and adults of lepidopteran pests, which require higher doses [40].

Some other applications such as heat, cold, and fumigation are used to disinfest host commodities before exporting them to pest free area. However, the treatment process other than irradiation requires balancing between the adverse effects and killing the pest insects to preserve commodity quality [41] since radiation treatments target pest insects without damaging the fruit or vegetable host [36]. For example, radiation prevents the temperature increase in commodities. International standard institutes approved that radiation is valid for all fruits and vegetables that are hosts for the given pests [42, 43].

Expanding the application spectrum of the generic irradiation treatments in the family or order level in other taxa would be practical, would easily promote international trade in agricultural products, and would supply an alternative treatment for infested commodities in cross-country transportation [44].

## 8. Integrated pest management programs

The process of pest control is becoming more complex and requires new solutions in the course of time due to the emergence of new pest population, strict regulation in international trade, insecticide resistance, and residue problems. These new problems made it necessary to develop new and cleverly designed pest control techniques. Integrated pest management (IPM) is largely accepted as a powerful and environmentally sensitive method in managing pest insects that relies on a combination of commonsense practices [45]. In IPM strategies, comprehensive information on the life cycles of pests and their interaction with the environment is used. This method, in combination with available pest control tactics, is applied to manage pest population damage with the least possible hazard to people, property, and environment [45].

As a part of an area-wide integrated pest management (AW-IPM) approach, the sterile insect technique (SIT) is regarded as a vigorous control strategy for establishing pest free areas. The development of more competitive moths may improve the effectiveness of AW-IPM programs integrated with SIT technique [46, 47]. Species-specific nature and compatibility with existing control methods (biological control, mating disruption, cultural control, and use of biorational pesticides) make SIT an indispensable part of AW-IPM application and also make it superior to other control methods [48-50]. There are a number of successful models in terms of integrating the SIT in AW-IPM programs against many important lepidopteran pests [51, 52].

Based on herein and other numerous literature, it may be said that SIT is a very convenient method as part of AW-IPM programs and can be further developed by decreasing the production costs, improving the effectiveness of released sterile moths and combining to other effective control tactics [53] .

## 9. Principles and practices of sterile insect technique

The idea that populations of economically important insect species might be controlled, managed, or eradicated through genetic manipulation was supposed by Knipling in 1930s. A similar concept was published independently by Serebrovsky [54]. In the late 1930s, Knipling recommended that if there could be a way to genetically sterilize male insects without affecting their ability to mate and competitiveness, then subsequent to their release and mating with wild females, the fertility of a target population could be reduced. The sterile insect technique is an environmentally innocuous and target-specific control tactic in suppressing the pest population [49]. With the development of modern genetic methods, this method will become a promising technique in the near future in controlling many important pest populations [55, 56].

The first applications were performed on the New World screwworm *Cochliomyia hominivorax* to evaluate this procedure [57]. The induction of sterility in this species by X-rays was the first small step on the way to the eradication of the serious livestock pest from Southern America and now from most countries of Central America [57]. This long-term and successful program has demonstrated that radiation-induced mutations can play an important role in developing environmentally acceptable, area-wide, and pest intervention strategies.

Although open to scientific criticism, the eradication process has been processed across the southern parts of the United States. With the help of this program, which began in Florida in 1957, the entire population of the pest is eradicated in the United States within a period of 10 years. Due to the reinfestations of migrating flies from neighboring Mexico, the program has been compromised, and the United States–Mexico joint program has become a necessity in 1972. With the success of the program, Mexico in 1991, Belize and Guatemala in 1994, and El Salvador in 1995 officially declared that they are free of screwworms. Because no flies have been detected since January 1995, Honduras was technically considered as free of screwworms. Eventually, the United States–Central America project proposed to maintain a sterile insect barrier at the Darien Gap in Panama starting in 1997. By the implementation of this program, billions of dollars was saved in livestock and wildlife loses [57]. Screwworm is an obligatory parasite of warm-blooded animals infesting livestock and mammals, including humans. Female flies lay their eggs on the wounded inflammatory region of the body. Larvae hatching from the eggs feed on the flesh. Because these flies were easy to rear, the program was composed of a small-scale wild adult population. The flies tend to mate only once the screwworm was a good candidate for SIT program. These factors were optimum to achieve high sterile/fertile ratios fort this pest [58].

Although not as successful as the screwworm eradication program, SIT has been implemented for some other pest populations such as the tephritid fruit flies, including *Ceratitis capitata* Wiedemann, *Pectinophora gossypiella* Saunders, and *Cydia pomonella* L. in many parts of the world [59].

Pests of agricultural, veterinary, and medical importance can be specifically controlled using the SIT method, the integral component of AW-IPM. The sterile insect technique (SIT) is a specific control method that may be applied in the area-wide integrated pest management of insects. It is important to release only the sterile males to implement this method effectively [60]. As the next generation is to be established by wild females, the removal of wild males is essential for reducing the size of target population [61]. Infertility in the wild population can only be achieved with the help of sterile males. Thus, this method was initially named as the sterile-male method [62]. At the first application, both sexes were released in controlling the New World screwworm *C. hominivorax*. However, the benefit of this bisexual releases was determined to be limited for the Mediterranean fruit fly [63-65]. In such a design, released sterile males and females tend to mate with each other. This inclination reduces the mating potential of sterile males with wild females, and less sterility is introduced into the wild population. Only sterile male release reduces mass rearing costs for both production and postproduction stages. In the postproduction stage, considerable reductions can be achieved in the cost of workload, marking, irradiation, transport, release, and monitoring [66]. In many cases, releasing sterile females is not an easy process and brings about further negative effects. For example, females of fruit fly may cause extra damage in some fruits, females of biting flies result in reducing livestock meat production, and females of blood-sucking species may transmit disease [67]. However, it is not easy to make sex separation in such large populations. Therefore, to overcome this problem, it is obligatory to develop new specific strains. To date, Mendelian genetics, chromosome rearrangements, and specific mutations can successfully be used to develop new strains. When the sterile insect technique is compared with pathogenic biological entities and toxic chemicals, it is noninvasive. Therefore, the environmental risks of SIT application are exceptionally very low [68, 69]. This method is also compatible with the food chain in terms of integrating ecosystems with living but nonreproductive organisms. When considering all these situations, the hazard of the SIT to the environment is negligible.

There are some components that make the sterilization techniques successful [70], and the principles of sterility have not changed significantly since E. F. Knipling's formulation:

1. Mass rearing of target insects should be easy and applicable (rearing component).

2. Large numbers of the target insect should be possible to sterilize (treatment component).

3. Following sterilization, fairly competitive insects should be released (competitiveness component).

4. Release and distribution of sterile insects into fields should be cost effective (release component).

5. Before and after the release of the treated insects, population should be assessed accurately by using special tools (evaluation component).

6. The treated area should be well isolated to prevent inseminated females from entering the field (reinfestation component).

## 10. Improvement of the sterile insect technique thorough genetic engineering technology

The sterile insect techniques (SIT) are considered as releasing sterile males in area-wide pest management. In this context, with the use of genetic methods, infertile matings were enhanced utilizing the release of mass-reared sterile insects [49, 71-73]. Therefore, by genetic means, new insect strains developed for improving SIT activity or avoiding potential adverse effects of such releases. Two categories of genetic methods for strain development are considered as conventional genetics and transgenesis [74, 75]. Using these methods, the development of an efficient and cost-effective SIT program would have a great importance in eliminating females from the released population. In this respect, the sterile insect technique may possibly be improved and extended using modern molecular tools. For example, SIT programs are improved by releasing unirradiated but instead homozygous insects with dominant lethal (RIDLs) constructs that are repressible during mass production [56, 76, 77].

A female-lethal version of RIDL, with insects homozygous for one or more female-specific dominant lethal genetic constructs, has been created in *C. capitata* and offered for many other species [78] . This approach is also known as autocidal biological control [79]. The identification of alternative and more promiscuous transposable elements as *hermes, hobo, minos, mosl,* and *piggyBac* and novel gene delivery systems such as microinjection, electroporation, sonoporation, lipofection, and biolistics prompted studies on genetic manipulation of many insects of agricultural importance for various purposes [76, 80, 81].

The nature and timing of lethality is one of the most important potential advantages of genetic methods over radiation-based SIT programs. The transmission of transgenic SIT methods to insects of agricultural importance is now applicable through the development of sophisticated vectors incorporating the *piggyBac* transposable element [82, 83] and transformation markers based on improved green fluorescent protein (EGFP) variants [84, 85]. This technology was supposed to facilitate the development of transgenic insects for sustainable tactics to control pest populations or disease vectors [86, 87]. In addition, the use of systems for marking transgenic sperm in SIT programs is one of the other significant improvements for addressing the lack of efficient and reliable methods in field monitoring of insects. In SIT programs, producing male-only sexing strains for converting female insects into males thorough genetic manipulations in sex determining pathways can be another strategy [88, 89]. In the medfly, such a phenomenon has been shown to conditionally express a transgene that interferes with the expression of female-specific *tra* gene expression. The resulting population was reported to comprise 95% males and 5% intersexes [89].

# 11. Radiation-induced $F_1$ sterility in lepidopteran pests

If the parental generation of these insects was irradiated with substerilizing doses of gamma radiation, the degree of sterility would be higher than that of parental generation, and this circumstance is known as radiation-induced $F_1$ sterility [15, 90]. Because the pest insects from Lepidoptera are radioresistant species, high doses are required to achieve complete sterility when compared to other pest insects from different orders [4, 91]. Despite continued for several generations, radiation-induced detrimental effects are most pronounced in the $F_1$ generation. Inherited sterility is also referred as inherited partial sterility, partial sterility, delayed sterility, semisterility, and $F_1$ sterility [52]. Mutagenic chemical substances (chemosterilants) can be used to induce sterility as an alternative to radiation, but due to human health and environmental concerns, chemicals are not preferable for obtaining sterilized mass-reared insects today [92, 93].

Inherited sterility has been shown for the first time on silkworm *Bombyx mori* (L.) [94]. Early investigations related to this topic were revised and discussed in terms of its pest control potential and genetic aspects [15, 90]. Experiments of the Proverb [95] showed that the $F_1$ generation of insects was sterile when their parents irradiated with gamma radiation. This first application has opened up new horizons and given impetus to research on $F_1$ sterility [96]. Knipling [97] and LaChance [90] recommended the use of $F_1$ sterility as the potential component of area-wide integrated management of lepidopteran pests. The validity and efficacy of the method has been indicated on various pests in a number of laboratory studies [90]. However, a high dose of radiation adversely affects some important traits of the pest population as mating ability and longevity and causes reduction in the competitiveness of the sterile insects against the wild population [98]. This control tactic represents an environmentally friendly alternative and provides facilities for control of many important pest species. The superiority of this method over the completely sterile insect is discussed by many authors [52, 99-102].

For the high radio resistance in lepidopteran insects, the presence of possible DNA repair mechanisms and an inducible cell recovery system was proposed [91]. The radio-tolerant talent of these insects has also been attributed to the holokinetic nature of their chromosomes [103]. Radiation-induced sterility is generally a result of dominant lethal mutations (DLMs) in insects other than lepidopterans and is expressed during early cell proliferation in embryogenesis [52, 104]. However, the frequency of DLMs is much lower in Lepidoptera than that of other pest orders and is seen toward the end of embryonic development [105].

Since sterile $F_1$ progeny are produced under field conditions, releasing partially sterile males with fully sterile females is more compatible with other tactics [106]. A significant amount of the unfertilized eggs or early embryonic mortality was observed for different lepidopteran pests in treated males mated with the females as in the case of *Manduca sexta* (L.), *Ephestia kuehniella* Zeller, and *Spodoptera litura* (F.) [107-109]. It can be inferred that the most important cause of male sterility results from physiological impairments, including failed mating and inability to complete sperm transfer [52].

Males of Lepidoptera are more radio resistant than females. Several authors indicated that in different species of Lepidoptera, the sex ratio was biased toward the males [15, 108-111]. This difference is attributed to the gametes at the time of irradiation. Radiation is generally applied to mature pupa or newly emerged adults of Lepidoptera. Euprene sperm production is completed at the time of irradiation, and dividing cell reaches to interphase. However, the oocytes are stalled in metaphase I, and the process could not be completed up to the oviposition [112]. Thus, radiation disrupts the normal course of meiosis. The secondary harmful effect seen in the oocytes is the degradation of the cytoplasmic components. The treated oocytes have large amounts of cytoplasm than that of cytoplasm-free sperm, and this cytoplasm contains many components required for embryogenesis [52].

The higher sterility level in $F_1$ male progeny was attributed to three factors by Tothová and Marec [113]:

1. Despite large inherited chromosomal breaks, $F_1$ males continue to survive, and the frequency of the chromosome breaks indicates a positive correlation depending on increasing doses. However, this correlation is not seen in $F_1$ females. The differences result from the large number of chromosomal breaks inherited by $F_1$, and higher radiation doses might cause increasing damage rate on sex chromosome (Z). F1 females might be affected of recessive lethal mutations, but not males.

2. Crossing-over process during spermatogenesis

3. Radiation-induced deleterious effects on the fertility of $F_1$ males

Genetic sexing system was suggested to introduce lethal mutations in the wild population firstly in *Bombix mori* by Strunnikov [114] and subsequently developed in Mediterranean flour moth *E. kuehniella* by Marec [115], Marec and Mirchi [116], and Marec [117]. Almost all $F_1$ generations consist of male progeny due to the inheritance of one of the lethal mutations from their father when BL-2 males are mated to wild-type females. Females are hemizygous with regard to sex-linked recessive mutations (*sl-2 and sl-15*) and die during embryogenesis. For introducing lethal mutations into the wild population, balanced BL-2 males could be released directly into nature or could be reared in laboratory conditions to generate male mutant strains [108, 118]. The combination of $F_1$ sterility with male-only colonies would be useful for reducing rearing costs and enhancing population suppression. Despite these advantages of genetic sexing system in $F_1$ sterility applications against lepidopteran pests, lack of suitable markers for constructing mutant strains, difficulties in sex separation under mass-rearing conditions, and constantly checking requirements of mutant strains to keep its genetic structure through genetic recombination or colony contamination are some of the significant drawbacks that still need to be overcome [52, 119].

$F_1$ sterility can effectively be combined with other control tactics, such as pheromone disruption [120-123], host plant resistance [122], and natural enemies [123]. The production of sterile $F_1$

larvae should be considered as an opportunity for producing natural enemies and sterile moths in field conditions [124]. This would be an additional advantage ensured from this method. The extra eggs of sterile population will constitute additional host material for the egg parasitoids [111, 125]. These sterile eggs do not affect parasitoid preference adversely [126], and this tactic could also be a suitable way for combining SIT and augmentative release.

The benefits of the radiation-induced $F_1$ sterility can be summarized as follows.

- Reduced egg hatch and highly sterile and predominant $F_1$ male progeny
- Lower doses are adequate to induce $F_1$ sterility and hence to increase the quality and competitiveness of the released insects [15]
- Dispersal ability improvement following release [120]
- Increase in mating competitiveness [127]
- Improved sperm competitiveness [127]
- Sterile $F_1$ progeny production in the field
- Supplementing extra host material for the egg parasitoids [111, 125]
- For increasing the natural enemy population, $F_1$ eggs, larvae, and pupae of the pest insects can also be used as host [128]

## 12. International database on insect disinfestation and sterilization (IDIDAS)

The International Database on Insect Disinfestation and Sterilization (IDIDAS) is a data bank collecting the radiation doses applied to important pest arthropods, which are important in terms of veterinary, medicine, and agriculture. Data collection and share about radiation doses are the main purpose of this database for disinfesting and reproducing sterile pests by comparative analysis and quality check [129]. This data bank can be accessed from the website of IDIDAS [21].

## 13. Conclusions and recommendations for future research

The future trends in controlling important pest population would predominantly be directed to biological control methods as SIT treatments. As an integrated part of area-wide pest management programs, the applications of SIT treatments will continue to increase and be desired by all sectors as farmers, commercial companies, and consumers. Cooperation and contributions of all stakeholders are essential to ensure effective implementation of these technologies. The development and the applicability of the proposed methods are required to be inexpensive and environmentally sensitive. For the mass rearing of biological control agents and improving their transport facilities, various studies are carried out with great efforts. In this respect, with the utility of novel and innovative methods, the cost-effective augmentation of natural enemies in field conditions will be possible. The use of modern biotechnology and molecular methods for the manipulation of many insects of agricultural importance for increasing the competitiveness of released male-only population in the field, release of insects carrying a dominant lethal, and timing of lethality will contribute to radiation-induced sterility. Thus, sterile insect populations with highly competitive and desired properties can be achieved for protective purposes.

## Author details

Abdurrahman Ayvaz[1*] and Semih Yilmaz[2]

*Address all correspondence to: ayvaza@erciyes.edu.tr

1 Faculty of Science, Department of Biology, Erciyes University, Kayseri, Turkey

2 Faculty of Agriculture, Department of Agricultural Biotechnology, Erciyes University, Kayseri, Turkey

## References

[1] Ahmed MSH. Irradiation disinfestation and packaging of dates. Insect disinfestation of food and agricultural products by irradiation. Proceedings of the final research co-ordination meeting; 25-29 MAY 1987; Beijing. China. Vienna: IAEA;1991. p. 7-26

[2] Marcotte M: United Nations Environment programme Methyl Bromide Technical Options Committee. Food Irrad. Newslett. 1993; 17: 27- 32

[3] Brower JH, Tilton EW. The potential of irradiation as commodities. In: Proceedings of Radiation disinfestation of food an agricultural products conference; Hawaii: Institute of Tropical Agricultural and Human Research, University of Hawaii, 1983. p. 75-86

[4] Bakri A, Mehta K, Lance DR. Sterilizing insects with ionizing radiation. In: Dyck VA, Hendrichs J, Robinson AS., editors. Sterile Insect Technique: Principles and Practice in Area-Wide Integrated Pest Management. Dordrecht, The Netherlands: Springer; 2005. p. 233—268.

[5] Loaharanu P, Mainuddin A. Advantages and disadvantages of the use of irradiation for food preservation. Journal of Agricultural and Environmental Ethics. 1991; 4(1): 14-30.

[6] Follett PA, Wall MM. Phytosanitary irradiation for export of fresh produce: commercial adoption in Hawaii and current issues. J Radioanal Nucl Chem. 2013; 296:517-522

[7] Wikilectures, [Internet]. Available from http://www.wikilectures.eu/index.php/Ionizing_ Radiation

[8] UNSCEAR 2010. Sources and effects of ionising radiation. United Nations Scientific Committee on the effects of Atomic Radiation. UNSCEAR 2008 Report to the general assembly with scientific annexes. Volume I. United Nations, New York 2010

[9] International Atomic Energy Agency. Natural and Induced Radioactivity in Food. IAEA-TECDOC-1287. IAEA, Vienna, Austria; 2002b

[10] Bushland RC, Hopkins DE. Sterilization of screw-worm flies with X-rays and gamma rays. Journal of Economic Entomology. 1953; 46(4):648–656

[11] Lindquist AW The use of gamma radiation for the control or eradication of the screwworm Journal of Economic Entomology 1955; 48(4):467–469

[12] Baumhover AH, Graham AJ, Bitter BA, Hopkins DE, New WD, Dudley FH, Bushland RC. Screwworm control through release of sterilized flies. Journal of Economic Entomology. 1955;48(4):462–466

[13] Hooper GHS. Sterilization and competitiveness of the Mediterranean fruit fly after irradiation of pupae with fast neutrons. Journal of Economic Entomology. 1971; 64(6): 1369 1372

[14] Offori ED, Czock KH. The use of 'fast' neutrons and gamma radiation to sterilize the tsetse fly Glossina tachinoides Westw. The International Journal of Applied Radiation and Isotopes. 1975;26 (5):257–260

[15] North DT Inherited sterility in Lepidoptera Annual Review of Entomology197520167–182

[16] Elias PS, Cohen AJ, editors. Radiation chemistry of major food components: its relevance to the assessment of the wholesomeness of irradiated foods. Amsterdam, The Netherlands: Elsevier; 1977. 220 p

[17] FAO/IAEA/WHO, Food and Agriculture Organization of the United Nations/International Atomic Energy Agency/World Health Organization. 1999. High-dose irradiation: wholesomeness of food irradiated with doses above 10 kGy. Joint FAO/IAEA/WHO Study Group, Technical Report Series 890. World Health Organization, Geneva, Switzerland

[18] International Atomic Energy Agency, Dosimetry for Food Irradiation. International Atomic Energy Agency. Technical Reports Series Number 409. IAEA, Vienna, Austria; 2002a

[19] [Internet]. Available from http://www.slideshare.net/brucelee55/module-7-irradiators-and-sealed-sources

[20] Hallman GJ. Ionizing radiation quarantine treatments. Anais da Sociedade Entomológica do Brasil. 1998; 27(3):313-323

[21] IDIDAS. [Internet]. 2004. Available from: http://www-ididas.iaea.org/ididas/ [Accessed: 2015-03-08]

[22] Cavallaro R, Delrio G. Sterilization of Dacus oleae Gmel. and Ceratitis capitata Wied with gamma rays and fast neutrons. Redia. 1974; 55(3):373–392

[23] Smittle BJ. Irradiation of Anastrepha suspensa (Diptera: Tephritidae): new irradiation facility. Florida Entomologist. 1993; 76(2):224–227

[24] EBFRF. Electron Beam Food Research Facility. Institute of Food Science and Engineering, Texas A&M University [Internet]. 2004. Available from: http://ifse.tamu.edu/centers/ebeam.html

[25] Farrell JP, Seltzer SM, Sillverman J. Bremsstrahlung generators for radiation processing. Radiation Physics and Chemistry. 1983; 22(3-5):469–478

[26] Parker A, Mehta K. Sterile insect technique: A model for dose optimization for improved sterile insect quality. Florida Entomologist. 2007; 90(1):88-95

[27] International Organization for Standardization/American Society for Testing and Materials (ISO/ASTM). (2004a). Standard guide for dosimetry for sterile insect release programs, ISO/ASTM 51940. Annual book of ASTM standards 12.02. ASTM International, Philadelphia, PA, USA

[28] Robinson AS. Mutations and their use in insect control. Mutation Research. 2002; 511(2):113–132

[29] International Organization for Standardization/American Society for Testing and Materials (ISO/ASTM). (2004b). Standard guide for selection and calibration of dosimetry systems for radiation processing. ISO/ASTM 51261. Annual book of ASTM standards 12.02. ASTM International, Philadelphia, PA, USA

[30] International Organization for Standardization/American Society for Testing and Materials (ISO/ASTM). (2004c). Standard guide for dosimetry in radiation research on food and agricultural products. ISO/ASTM 51900. Annual book of ASTM standards 12.02. ASTM International, Philadelphia, PA, USA

[31] International Organization for Standardization/American Society for Testing and Materials (ISO/ASTM). (2004d). Standard practice for dosimetry for a self-contained dry-storage gamma-ray irradiator. ISO/ASTM 52116. Annual book of ASTM standards 12.02. ASTM International, Philadelphia, PA, USA

[32] International Atomic Energy Agency. Irradiation as a phytosanitary treatment of food and agricultural commodities: proceedings of a final research coordination meeting organized by the Joint FAO/IAEA Division of Nuclear Techniques in Food and Agriculture 2002. Vienna, Austria: International Atomic Energy Agency (IAEA); 2004. 181 p

[33] Jarrett RD. (1982). Isotope radiation sources. In: Josephson, E.S., Peterson, M.S.(Eds) Preservation of Food by Ionizing Radiation. CRC Press, Boca rato′ n. pp. 137-163

[34] Tiryaki O. Inhibition of *Penicilllium expansum, Botrytis cinerea, Rhizopus stolonifer*, and *Alternaria tenuissima*, which were isolated from Ankara pears, by gamma irradiation. The Journal of Turkish Phytopathology. 1990; 19(3):133-140

[35] IAEA, International Atomic Energy Agency. 1986. Report of a Task Force Meeting on Irradiation as a Quarantine Treatment, Chiang Mai, Thailand, 17-28 February 1986. Vienna: International Atomic Energy Agency

[36] Follett PA: Generic Radiation Quarantine Treatments: The Next Steps J. Econ. Entomol. 2009; 102(4):1399-1406

[37] Hallman GJ, Levang-Brilz NM, Zettler JL, Winborne IC. Factors affecting ionizing radiation phytosanitary treatments, and implications for research and generic treatments. Journal of Economic Entomology. 2010; 103(6):1950-1963

[38] International Consultative Group on Food Irradiation (ICGFI). 1991. Irradiation as a quarantine treatment of fresh fruits and vegetables. ICGFI Document 13. International Atomic Energy Agency, Vienna, Austria

[39] Hallman GJ. Control of stored product pests by ionizing radiation. Journal of Stored Products Research. 2013; 52:36-41

[40] Follett PA, Weinert ED. Comparative radiation dose mapping of single fruit type and mixed-fruit boxes for export from Hawaii. J. Food Process. Preserv. 2009; 33:231–244

[41] Paull RE: Response of tropical horticultural commodities to insect disinfestation treatments. Hort Science. 1994; 29:988-996

[42] Food and Agricultural Organization (FAO). 2000. Phytosanitary treatments for regulated pests. ISPM #28. Food and Agricultural Organization, Rome, Italy

[43] Food and Agricultural Organization. 2010. Report of the 5th session of the Commission on Phytosanitary Measures. Food and Agricultural Organization, Rome, Italy

[44] Follett P, Armstrong JW. Revised irradiation doses to control melon fly, mediterranean fruit fly, and oriental fruit fly (Diptera: Tephritidae) and a generic dose for Tephritid Fruit Flies. Journal of Economic Entomology. 2004; 97(4):1254-1262

[45] EPA [Intrnet].2007. Available from: http://www.epa.gov/region1/assistance/univ/pdfs/bmps/CatholicUnivIPM1-8-07.pdf. [Accessed: 2015-03-08]

[46] Hendrichs J, Kenmore P, Robinson AS, Vreysen MJB. Area-wide integrated pest management (AW-IPM): principles, practice and prospects. In: Vreysen MJB, Robinson AS, Hendrichs J, editors. Areawide Control of Insect Pests. from Research to Field. Dordrecht, The Netherlands: Springer; 2007. p. 3–33

[47] Pimentel D. Area-wide pest management: environmental, economic and food issues. In: Vreysen MJB, Robinson AS, Hendrichs J, editors. Area-Wide Control of Insect

Pests. from Research to Field Implementation. Dordrecht, The Netherlands: Springer; 2007. p. 35–47

[48] Carpenter JE. Area-wide integration of lepidopteran F1 sterility and augmentative biological control. In: Area-wide control of fruit flies and other insect pests. In: Tan KH, editor. Proceedings International Conference on AreaWide Control of Insect Pests, and the 5th International Symposium on Fruit Flies of Economic Importance; 28 May-5 June 1998; Penang, Malaysia. Pulau Pinang, Malaysia: Penerbit Universiti Sains; 2000. p. 193–200

[49] Dyck VA, Hendrichs J, Robinson AS, editors. Sterile Insect Technique. In: Principles And Practice In Area Wide Integrated Pest Management. Dordrecht, The Netherlands: Springer; 2005. 723 p

[50] Schuster DJ, Stansly PA. Biorational insecticides for integrated pest management in tomatoes. Crop Protection. 2005; 64:88-92. DOI: doi:10.1016/j.cropro.2014.06.011

[51] Bloem KA, Bloem S, Carpenter JE. Impact of moth suppression/eradication programmes using the sterile insect technique or inherited sterility. In: Dyck VA, Hendrichs J, Robinson AS, editors. Sterile Insect Technique. Principles And Practice In Area-Wide Integrated Pest Management. Dordrecht, The Netherlands: Springer; 2005. p. 677–700

[52] Carpenter JE, Bloem S, Marec F. Inherited sterility in insects. In: Sterile Insect Technique. Principles And Practice In Area-Wide Integrated Pest Management. Dyck VA, Hendrichs J, Robinson AS ed. Dordrecht, The Netherlands: Springer; 2005. p. 115–146

[53] Simmons GS, Suckling DM, Carpenter JE, Addison MF, Dyck VA, Vreysen MJB. Improved quality management to enhance the efficacy of the sterile insect technique for lepidopteran pests. Journal of Applied Entomology. 2010; 134(3):261–273. DOI: DOI: 10.1111/j.1439-0418.2009.01438.x

[54] Serebrovsky AS. On the possibility of a new method for the control of insect pests.. Zoologichesky Zhurnal. 1940; 19:618-630

[55] Gong P, Epton M, Fu G, Scaife S, Hiscox A, Condon K, Condon G, Morrison N, Kelly D, Dafa'alla T, Coleman P, Alphey L. A dominant lethal genetic system for autocidal control of the Mediterranean fruitfly. Nature Biotechnology. 2005; 23(4):453-456

[56] Alphey L. Engineering insects for the sterile insect technique. In: Vreysen, M., editors. Area-Wide Control of Insect Pests: from Research to Field Implementation.. Dordrecht, The Netherlands: Springer; 2007. p. 51–60

[57] Wyss JH. 2000. Screw-worm eradication in the Americas — overview, pp. 79–86. In K. H. Tan (ed.), Proceedings: Area-Wide Control of Fruit Flies and Other Insect Pests. International Conference on Area-Wide Control of Insect Pests, and the 5th International Symposium on Fruit Flies of Economic Importance, 28 May–5 June 1998, Penang, Malaysia. Penerbit Universiti Sains Malaysia, Pulau Pinang, Malaysia

[58] Bushland RC, Hopkins DE. Experiments with screwworm flies sterilized by X-rays. Journal of Economic Entomology. 1951; 44(5):725–731

[59] Alphey L, Benedict M, Bellini R, Clark GG, Dame DA, Service MV, Dobson SL. Sterile-insect methods for control of mosquito-borne diseases: an analysis. Vector-Borne and Zoonotic Diseases. 2010; 10(3):295-311

[60] Franz G. Genetic sexing strains in Mediterranean Fruit Fly, an example for other species amenable to large-scale rearing for the sterile insect technique. In: Dyck VA, Hendrichs J, Robinson AS, editors. Sterile Insect Technique: Principles And Practice In Area-Wide Integrated Pest Management. Dordrecht, The Netherlands: Springer; 2005. p. 427-449

[61] Koyama J, Teruya T, Tanaka K: Eradication of the oriental fruit fly (Diptera: Tephritidae) from the Okinawa Islands by a male annihilation method. Journal of Economic Entomology. 1984; 77:468-472

[62] Knipling EF: Sterile-male method of population control. Science. 1959; 130:902-904

[63] McInnis DO, Tam S, Grace C, Miyashita D: Population suppression and sterility rates induced by variable sex ratio, sterile insect releases of Ceratitis capitata (Diptera: Tephritidae). Annals of the Entomological Society of America. 1994; 87:231-240

[64] Rendón P, McInnis D, Lance D, Stewart J. 2000. Comparison of medfly male-only and bisexual releases in large scale field trials, pp. 517–525. *In* K. H. Tan (ed.), Proceedings: Area-Wide Control of Fruit Flies and Other Insect Pests. International Conference on Area-Wide Control of Insect Pests, and the 5th International Symposium on Fruit Flies of Economic Importance, 28 May–5 June 1998, Penang, Malaysia. Penerbit Universiti Sains Malaysia, Pulau Pinang, Malaysia

[65] Rendón P, McInnis D, Lance D, Stewart J. Medfly (Diptera: Tephritidae) genetic sexing: large-scale field comparison of males-only and bisexual sterile fly releases in Guatemala. Journal of Economic Entomology. 2004; 97: 1547–1553

[66] Epsky ND, Hendrichs J, Katsoyannos BI, Vasquez LA, Ros JP, Zumreoglu A, Pereira R, Bakri A, Seewooruthun SI, Heath RR: Field evaluation of female targeted trapping systems for Ceratitis capitata (Diptera: Tephritidae) in seven countries. Journal of Economic Entomology. 1999; 92:156–164

[67] Lance DR, McInnis DO. Biological Basis of the Sterile Insect Technique. In: Dyck VA, Hendrichs J, Robinson AS, editors. Sterile Insect Technique: Principles and Practice in Area-Wide Integrated Pest Management. Dordrecht, The Netherlands: Springer; 2005. p. 69-85

[68] Müller P, and Nagel P: Tsetse fly eradication on Zanzibar Island; environmental surveys. End of Mission Report, URT/5/016-07-01. International Atomic Energy Agency, Vienna, Austria. 1994. Institut für Biogeographie, Universität des Saarlandes, Saarbrücken, Germany

[69] Hendrichs J. The sterile insect technique world wide. In: Proceedings of Seminar, Madeira Med: Sterile Insect Technique as an Environmentally Friendly and Effective Insect Control System. Região Autónoma da Madeira and European Union; 12-13 November 1999; Funchal, Madeira. Madeira Regional Direction of Agriculture, Portugal; 2001. p. 25-53

[70] Bartlett, AC.. Insect sterility, insect genetics, and insect control. In: D. Pimentel, editor. CRC Handbook of Pest Management in Agriculture, Vol. II. Boca Raton, FL: CRC Press; 1990. p. 279-287

[71] Knipling EF. Possibilities of insect control or eradication through the use of sexually sterile males. Journal of Economic Entomology. 1955; 48:459–462

[72] Krafsur ES. Sterile insect technique for suppressing and eradicating insect population: 55 years and counting. Journal of Agricultural Entomology. 1998; 15(4):303–317

[73] Klassen W, Curtis CF. History of the sterile insect technique. In: Dyck VA, Hendrichs J, Robinson AS, editors. Sterile Insect Technique. Principles and Practice in Area-Wide Integrated Pest Management. Dordrecht, The Netherlands: Springer; 2005. p. 3–36

[74] Alphey L, Baker P, Burton RS, Condon GC, Condon KC, Dafa'alla TH, Epton MJ, Fu G, Gong P, Jin L, Labbé G, Morrison NI, Nimmo DD, O'Connell S, Phillips CE, Plackett A, Scaife S, Woods A.. Genetic technologies to enhance the Sterile Insect Technique (SIT). In: Proceedings of the 7th International Symposium on Fruit Flies of Economic Importance; 10-15 September 2006; Salvador, Brazil. p. 319-326

[75] Morrison NI, Franz G, Koukidou M, Miller TA, Saccone G, Alphey LS, Beech CJ, Nagaraju J, Simmons GS, Polito LC. Genetic improvements to the sterile insect technique for agricultural pests. Asia-Pacific Journal of Molecular Biology and Biotechnology. 2010; 18(2):275-295

[76] Thomas DD, Donnelly CA, Wood RJ, Alphey LS. Insect population control using a dominant, repressible, lethal genetic system. Science. 2000; 287(5462):2474–2476. DOI: DOI: 10.1126/science.287.5462.2474

[77] Alphey L, Nimmo D, O'Connell S, Alphey N.. Insect population suppression using engineered insects.. In: Aksoy S., editor. Transgenesis and the Management of Vector-Borne Disease. Austin, Texas: Landes Bioscience; 2007a

[78] Fu G, Condon KC, Epton MJ, Gong P, Jin L, Condon GC, Morrison NI, Dafa'Alla TH, Alphey L. Female-specific insect lethality engineered using alter- native splicing. Nature Biotechnology. 2007; 25(3):353–357

[79] Fryxell KJ, Miller TA. Autocidal biological control: a general strategy for insect control based on genetic transformation with a highly conserved gene. Journal of Economic Entomology. 1995; 88(5):1221–1232

[80] Swartz M, Eberhart J, Mastick GS, Krull CE: Sparking new frontiers: using in vivo electroporation for genetic manipulations. Dev Biol. 2001; 233, 13-21

[81] Asokan R: Genetic engineering of insects. Resonance. 2007; 21: 47-56

[82] Handler AM, Harrel RA. Germline transformation of Drosophila melanogaster with the piggybac transposon vector. Insect Molecular Biology. 1999; 8(4):449-457

[83] Handler AM, Harrel RA. Transformation of the Caribbean fruit fly, Anastrepha suspensa, with a piggybac vector marked with polyubiquitin-regulated GFP. Insect Biochemistry and Molecular Biology. 2001; 31(2):199-205

[84] Horn C, Jaunich B, Wimmer EA. Highly sensitive, fluorescent transformation marker for Drosophila transgenesis. Development Genes and Evolution. 2000; 210(12): 623-629

[85] Horn C, Schmid BGM, Pogoda FS, Wimmer EA. Fluorescent transformation markers for insect transgenesis. Insect Biochemistry and Molecular Biology. 2002; 32(10): 1221-1235

[86] O'Brochta DA, Atkinson PW: Building the better bug. Sci Am.1998; 279:90-95

[87] Schetelig MF, Wimmer EA. Insect transgenesis and the sterile insect technique. In: Vilcinskas A, editor. Insect Biotechnology, Biologically-Inspired Systems 2. Springer Science+Business Media B.V; 2011. p. 169-194. DOI: DOI 10.1007/978-90-481-9641-8

[88] Pane A, Salvemini M, Bovi PD, Polito LC, Saccone G. The transformer gene in Ceratitis capitata provides a genetic basis for selecting and remembering the sexual fate. Development. 2002; 129(15):3715-3725

[89] Saccone G, Pane A, De Simone A, Salvemini M, Milano A. New sexing strains for Mediterranean fruit fly Ceratitis capitata: transforming females into males. In: Vreysen MJB, Robinson AS, Hendrichs J, editors. Area-wide Control of Insect Pests. From Research to Field Implementation. Dordrecht, The Netherlands: Springer; 2007. p. 95-102

[90] LaChance LE. (1985). Genetic Methods fort he control of lepidopteran species: status and potential, ARS-28. USDA, Washington, DC, USA

[91] LaChance LE, Graham CK. Insect radiosensitivity: dose curves and dose-fractionation studies of dominant lethal mutations in the mature sperm of 4 insect species. Mutation Research. 1984; 127(1):49–59

[92] Knipling EF. The Basic Principles of Insect Population Suppression and Management. Washington, D. C: U.S. Dept. of Agriculture; 1979. 659 p

[93] Bartlett AC, Staten RT.. The sterile insect release method and other genetic control strategies. [Internet]. 1996. Available from: http://ipmworld.umn.edu/chapters/bartlett.htm

[94] Astaurov BI, Frolova SL.. Artificial mutations in the silkworm (Bombyx mori L.). V.Sterility and spermatogenic anomalies in the progeny of irradiated moths concern-

ing some questions of general biological and mutagenic action of X-rays. Biologicheskii Zhurnal. 1935; 4:861–894

[95] Proverbs, M. D. Some effects of gamma radiation on the reproductive potential of the codling moth, *Carpocapsa pomonella* (L.) (Lepidoptera: Olethreutidae). Canadian Entomologist 1962; 94:1162–1170

[96] Genchev NP, Gencheva EM. Inheritance of radiation induced sterility in males of the oriental fruit moth, Grapholitha Molesta Busck., (Lepidoptera: Tortricidae). Bulgarian Journal of Agricultural Science. 2006; 12:489-499

[97] Knipling EF: Suppression of pest Lepidoptera by releasing partially sterile males: a theoretical appraisal. BioScience. 1970; 20:456-4

[98] Suckling DM, Stringer LD, Mitchell VJ, Sullivan TES, Barrington AM, El-Sayed AM: Comparative Ştness of irradiated sterile light brown apple moths (Lepidoptera: Tortricidae) in a wind tunnel, hedgerow and vineyard. J. Econ. Entomol. 2011; 104: 1301-1308

[99] Makee H, Saour G. Efficiency of inherited sterility technique against Phthorimaea operculella Zeller (Lepidoptera: Gelechiidae) as affected by irradiation of females. Journal of Vegetable Crop Production. 2004; 10(1):11–22

[100] Tate CD, Carpenter JE, Bloem S. Influence of radiation dose on the level of F1 sterility in the cactus moth, Cactoblastis cactorum (Lepidoptera: Pyralidae).. Florida Entomologist. 2007; 90(3):537–544. DOI: http://dx.doi.org/10.1653/0015-4040(2007)90[537:IORDOT]2.0.CO;2

[101] Soopaya R, Stringer LD, Woods B, Stephens AEA, Butler RC, Lacey I, Kaur A, Suckling DM. Radiation biology and inherited sterility of light brown apple moth (Lepidoptera: Tortricidae): developing a sterile insect release program. Journal of Economic Entomology. 2011; 104(6):1999–2008. DOI: http://dx.doi.org/10.1603/EC11049

[102] Jang EB, McInnis DO, Kurashima R, Woods B, Suckling DM. Irradiation of adult Epiphyas postvittana (Lepidoptera: Tortricidae): egg sterility in parental and F1 generations. Journal of Economic Entomology. 2012; 105(1):54–61

[103] LaChance LE, Schmidt CH, Bushland RC. Radiation induced sterilization. In: Kilgore WW, Doutt RL, editors. Pest Control: Biological, Physical and Selected Chemical Methods. New York, NY, USA: New York Academic Press; 1967. p. 147–196

[104] LaChance LE. The induction of dominant lethal mutations in insects by ionizing radiation and chemicals - as related to the sterile-male technique of insect control. In: Wright JW, Pal R, editors. Genetics of Insect Vectors of Disease. Amsterdam, The Netherlands: Elsevier; 1967. p. 617–650

[105] LaChance LE. Dominant lethal mutations in insects with holokinetic chromosomes. 2. irradiation of sperm of cabbage looper. Annals of the Entomological Society of America. 1974; 67(1):35-39

[106] Carpenter JE, Layton RC. Computer model for predicting the effect of inheritedsterility on population growth. In: Proceedings: Radiation Induced F1 Sterility in Lepidoptera for Area-Wide Control. Final Research Co-ordination Meeting, Joint FAO/IAEA Division of Nuclear Techniques in Food and Agriculture; 9–13 September 1991; Phoenix, AZ, USA. Vienna, Austria: IAEA; 1993. p. 49–55

[107] Seth RK, Reynolds SE: Induction of inherited sterility in the tobacco hornworm Manduca sexta (Lepidoptera: Sphingidae) by substerilizing doses of ionizing radiation. Bulletin of Entomological Research. 1993; 83:227-235

[108] Marec F, Kollárová I, Pavelka J. Radiation induced inherited sterility combined with a genetic sexing system in Ephestia kuehniella (Lepidoptera: Pyralidae). Annals of the Entomological Society of America. 1999; 92(2):250–259

[109] Seth RK, Sharma VP. Inherited sterility by substerilizing radiation in Spodoptera litura (Lepidoptera: Noctuidae): bio-efficacy and potential for pest suppression. Florida Entomologist. 2001; 84(2):183–193

[110] Brower JH. Reproductive performance of inbred or outbred F1 and F2 progeny of Indian meal moth females or males x females partially sterilized by gamma radiation. Annals of the Entomological Society of America. 1981; 74(1):108–113

[111] Ayvaz A, Albayrak S, Tuncbilek AS.. Inherited sterility in Mediterranean flour moth Ephestia kuehniella Zeller (Lepidoptera: Pyralidae): Effect of gamma radiation doses on insect fecundity fertility and developmental period. Journal of Stored Product Research. 2007; 43:234–239

[112] Traut W. A study of recombination, formation of chiasmata and synaptonemal complexes in female and male meiosis of Ephestia kuehniella (Lepidoptera). Genetica. 1977; 47(2):135–142

[113] Tothová A, Marec F: Chromosomal principle of radiation-induced F1 sterility in *Ephestia kuehniella* (Lepidoptera: Pyralidae). Genome. 2001; 44:172-184

[114] Strunnikov VA: Sex control in silkworms. Nature (Lond.). 1975; 255:111-113

[115] Marec F: Genetic control of pest Lepidoptera: induction of sex-linked recessive lethal mutations in Ephestia kuehniella (Pyralidae). Acta Entomologica Bohemoslovaca. 1990; 87:445-458

[116] Marec F, Mirchi R: Genetic control of the pest Lepidoptera: gamma-ray induction of translocations between sex chromosomes of *Ephestia kuehniella* Zeller (Lepidoptera: Pyralidae). Journal of Stored Products Research. 1990; 26:109-116

[117] Marec F. Genetic control of pest Lepidoptera: construction of a balanced lethal strain in Ephestia kuehniella. Entomologia Experimentalis et Applicata. 1991; 61(3):271–283

[118] Marec F: Synaptonemal complexes in insects. International Journal of Insect Morphology and Embryology. 1996; 25:205-233

[119] Marec F. Genetic control of pest Lepidoptera: construction of a balanced lethal strain in *Ephestia kuehniella*. Entomologia Experimentalis et Applicata. 1991; 61(3):271–283

[120] Bloem S, Carpenter JE. Evaluation of population suppression by irradiated pidoptera and their progeny. Florida Entomologist. 2001; 84(2):165–171

[121] Hamm JJ, Carpenter JE. Compatibility of nuclear polyhedrosis viruses and inherited sterility for control of corn earworm and fall armyworm (Lepidoptera: Noctuidae). Journal of Entomological Science. 1997; 32(2):48–53

[122] Carpenter JE, Wiseman BR. Spodoptera frugiperda (Lepidoptera: Noctuidae) development and damage potential as affected by inherited sterility and host plant resistance. Environmental Entomology. 1992; 21(1):57-60

[123] Carpenter JE, Young JR, Sparks AN. Fall armyworm (Lepidoptera: Noctuidae): comparison of inherited deleterious effects in progeny from irradiated males and females. Journal of Economic Entomology. 1986; 79(1):47–49

[124] Carpenter JE, Bloem S, Marec F. Inherited sterility in insects. In: Sterile Insect Technique. Principles And Practice In Area-Wide Integrated Pest Management. Dyck VA, Hendrichs J, Robinson AS ed. Dordrecht, The Netherlands: Springer; 2005. p. 115–146

[125] Bloem S, Carpenter JE, Hofmeyr JH. Radiation biology and inherited sterility in false codling moth (Lepidoptera: Tortricidae). Journal of Economic Entomology. 2003; 96(6):1724– 1731

[126] Ayvaz A, Karasu E, Karabörklü S, Tunçbilek AS.. Effects of cold storage, rearing temperature, parasitoid age and irradiation on the performance of Trichogramma evanescens Westwood (Hymenoptera: Trichogrammatidae). Journal of Stored Products Research. 2008; 44:232-240

[127] Carpenter JE, Sparks AN, HL Cromroy: Corn earworm (Lepidoptera. Noctuidae): influence of irradiation and mating history on the mating propensity of females. J. Econ. Entomol. 1987; 80:1233-1237

[128] Proshold FI, Gross HR, Carpenter JE. Inundative release of Archytas marmoratus (Diptera: Tachinidae) against the corn earworm and fall armyworm (Lepidoptera: Noctuidae) in whorl-stage corn. Journal of Entomological Science. 1998; 33:241–255

[129] Bakri A, Mehta K, Lance DR.. Sterilizing Insects With Ionizing Radiation. In: Dyck VA, Hendrichs J, Robinson AS., editors. Sterile Insect Technique: Principles And Practice In Area-Wide Integrated Pest Management. Dordrecht, The Netherlands: Springer; 2005. p. 233-268

Chapter 11

# Gamma Radiation as a Recycling Tool for Waste Materials Used in Concrete

Gonzalo Martínez-Barrera, Liliana Ivette Ávila-Córdoba, Miguel Martínez-López, Eduardo Sadot Herrera-Sosa, Enrique Vigueras-Santiago, Carlos Eduardo Barrera-Díaz, Fernando Ureña-Nuñez and Nelly González-Rivas

Additional information is available at the end of the chapter

http://dx.doi.org/10.5772/60435

**Abstract**

Over the course of the last 50 years, a large number of major technological advances have contributed to the development of higher-strength, high-performance materials that provide excellent benefits. Nevertheless, in most cases, after a very short useful life, these products become waste material and contribute to environmental degradation. This situation has created an environmental crisis that has reached global proportions. In efforts to combat this issue and to promote sustainable development and reduce environmental pollution, some investigations have focused on recycling using innovative and clean technologies, such as gamma radiation, as an alternative to conventional mechanical and chemical recycling procedures. In this context, the reuse and recycling of waste materials and the use of gamma radiation are useful tools for improving the mechanical properties of concrete; for example, the compressive strength and modulus of elasticity are improved by the addition of waste particles and application of gamma radiation. In this chapter, we propose the use of gamma radiation as a method for modifying waste materials; for instance, polyethylene terephthalate plastic bottles, automotive tire rubber, and the cellulose in Tetra Pak containers, and their reuse to enhance the properties of concrete.

**Keywords:** Recycling, Waste materials, Gamma radiation, Concrete, Mechanical properties

## 1. Introduction

Concrete is the most widely used structural material in the world, due to its easy preparation and low cost. Nevertheless, it has some disadvantages: a) pores, which can become entrance points for water, water vapor, gases and chemical substances that might damage concrete; b) the rapid deterioration in roughness of the concrete surface because of its high abrasion; c) poor resistance to aggressive substances and salty water; and d) low resistance to heating. One alternative for remediating these problems is the incorporation of polymeric materials, which bind well with mineral aggregates that are the main components of concrete.

Currently, because of the more stringent legislation regarding the environment and the market demand for environmentally friendly products, manufacturers are interested in developing approaches aimed at reducing the environmental impact of industrial processes through reductions in the amount of residues produced or by treating those that are inevitably generated. The environmental damage caused during the extraction of raw materials, as well as the high cost of extraction methods, provides good motivation for the use of industrial and domestic residues as substitutes or complements of fresh materials in several areas of manufacturing. The depletion of reliable, secure raw material reserves and conservation of non-renewable sources are also incentives to develop ways to reuse waste materials.

In recent years, various tools and strategies have been proposed to meet environmental challenges within the building industry, including a) increasing the use of waste materials, especially those that are by-products of industrial processes; b) using recycled materials instead of natural resources, which will make the industry more sustainable; and c) improving durability as well as mechanical and other properties, thus reducing the volume of construction replacement materials for structures that are damaged or destroyed.

In principle, the molecular structure of composite materials can be modified using gamma radiation. Cross-linking and polymer degradation (by chain scission) can occur with radiation; the chemical composition of the polymer is the key factor determining the extent to which these processes occur. Materials with superior properties can be obtained from recovered scrap polymer cross-linked by gamma radiation. The application of radiation technology in the recycling of polymers is a good option from both an economic and ecological point of view.

The purpose of this chapter is to show how the combination of gamma radiation with waste and recycling materials can provide alternative tools for improving the physical and chemical properties of concrete. Waste materials such as polyethylene terephthalate (PET) bottles, tire rubber, and cellulose in Tetra Pak containers are discussed in terms of their physicochemical modification by radiation and their use in enhancing the properties of concrete. Such information is focused on contributions to improving the care of the environment.

## 2. Recycled and waste materials used in concrete

In light of the growing awareness of environmental concerns, the use of waste materials in industrial processes is an attractive area of opportunity. The recovery and recycling of solid waste has long been the subject of research. Its use in building, road construction and paving materials is beneficial in helping to reduce environmental pollution and as a solution to waste disposal issues [1, 2].

Solid waste is classified by its chemical nature as organic and inorganic. Glass, ceramics and metals such as aluminum used in packaging materials are the main components of inorganic solid waste; others include zinc, copper and iron [3]. In the case of organic solid waste, one of the most representative components is polyethylene terephthalate (PET). In 2007, the world's annual PET consumption comprised 250,000 million bottles (10 million tons of waste). In the United States 50,000 million bottles are discarded in landfills each year. Since PET waste is not biodegradable, it can remain in the environment for hundreds of years. PET waste can be used to produce an unsaturated polyester resin (UPR) in the presence of glycols and dibasic acid. This material can serve as a binder to produce polymer concrete (PC) with high compressive strength. With a PET/glycol ratio of 2:1, higher compressive strength of polymer concrete can be obtained [4].

Due to the increasing number of cars worldwide, the accumulation of huge volumes of discarded tires has become a major waste management problem. In 2002, approximately 275 million of scrap tires were generated in the U.S., 110 million in Japan, and 37 million in the UK. Over 100,000 tons of waste tires are generated annually in Taiwan. The final disposal of used tires is a major environmental concern; landfills where tires are discarded represent a severe fire and health hazard. Burning tire scrap to provide energy for the production of vapor or electricity is one of the most common methods for eliminating tire waste [5]. The use of waste tires as alternative fuel in cement furnaces has been established across the U.S. and Europe. Applications utilizing waste tires include the bituminous hot mixing of pneumatic dust for agglutinative modification in asphalt pavement [6, 7]. Another alternative is its use as a substitute of fine or coarse aggregates in concrete. Its characteristics can improve the mechanical properties of concrete such as strength and modulus of elasticity over those achieved by sand or stone.

Recycling of waste tires in the construction industry can aid in preventing environmental pollution and in the design of more economically efficient buildings. In this respect, the use of waste tire rubber in ready-mixed concrete has become increasingly popular globally, generating significant research interest in the last two decades. A modest quantity of unprocessed scrap tires is used to provide shock protection for marine platforms against the impact of waves or ships. In some regions of the world, people still resort to burning tires, which produces unacceptable levels of pollution. As such, new and innovative techniques to promote recycling are important. Many countries avoid/forbid the stockpiling or landfill of waste tires, providing a significant incentive for investigating recycling strategies. One of such strategies involves the transformation of scrap tires into alternative aggregates, generating increased economic value while reducing aggregate consumption [8].

Materials from tires are used in a variety of elastomers and plastic products, as well as for asphalt rubber (AR) pavement. Oxychloride cement is a binder for rubberized concrete mixtures. In a recent study, asphalt rubber was prepared in two ways, one with a gap-graded design and the other using open gradation. The results showed satisfactory performance and the potential for household use. The wet process is the most suitable for normal asphalt mixtures with ground tire rubber (GTR). It is worth mentioning that rubber asphalt mixtures meet ASTM International specifications. Through the use of different concentrations of AR and GTR, modified asphalt can represent a superior alternative to conventional mixtures for use of local materials and paving techniques [7].

In a study,, the mechanical properties of polymer concrete made from reinforced epoxy powder tire rubber were studied. Mixtures were optimized using direct neural modeling and reverse neuronal modeling at minimal cost; in this case, the most important cost variable is resin content. Direct neural modeling gave the optimum composition for obtaining maximum values of compressive, flexural and tensile strength. Reverse neural modeling was used to analyze the maximum values of mechanical properties obtained with varying concentrations of the epoxy resin powder. The results show a high resistance to compression for composition of 0.215 (weight fraction) for epoxy resin and 0.3 (weight fraction) of tire powder. The maximum flexural strength of 0.23 was obtained with 0.17 resin tire powder epoxy, and maximum tensile strength for the 0.24 and 0.17 resin [9].

The use of tire rubber as aggregate reduces the compressive strength of the concrete, which may limit its usefulness in some structural applications. Nevertheless, it has desirable characteristics including lower density, higher impact resistance and toughness, higher ductility, and better sound insulation properties. These features may be advantageous for a variety of construction applications, such as access roads. A significant reduction in used tire waste could be accomplished by using scrap tires for concrete-coated tire rubber particles. The use of magnesium oxychloride makes it possible to produce high-strength concrete with better elastomer adhesion characteristics and with significantly improved performance. Moreover, the adhesion between tire rubber particles and other constituent concrete materials may be improved by pretreatment of the aggregates of magnesium oxychloride tire rubber. Adhesion depends on several factors, including size and concentration of tire particles, type of cement, the use of chemical and mineral additives, and methods of pretreating tire rubber particles. In terms of size, it is possible to use tire powder in both mortars and concrete [10]. Additionally, higher amounts of textile fibers (from used tires in plasters and plasterboards of pressed gypsum) cause less resistance reduction compared to plaster without additives.

Composites incorporate various waste materials, including granulated cork, cellulose fibers from waste paper, and fibers from the recycling of used tires. Several studies have concentrated on developing new composite materials through the use of different processes for composite production, including simple molding or pressing.

The main components of natural fibers are cellulose, hemi-cellulose and lignin, with minor concentrations of pectin, waxes and water-soluble substances. Linear cellulose molecules are linked laterally by hydrogen bonds to form linear bundles, giving rise to a crystalline structure. The degree of crystallinity is one of the most important structural parameters of cellulose. The rigidity of cellulose fibers increases, while flexibility decreases, with an increasing ratio of crystalline to amorphous regions. Moreover, the addition of cellulose fibers improves the bending behavior of the composites [11].

Some of the most important waste materials are those containing cellulose, for example, Tetra Pak containers. Such packaging is made from three raw materials: paper (about 75 %), low-density polyethylene (about 20 %) and aluminum (about 5 %). Discarded containers are recycled through a simple, well-established process called hydropulping. In this process, the cellulosic fibers are separated from thin layers of polyethylene and aluminum. Most of the waste from the paper industry is known as paper sludge (PS), which is burnt and becomes PS ash. It is used as a soil improvement material and raw material for cement. PS ash increases the strength of extremely stiff concrete with its high water absorption capacity. It can be added to concrete, and undergoes a pozzolan reaction with calcium hydroxide due to the hydration of cement, resulting in an obtained material with increased compressive strength relative to concrete without PS ash. The material contains 38.1 % silica ($SiO_2$), 21.4 % alumina ($Al_2O_3$) and 28.9 % CaO [11]. SEM images of PS ash show particles with a rough shape, but no spherical particles are present. Typical concentrations are 200 kg/m$^3$ of cement and between 100 and 300 kg/m$^3$ of PS ash. Plant fibers and "man-made" cellulose fibers are used as substitutes for asbestos fibers in cement matrices; they show comparable properties at lower cost, with values essentially dependent on the properties of the fiber and the adhesion between fiber and matrix [12].

For the preparation of composites, paper recovered from packaging has been utilized, with the pulped fibers composed of 40 % resinous wood, 35 % Alfa grass (*Stipa tenacissima* L.) and 25 % leafy wood. The fiber sizes are classified as fine for values < 1.25 mm and coarse for values of > 1.25 mm to < 5 mm. The results show that compressive strength decreases as pulped fiber content increases, largely due to the fact that increasing fiber content induces more voids that reduce weight and weaken the composite. When waste fibers are added to cement, the amount of water for the preparation increases to compensate for the water absorbed by the fibers; thus it is necessary to calculate the water/cement ratio (W/C). For a composite with 10 % fibers (W/C = 0.56), SEM microscopy images show agglomerations of fibers in non-homogeneous dispersion. When an additional water quantity is added (W/C = 0.64), better dispersion of fibers is observed, but strength decreases because of the voids formed by the added water.

Thermal conductivity, $\kappa$, is a measure of thermal insulating efficiency of materials; when cellulose fibers are added to composites (2–16 % by weight), thermal conductivity values diminish, and consequently energy is saved. The fibers are thus used as cement replacement. This behavior is due to the porosity that occurs in the packing of fibers that is induced by bubbles of air formed during the mixing operation, and to the insulation properties of the fibers themselves. When more voids are in the mix, a lighter composite specimen is obtained and its thermal conductivity is diminished.

Cellulose fibers have been used as cement replacement in lightweight concrete; the fibers were recycled from packages and mixed at concentrations up to 16 % by weight. Results of studies showed that an increase in the fiber content led to a reduction in the compressive strength of concrete and an improvement in thermal insulation properties, along with a homogeneous distribution of fibers in the matrix, when an appropriate water–cement ratio was used. Better thermal insulation of the cement matrix and low density provide for a lightweight construction material. This type of lightweight concrete is used for the construction of partition walls (compressive strength 8.6 MPa), partitions, ceilings and roofs [13].

One important alternative for recycling PET materials is their use as concrete aggregate substitutes. Given the technological demands in the construction area, studies are exploring the possibility of generating alternative materials with increasing functionality, lower cost, and better physical, chemical and mechanical properties than those of conventional materials [14, 15].

In the last two decades, virgin polymers used in road surfaces have shown advantages by virtue of certain improved characteristics of these materials. Researchers have used different polymers which, when properly mixed with asphalt, have resulted in improved road surface yield and lifespan. However, waste polymers can be dangerous and remain in the environment; and thus it is important that they would be recycled or reused effectively.

Road surface yield can be improved through modification of the asphalt with various substances, most of which are virgin materials that are scarce and costly. An alternative is the use of waste materials, such as plastic bottles, which can help reduce waste material and potentially improve its yield [16]. To improve concrete ductility, PET fibers from plastic bottles have been used. Results show that the addition of only a few fibers has a considerable influence on the concrete post-cracking. Both type lamellar and type O fibers improve concrete hardness. The latter helps to join together the concrete of each cracked section side [17].

Various studies have predicted the long-term creep of polymer concrete containing $CaCO_3$ and fly ash particles, as well as recycled PET resin, through short-term creep experiments. Results have shown more rapid creep deformation of early-age concrete with PET in comparison to ordinary concrete deformation. More than 20 % of the long-term creep occurs during the first two days, and 50 % during the first 20 days. Furthermore, creep deformation of polymer concrete without reinforcement is greater than that for concrete with $CaCO_3$, due to the higher surface area of $CaCO_3$ particles. Reinforcement plays an important role in reducing polymeric concrete deformation. Creep values increase with an increase in applied effort, although the increases are not proportional, due to the viscoelastic, non-linear behavior of polymeric concrete with recycled PET [18].

Concrete has been manufactured with up to 3 % recycled PET bottle fibers. The main concern in the development of PET fiber is its alkali strength; however, research has found that this is not an issue for fiber used in concrete. PET fiber has been used for tunnel pulverization and covering, including motorbike tunnels. Future applications include underground structures found in hostile environments, for example, near the coast or in the sea. Moreover, it can be considered use as pavement in narrow, winding and steep roads. In a comparison study of PET with other fibers, moisture levels of PET fibers were found to be lower than those of polyvinyl alcohol (PVA) fibers but higher than polypropylene (PP) fibers [19].

To reduce cracks in concrete, PET particles obtained from recycled bottles, with lengths of 10, 15 and 20 mm and concentrations of 0.05, 0.18 and 0.30 % by volume, were added. Bending and impact tests were carried out at 28 and 150 days. Significant effects on compression strength values were observed with the addition of fiber. Moreover, Young modulus values were reduced with higher fiber content, where surface changes occurred according to the increment of the fiber concentration.

Compressive strength of concrete is dependent on PET concentrations. Such behavior can be explained in terms of the surface characteristics of concrete with PET particles (Figure 1). For concrete without PET particles, dispersed particles of mineral aggregates (sand and gravel) show rough surfaces (0 % PET). At lower concentrations, PET particles cover the mineral aggregates, and more rough surfaces are detected (1.5 % PET). Concrete surface morphology changes with increased concentration of PET particles, and the mixture produces a more homogeneous surface, with some compact regions (2.5 % PET). However, when PET particle concentration is further increased (5.0 % PET), regions with some cracks are observed.

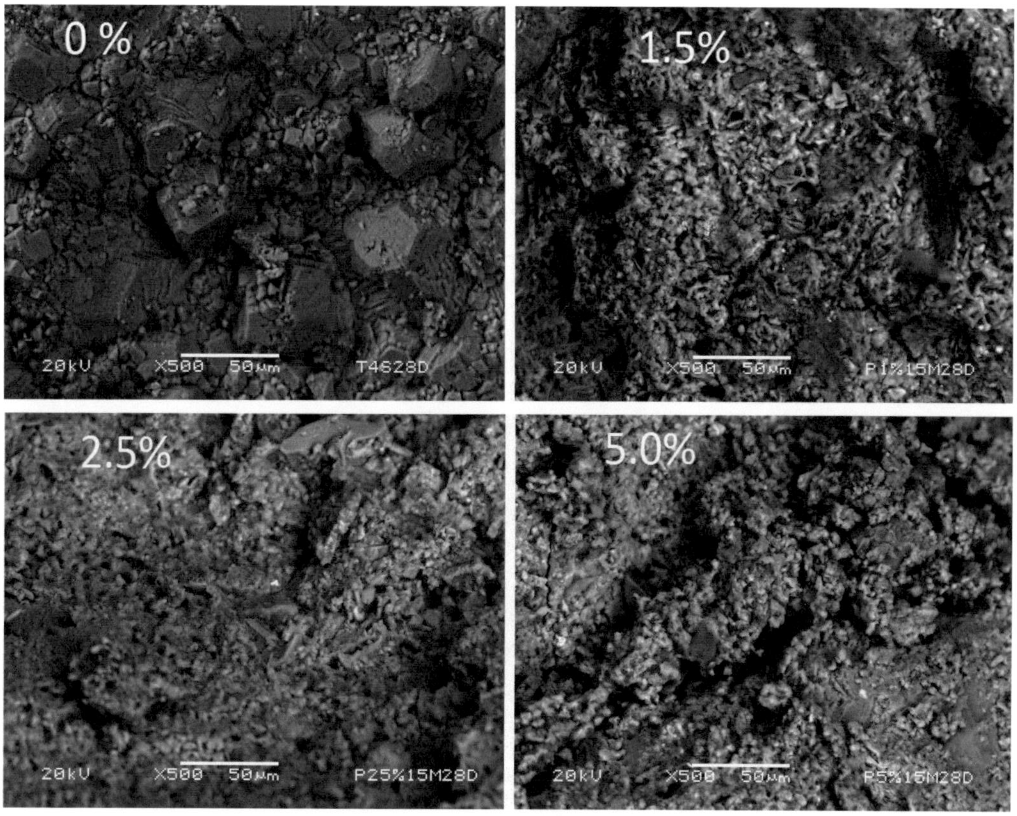


**Figure 1.** Concrete with different concentrations of PET particles

In another study investigating curing time, it was observed that, at 28 days, the flexural, impact and tensile strength were increased with the presence of fibers. However, at 150 days, this improvement was no longer present, as a result of fiber fragilization and degradation in the alkaline concrete environment. After a year, porosity was increased in concrete with fibers. In view of the important aspects of sustainability, such as the use of recycled materials in construction, fibers obtained from recycled PET bottles are an alternative in reinforced concrete [14].

## 3. Structural modification of waste materials using gamma radiation

Gamma rays are produced in the disintegration of radioactive atomic nuclei and in the decay of certain subatomic particles. The commonly accepted definitions of the gamma ray region in the electromagnetic spectrum include some wavelength overlap. Gamma ray radiation has wavelengths that are generally shorter than a few tenths of an angstrom, and gamma ray photons having energies greater than tens of thousands of electron volts [20].

The effects of gamma radiation on polymers are usually evaluated through changes in their chemical structure and mechanical behavior. These modifications occur as a result of reorganization of chemical bonds, which allows an increase in the degree of polymerization or structural reticulation. Polymers have been modified with the purpose of optimizing properties and increasing their compatibility in composite materials [21].

Gamma radiation is being used successfully today for post-consumer plastics recycling. Such technology is feasible from both an ecologic and economic point of view. Among the most important benefits of this application are the following: a) improvement in mechanical properties and performance of recovering polymers or polymer mixtures, mainly through cross-linking or modification of several combined-phase surfaces; b) more rapid polymer decomposition, particularly by chain scission, which produces low molecular masses that can be used as additives or raw materials in several processes; and c) advanced polymeric materials production, designed specifically to be environmentally compatible [22].

The effects of gamma radiation on PET have been evaluated in several studies. For example, the processes involved in PET degradation induced by radiation were assessed through the use of electron spin resonance (ESR) and optical absorption spectroscopy. PET films were irradiated at a temperature of −196 °C in darkness. Upon irradiation, the film changed to reddish purple in color, which enabled the detection of PET radical ionic species by ESR [23]. In another work, a photosensitization process through gamma radiation was carried out; changes were followed by infrared spectroscopy and reversed-phase high-performance liquid chromatography. PET break zones were observed as well as the formation of terephthalic acid as a result of radiolysis [24].

Results of studies on the effects of gamma radiation on packaging PET films in the 0–200 kGy dose range demonstrate that diethylene glycol content increases at low doses (5–10 kGy) but decreases at high doses (30–200 kGy). While molecular mass, intrinsic viscosity and terminal carboxyl groups decrease slightly at doses greater than 60 kGy, permeability, thermal properties, color, and surface resistivity are not significantly affected at any dose [25].

The morphology of the surfaces of recycled PET particles were evaluated by scanning electron microscopy (SEM); particles varied in size from 0.5 to 3.0 mm and were obtained following a cutting process of PET bottles. After irradiation, several changes on the surfaces were observed, as shown in Figure 2.

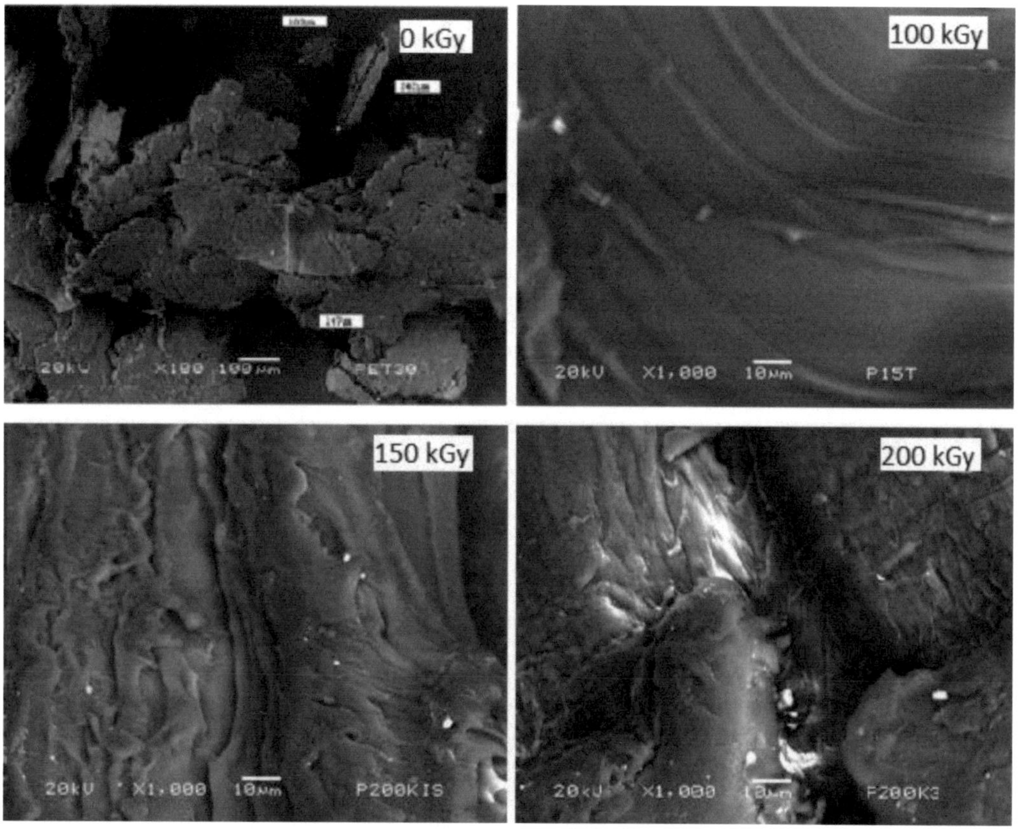


**Figure 2.** SEM images of non-irradiated and irradiated recycled PET particles

The thermal behavior of gamma-irradiated amorphous PET films under environmental conditions was studied for doses up to 3.5 MGy at a rate of 28 kGy/h. Differential scanning calorimetry (DSC) was used to determine the glass transition temperatures (Tg) and degrees of crystallinity. The results showed that both Tg and heat capacity decreased as the dose was increased, which was due to the breaking processes in polymeric chains. It is possible to conclude that Tg could be used as an indicator of dose absorption ratios in PET [26].

DSC, X-ray photoelectron spectroscopy, SEM, and molecular mass determination were used to evaluate the effects of gamma radiation on PET in varying doses up to 15 MGy at a ratio of 1.65 MGy/h. A decrease in molecular mass was observed at a dose of 5 MGy, which was attributed to polymer chain scission; however, molar mass increased at doses greater than 5 MGy, due mainly to recombination and branch formation [27].

X-ray diffraction and UV spectroscopy were used to evaluate the optical and structural properties of irradiated PET in the 0–2 MGy dose range. The diffraction pattern revealed the

PET semicrystalline nature, with crystallinity increased as the radiation dose increases. UV analysis revealed an increment in both activation and absorption energies, but the forbidden band decreased at a higher applied dose [28]. Another study in which PET bottles were irradiated at doses from 0 to 670 kGy, the results showed an increase in both crystallinity and crystal size in the formed particles after irradiation [29].

The effects of gamma radiation on the mechanical and thermal properties of recycled PET mixtures with low-density polyethylene (LDPE) and ethylene vinyl acetate (EVA) were studied using applied doses of 25, 50, and 100 kGy. The results showed maximum cross-linking chains for 10 % of recycled PET irradiated at 100 kGy [30].

An ethylene–methyl acrylate–glycidyl methacrylate monomer was grafted into PET through gamma radiation. The formed elastomer shown a 30 % increase in impact strength, with only 0.1 % terpolymer mass, compared with the non-irradiated mixture. From these observations it can be concluded that gamma radiation is a very adequate technique (on-site) for improving the compatibility of polymers in composites [31].

A study of thermoplastic aromatic polyesters (used for their electrical insulating capacity) showed stable polymeric chains due to the presence of benzene rings upon irradiation with doses up to 1 MGy. For higher doses (5 MGy), irradiated PET samples showed diminution of molecular mass due to chain scissions [13].

Some investigations focusing on the effects of gamma radiation on the physicochemical properties of cellulose have been described; the results show that an increase of 25 kGy (on average) caused a loss of 1 % in cellulose crystallinity in a dose range of 0–1 MGy. Cellulose shows degradation (from 6 to 12 %) at up to 31.6 kGy, and the degree of crystallinity is unchanged up to 300 kGy. The degree of polymerization (DP) is obtained up to 1 kGy; this decreases above 10 kGy. Moreover, changes in specific gravity and lattice constant are observed up to 1 MGy, with complete degradation of cellulose at 6.55 MGy.

In cellulose, there are amorphous zones along the microfibril length in which the crystallinity is interrupted. These zones allow the penetration of chemicals into the microfibrils. Furthermore, gamma radiation causes the breakdown of cellulose into shorter chains, which are water-soluble, and also leads to an "opening of additional micro-cracks" that are easily penetrable by water molecules.

Figure 3 shows SEM images of irradiated recycled cellulose. For non-irradiated cellulose, smooth and homogeneous surfaces are observed, and some particles are present. When the radiation dose is increased to 50 kGy, more dispersed particles and some cracks are observed; for higher doses, more space between cellulose surfaces appears, together with small voids. Such modifications can be attributed to the main effects produced by gamma radiation: scission and cross-linking of molecular chains in cellulose.

In Figure 4, the surface characteristics of the recycled tire particles are shown. Non-irradiated particles show a homogenous surface, while particles irradiated at 200 kGy show roughness and voids on the fiber surfaces. Incremental doses of gamma radiation provoke more damage to the surface, and voids are formed (greater than 100 μm). Finally, with irradiation at 300 kGy,

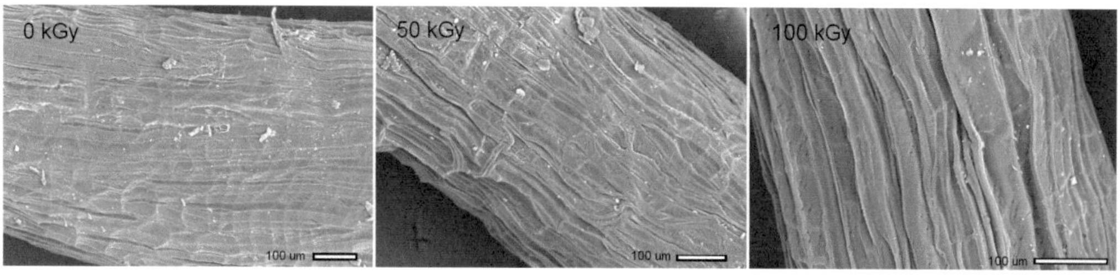


**Figure 3.** SEM images of non-irradiated and irradiated cellulose fibers

the surface damage is more prominent, showing large cracks as a consequence of gamma radiation.

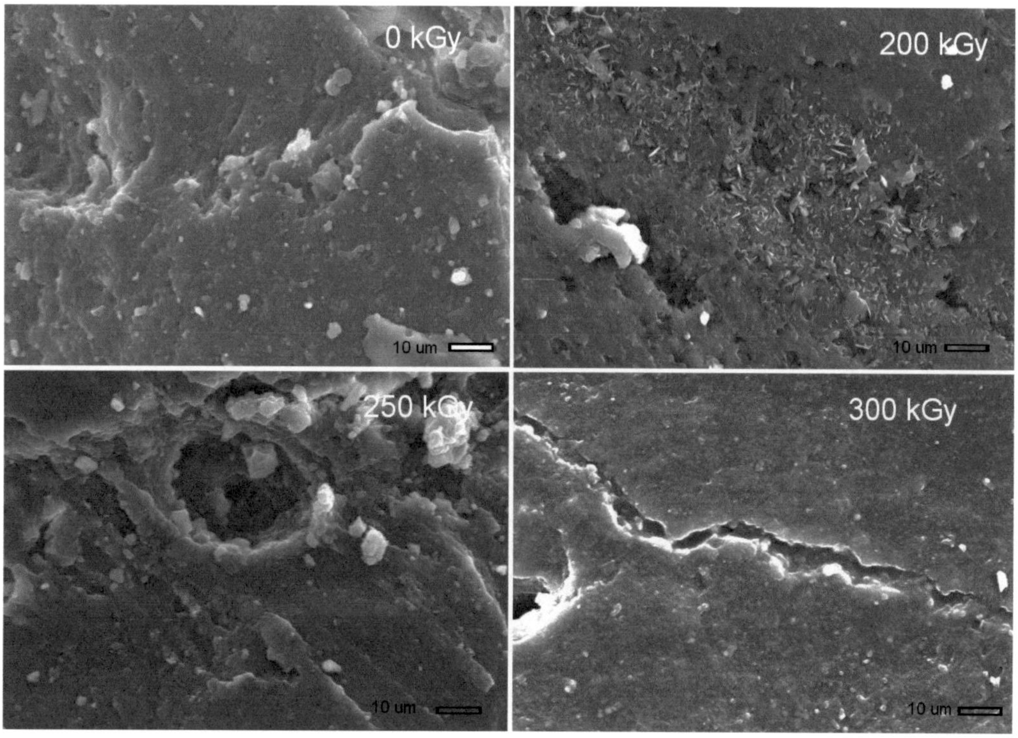


**Figure 4.** SEM images of non- irradiated and irradiated tire rubber particles

## 4. Concrete with waste and recycled materials: Effects of gamma irradiation

The use of gamma radiation presents significant advantages for PET recycling and for improving the mechanical properties of concrete, which can be explained by changes in the chemical structure of the surface. Gamma radiation accelerates the initiation of polymerization of a monomer into the ceramic matrix, and can provide considerable benefits, the most important of which is better adhesion between the fibers and matrix [32]. The main mechanical properties analyzed in concrete are strain, compression, and impact strength, deformation in the yield point and breakdown, as well as deformation values and elasticity modulus.

Some studies have investigated the effects of ionizing radiation on polymer–ceramic composite materials. For example, in gypsum/poly(methyl acrylate) composites, polymerization yield increased with increasing radiation dose. A yield of 87-88 % was obtained at doses of 3–4 kGy. Since this process is carried out at room temperature, there is substantial economy of heat energy in addition to reduction in cost of keeping the composite under pressure. The pressure allows much of the monomer (usually with high volatility) to fill the interstices of the ceramic matrix during the conversion to ceramic–polymer composite.

There are a few studies on the effects of gamma radiation on concrete [33–35]. The effects produced can be controlled through the use of appropriate radiation doses. For instance, it is possible to modify the surface to obtain a rougher and more cracked material, allowing for greater compatibility with the cementitious material [34].

For concrete with irradiated waste tire particles, compressive strength values followed similar patterns. The values decreased as particle concentration increased, with values ranging from 7.4 to 17.5 MPa. The highest value was for concrete with 10 % of particles 2.8 mm in size; this value was 27 % lower relative to that of the control concrete. Concretes with particles 2.8 mm in size had larger values than those with 0.85 mm particles. The concretes with 20 or 30 % of particles had higher values in comparison to concrete with non-irradiated particles. Thus, the use of larger particle sizes is more efficient than using smaller particles.

The mechanical properties of concrete are dependent on the size and concentration of the waste tire particles. Compressive and tensile strength values are reduced by the presence of these particles, as they promote stress concentration zones and introduce tensile stresses into concrete, resulting in rapid cracking and concrete failure. Nevertheless, in some cases, improvement in mechanical properties is observed when gamma radiation is applied to the waste tire particles. It appears that the best results are obtained in concrete with 10 % irradiated particles 2.8 mm in size. Concrete with irradiated particles can receive up to 30 % of tire particles, helping to reduce the final cost of the concrete.

Changes in mechanical properties may be related to morphological changes that occur in the fracture zones of concrete specimens, as illustrated in Figure 5. For non-irradiated concrete, a rough surface can be seen (0 kGy); when a radiation dose of 200 kGy is applied, the dispersed particles are covered with irradiated tire fibers as a consequence of the scission of the polymer chains. At a higher dose, 250 kGy, the tire rubber fibers continue to cover the hydrated cement particles, as evidenced by the presence of cross-linked regions accompanied by larger quantities of ceramic particles. Strong bonds are progressively developed between tire particles and cement matrix with incremental irradiation.

Concrete acting as a binder mixed with crumb rubber can produce more flexible concrete blocks, thus providing a softer surface. The crumb rubber block produced for pedestrian pavement also performs quite well in terms of both skid and abrasion resistance. One of the

**Figure 5.** SEM images of gamma-irradiated concrete with tire rubber fibers

clearest effects of irradiation of thermoplastic/elastomer blends is the change in tensile strength due to their cross-linking and degradation. These values increase with doses up to 150 kGy and then decrease as the dose is increased to 250 kGy. At any dose, however, tensile strength values decrease as the ratio of ground tire rubber (GTR) into the blend increases. The GTR particles separate the molecules of ethylene propylene diene monomer (EPDM) and high-density polyethylene (HDPE), and hence retard the formation of cross-linking. Moreover, a higher ratio of GTR indicates the presence of more spaces in the polymeric network due to the incomplete cross-linking of GTR. Therefore, it may be concluded that the presence of GTR reduces the cross-linking of EPDM and HDPE and cannot protect the blend from deterioration, particularly at higher doses, thus acting only as filler [36].

Waste tire rubber has been used to produce a composite used in a multilayer plate subjected to the direct impact of a bullet, in which a sandwich of soft and hard materials is used to stop the ballistic force. The role of the soft layer, which is rubberized concrete, is to act as a cushion to absorb some of the total energy, thus reducing the impact force reaching the hard layer. This results in a delayed response time at the beginning of the impact event, decreasing the acceleration peak and lowering the vertical displacement of the center of mass [36].

The elongation of blends with different composition decreases when the dose is increased. With a larger dose, more cross-linking is produced in the sample, which prevents structural reorganization during drawing and reduces internal chain mobility and elongation [37].

In real applications under practical or engineering conditions, polymer-based materials such as blends are not stretched until they undergo rupture. Therefore, despite the improved mechanical properties, the tensile strength measurements are not adequate. The property that measures the resistance to a limited strain deformation of polymeric materials under practical applications is the tensile modulus at 100 % elongation. Radiation-induced cross-linking in polymer materials should be reflected by an increase in hardness. The hardness values of all blends increase slightly with doses up to 250 kGy, and hardness values decrease with increasing GTR content. The high degree of crystallinity of HDPE has a significant effect on hardness. Temperatures of the maximum rate of reaction ($T_{max}$) taken from the thermogravimetric analysis, thermogram show an increase with an increasing ratio of GTR up to 33 %, and then a decrease at higher ratios.

In general, the higher thermal stability of the composition containing GTR compared to the EPDM/HDPE blend is due to the oxidative degradation of the GTR and the formation of carbonyl groups with higher dissociation energy than that of CH groups. SEM images of blends with different composition show that EPDM and HDPE are non-compatible polymers. However, upon irradiation, the surface is homogeneous and smooth, and exhibits no indication of phase separation, due to the occurrence of cross-linking between the incompatible polymers. The appearance of white particles across the SEMs indicates the non-compatibility among EPDM, HDPE and GTR. Meanwhile, the presence of GTR does not affect the cross-linked polymer matrix. These features increase with the ratio of GTR, which may explain the decrease in tensile and hardness properties associated with the introduction of GTR [38].

Compressive strength values of concrete with waste cellulose were obtained. Concrete without waste cellulose at 28 days of curing had the highest compressive strength value, 21.7 MPa. Some general patterns were observed. The values gradually decreased as cellulose concentration was increased. Concrete with 3 wt% of waste cellulose had a minimal difference (5 %) relative to control concrete (without cellulose). This did not occur for concrete with 7 wt% of waste cellulose, as it had a 47 % reduction. The compressive strength values increased with longer curing time, no matter of the percentage of cellulose.

Such reductions in values can be explained in terms of waste cellulose added. The strength is dependent on the amount of waste cellulose and water cement ratio (w/c). Cellulose, which is hydrophobic in nature, can be substitute for up to 7 wt% of sand in the mixture, and thus a greater amount of water is available to interact with the surface of non-hydrated grains of cement particles. As a result, weak interfacial adhesion between cellulose and hydrated cement particles is obtained, and consequently, a reduction in compressive strength values is observed.

Recovery of these materials has long been the subject of research. Other characteristics, such as electric properties, have been studied for irradiated PET covering a dose range from 100 kGy to 2 MGy; both \conductivity and electric constant values increase with the increment of irradiation dose. This raises the possibility of using PET films in electronic components such as capacitors and resistors. With irradiation at low doses (8, 10 and 15 kGy), two types of laminated PET films showed improved physical and mechanical properties at 15 kGy [3].

A study of concrete reinforced with waste PET particles found that non-irradiated concrete followed typical behavioral patterns for compressive strain, increasing progressively with incremental PET particle concentration, but no such pattern was observed for compressive strength or elasticity modulus. Minimal value is obtained for compressive strength and maximal value for elasticity modulus when adding 2.5 wt% of PET. Both compressive strength and elasticity modulus values are maximal when adding 0.5 mm PET particles to concrete.

Different behaviors can be observed with irradiated versus non-irradiated concrete. When PET concentration is increased, the compressive strength values diminish, and a notable reduction in compressive strain is obtained. However, elasticity modulus exhibits the opposite behavior with non-irradiated concrete. In this study, at 2.5 % PET, a minimal value was observed. With regard to PET particle size, a similar behavior for non-irradiated concrete was observed:

maximal values for both compressive strength and elasticity modulus are obtained by adding a 0.5-mm particle. In general, irradiated concrete containing PET particles had similar elasticity modulus, higher compressive strength, and lower compressive strain values compared to non-irradiated concrete.

Since compressive strength of concrete is one of the key structural design parameters used by engineers, waste PET particles can provide suitable material for construction. A small amount of PET (5.0 %) substituted for fine aggregates in the mix design can increase strength as much as 23 % and diminish strain up to 26 %. Thus irradiation represents a useful tool and suitable method for recycling waste PET.

In one study, recycled PET was incorporated into hydraulic concrete as a substitute for sand, and the effects on mechanical properties (compressive strength, elasticity modulus and unitary deformation) were evaluated. The considered variables were particle size (0.5, 1.0 and 3.0 mm), volume PET concentration (1.0, 2.5 and 5.0 %) and gamma radiation dose (100, 150, and 200 kGy). Results showed that samples irradiated at a dose of 100 kGy exhibited greater compression strength (between 15 and 35 %) than non-irradiated specimens. In addition, compression strength decreased with increasing PET particle size, regardless of the percentage used [39].

In the case of samples irradiated at 150 and 200 kGy, a 50 % increase in mechanical strength was observed in comparison to samples irradiated at 100 kGy. However, no difference in strength was obtained for samples irradiated to 150 and 200 kGy with PET at any size or concentration. With regard to elasticity modulus, values were similar for both types of specimens, with an inverse relationship existing between mechanical property and PET particle size: the smaller the size, the greater the elasticity modulus. Finally, with respect to unitary deformation, the values obtained from irradiated specimens were between 20 and 70 % less than those of non-irradiated samples, as shown in Figure 6.

Reductions in compressive strain values are due to irradiation effects in both cement paste and PET particles. Irradiation causes chain scission and generation of free radicals, which can cause bonds to form in hydrated cement paste, and consequently produces a hard rather than ductile material. A SEM image of irradiated concrete with 1.0 % PET 0.5 mm in size shows a homogeneous distribution of PET particles; when PET particles are added (2.5 %), morphological changes in the homogeneous regions of hydrated cement with irradiated PET particles can be observed. With higher PET particle concentration, inhomogeneous surface areas with fewer hydrated regions are detected. These morphological changes are not enough to cause a significant difference in compressive strain values, as minimal differences are observed among them, independently of PET size and concentration.

In a current study, gamma radiation and waste cellulose were investigated as tools for improving the mechanical properties of cement concrete. Waste cellulose was obtained from Tetra Pak packages. A simple and inexpensive process was sought, as well as a contribution to environmental care. Prior to the preparation of concrete specimens, one set of waste cellulose particles with an average size of 0.5 mm was obtained from Tetra Pak containers, and was used in concentrations of 3, 5, and 7 wt%; these values were selected in order to avoid problems related to homogeneity and workability.

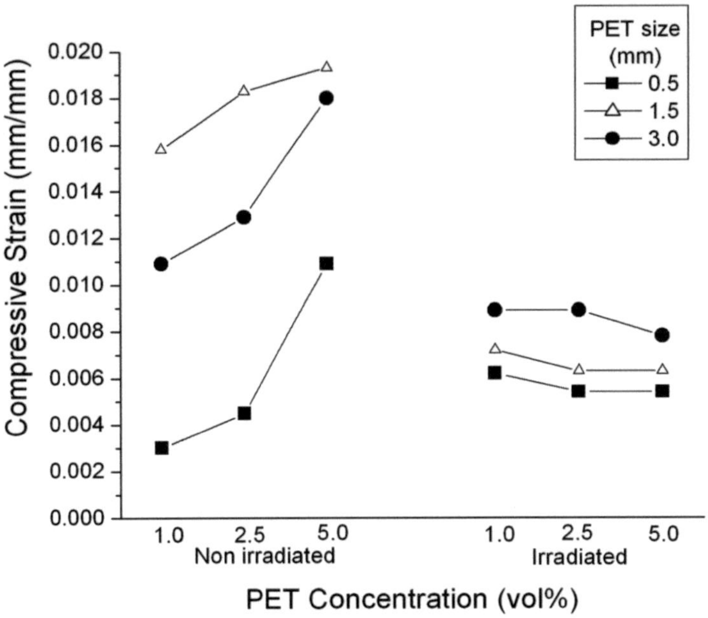


**Figure 6.** Compressive strain of non-irradiated and irradiated concrete with PET particles

Compressive strength values of non-irradiated concrete with waste cellulose demonstrated that concrete without waste cellulose after 28 days of curing had the highest compressive strength value. Some characteristic patterns were observed. The values gradually decreased as cellulose concentration was increased. Concrete with 3 wt% of waste cellulose had a minimal difference (5 %) relative to control concrete (without cellulose). This did not occur for concrete with 7 wt% of waste cellulose, as it had a 47 % reduction. Independent of cellulose percentage, the compressive strength values increased with longer curing time.

In the case of irradiated concrete, different behavior was observed. The values increased as the waste cellulose concentration increased. The highest value was observed for concrete after 28 days of curing and irradiated at 300 kGy, which was 47 % higher relative to control concrete. For each radiation dose, the values increased with longer curing time. In general terms, irradiated cellulose covered the sand particles, and thus the zone around them was affected by stress concentration. Therefore, if the distance between particles is sufficiently small, these zones intersect and form a percolation network, which generates good adhesion between cement matrix and cellulose, and thus an increment in the modulus of elasticity is obtained. The results can be attributed to the effects of gamma irradiation on the waste cellulose. Improvement in the modulus of elasticity values indicates a predominant domain of cross-linking of polymer chains in cellulose. However, some shorter chains are produced, which are water-soluble, and as a consequence, a solubility increment is reached.

Improvements in compressive strength can be explained in terms of the effects of gamma radiation on the concrete components and waste cellulose. As we know, many types of

http://dx.doi.org/10.5772/60435

chemical reactions take place during gamma irradiation of polymeric materials —cross-linking and degradation by chain scission, among others— but one or the other of these effects may be predominant in some materials.

The formation of cross-linking of the polymeric chains in the cellulose under the effects of the irradiation dose is highly significant, with impacts on the cement and water molecules. Cross-linking is the most important effect of polymer irradiation, as it generally improves the mechanical, thermal and chemical properties of concrete. Moreover, application of high-energy irradiation to cellulose creates free radicals by the scission of the weakest bonds; such radicals can react with certain molecules in the cement matrix. The interaction between calcium silicate hydrate (formed during the hydration process) and the cellulose present in the pores during irradiation polymerization enhances the interphase bonding, resulting in improved mechanical strength.

## 5. Conclusions

Waste or recycled materials and gamma radiation are both useful tools for improving the mechanical properties of concrete, where waste materials are substitute for gravel or sand. In particular, the compressive strength and modulus of elasticity values exhibit improvement with the addition of certain concentrations of waste materials and the application of a specific gamma radiation dose. In contrast, non-irradiated concrete possesses poor mechanical properties.

As concrete compressive strength is a key structural design parameter used by engineers, waste materials of different shapes (particles or fibers) may be suitable as construction material. A small amount of waste materials can be substitute for fine aggregates in the mix design to enhance the mechanical properties. In addition, gamma radiation can be a useful tool and suitable method for recycling waste materials. Properties of flexural and compressive strength are dependent upon waste concentrations. In general, mechanical properties are improved when the waste concentration is sufficient to decrease the negative effect of poor particle–matrix adherence. A more ductile material is obtained at the expense of flexural and compressive strength.

In the case of concrete with PET particles, for non-irradiated samples, compressive strain typically increases progressively as PET particle concentration increases, while compressive strength and elasticity modulus are not affected by changes in concentration. With regard to PET particle size, maximal values of both compressive strength and modulus of elasticity are dependent upon PET particle size. Different behaviors are observed for irradiated versus non-irradiated concrete. When PET concentration is increased, compressive strength values are reduced. More notable is the reduction in compressive strain. However, the elasticity modulus exhibits opposite behavior to that shown for non-irradiated concrete. Lastly, with SEM images, the influence of gamma radiation on waste materials and its effect on the mechanical properties of concrete is corroborated.

## Acknowledgements

Financial support of the Autonomous University of the State of Mexico (UAEM), Toluca, by grant UAEM 2015/3223FS is acknowledged.

## Author details

Gonzalo Martínez-Barrera[1*], Liliana Ivette Ávila-Córdoba[2], Miguel Martínez-López[1], Eduardo Sadot Herrera-Sosa[3], Enrique Vigueras-Santiago[1], Carlos Eduardo Barrera-Díaz[3], Fernando Ureña-Nuñez[4] and Nelly González-Rivas[3]

*Address all correspondence to: gonzomartinez02@yahoo.com.mx

1 Laboratorio de Investigación y Desarrollo de Materiales Avanzados (LIDMA), Facultad de Química, Universidad Autónoma del Estado de México, San Cayetano, México

2 Unidad Académica Profesional Tianguistenco, Universidad Autónoma del Estado de México, Tianguistenco, Estado de México, México

3 Centro Conjunto de Investigación en Química Sustentable, Universidad Autónoma del Estado de México – Universidad Nacional Autónoma de México (UAEM-UNAM), Carretera Toluca-Atlacomulco, Unidad El Rosedal, Toluca, Estado de México, México

4 InstitutoNacional de Investigaciones Nucleares, Carretera México-Toluca S/N, La Marquesa Ocoyoacac, México

## References

[1] Meyer C. The economics of recycling in the US construction industry. In: Chun Y.-M., Claisse P., Naik T.R., Ganjian E. (eds.) Sustainable construction materials and technologies. London: Taylor and Francis Group; 2007, p. 509–513.

[2] Meyer C. Recycled materials in concrete. In: Mindess S. (ed.) Developments in the formulation and reinforcement of concrete. Cambridge: Woodhead Publishing Limited; 2009, p. 208–230.

[3] Murathan A., Murathan A.S., Guru M., Balbas M. Manufacturing low density boards from waste cardboards containing aluminum. Materials & Design 2007; 28: 2215–2217.

[4] Pacheco-Torgal F., Ding Y., Jalali S. Properties and durability of concrete containing polymeric wastes (tyre rubber and polyethylene terephthalate bottles): an overview. Construction and Building Materials 2012; 30: 714–724.

[5] Brown K.M., Cumming R., Morzek J.R., Terrebonno P. Scrap tire disposal: three principles for policy of choice. Natural Resources Journal 2001; 41: 9–22.

[6] Navarro F.J., Partal P., Martinez-Boza F., Gallegos C. Influence of crumb rubber concentration on the rheological behavior of crumb rubber modified bitumen. Energy and Fuels 2005; 19: 1984–1990.

[7] Chiu C.-T. Use of ground tire rubber in asphalt pavements: field trial and evaluation in Taiwan. Resources, Conservation and Recycling 2008; 52: 522–532.

[8] Sukontasukkul P., Chaikaew C. Properties of concrete pedestrian block mixed with crumb rubber. Construction and Building Materials 2006; 20: 450–457.

[9] Diaconescua R.M., Barbutab M., Harja M. Prediction of properties of polymer concrete composite with tire rubber using neural networks. Materials Science and Engineering: B 2013; 178: 1259–1267.

[10] Siddique R., Naik T.R. Properties of concrete containing scrap-tire rubber—an overview. Waste Management 2004; 24: 563–569.

[11] Paranhos Gazineu M.H., dos Santos W.A., Hazin C.A., de Vasconcelos W.E., Dantas C.C. Production of polymer–plaster composite by gamma irradiation. Progress in Nuclear Energy 2011; 53: 1140–1144.

[12] Fujiwara H., Maruoka M., Koibuchi K., Fujita K. The application of paper sludge ash to extremely stiff consistency concrete product. In: Chun Y.-M., Claisse P., Naik T.R., Ganjian E. (eds.) Sustainable construction materials and technologies. London: Taylor and Francis Group; 2007, p. 303–311.

[13] Bentchikou M., Guidoum A., Scrivener K., Silhadi K., Hanini S. Effect of recycled cellulose fibres on the properties of lightweight cement composite matrix. Construction and Building Materials 2012; 34: 451–456.

[14] Pelisser F., Montedo O.R.K., Gleize P.J.P., Ramos Roman H. Mechanical properties of recycled PET fibers in concrete. Materials Research 2012; 15 (4): 679–686.

[15] Israngkura B., Ungkoon Y. Bond performance of fiber reinforced polymer (FRP) bars in autoclaved aerated concrete. Journal of Civil Engineering and Architecture 2010; 4: 41–44.

[16] Kalantar Z.N., Karim M.R., Mahrez A. A review of using waste and virgin polymer in pavement. Construction and Building Materials 2012; 33: 55–62.

[17] Foti D. Preliminary analysis of concrete reinforced with waste bottles PET fibers. Construction and Building Materials, 2011; 25: 1906–1915.

[18] Jo B.-W., Tae G.-H., Kim Ch.-H. Uniaxial creep behavior and prediction of recycled-PET polymer concrete. Construction and Building Materials 2007; 21: 1552–1559.

[19] Ochi T., Okubo S., Fukui K. Development of recycled PET fiber and its application as concrete-reinforcing fiber. Cement and Concrete Composites 2007; 29: 448–455.

[20] Bagher, A.M. Advantages of gamma radiation in science and industry. Journal of Advanced Physics 2014; 3 (2): 97–103.

[21] Martínez-Barrera G., Menchaca-Campos C., Barrera-Díaz C., Ávila-Córdoba L. Recent developments in polymer recycling. In: Bikit I. (ed.) Gamma rays: technology, applications, and health implications. New York: Nova Science Publishers, Inc.; 2013, p. 237–255.

[22] Burillo G., Clough R. L., Czvikovszky T., Guven O., Le Moel A., Liu W., Singh A., Yang J., Zaharescu T. Polymer recycling: potential application of radiation technology. Radiation Physics and Chemistry 2002; 6 (1): 41–51.

[23] Torikai A., Fueki K. Elementary processes in the radiation-induced degradation of poly(ethylene terephthalate)—a note. Polymer Degradation and Stability 1985; 13 (2): 183–189.

[24] Vilensky A.I., Zagorski D.L., Kabanov V.Y., Mchedlishvili B.V. UV- and $\gamma$-sensitization of latent tracks in polyethylene terephthalate. Radiation Measurements 2003; 36 (1–6): 131–135.

[25] Jeon D.H., Lee H.K., Park H.J. The effects of irradiation on physicochemical characteristics of PET packaging film. Radiation Physics and Chemistry 2004; 71: 1059–1064.

[26] Kattan, M. Thermal behavior of gamma-irradiated amorphous poly (ethylene terephthalate) films. Polymer Engineering and Science 2006; 46 (10): 1374–1377.

[27] Burillo G., Tenorio L., Bucio E., Adem E., Lopez G.P. Electron beam irradiation effects on poly(ethylene terephthalate). Radiation Physics and Chemistry 2007; 76 (11–12): 1728–1731.

[28] Siddhartha S.A., Kapil D., Suresh-Kumar R., Krishna J.B., Wahab M.A. Effect of gamma radiation on the structural and optical properties of polyethyleneterephthalate (PET) polymer. Radiation Physics and Chemistry 2012; 81(4): 458–462.

[29] Kumar V., Ali Y., Sonkawade R., Dhaliwal A. Effect of gamma irradiation on the properties of plastic bottle sheet. Nuclear Instruments and Methods in Physics Research Section B: Beam Interactions with Materials and Atoms 2012; 287: 10–14.

[30] Abdel-Tawab K., Ibrahim S., Magida M. The effect of gamma irradiation on mechanical, and thermal properties of recycling polyethylene terephthalate and low density polyethylene (R-PET/LDPE) blend compatibilized by ethylene vinyl acetate (EVA). Journal of Radioanalytical and Nuclear Chemistry 2012; 295 (2): 1313–1317.

[31] Liang, M., Mozhen, W., Xuewu G. Surface treatment of poly(ethylene terephthalate) by gamma-ray induced graft copolymerization of methyl acrylate and its toughening

effect on poly(ethylene terephthalate)/elastomer blend. Radiation Physics and Chemistry 2003; 90: 92–97.

[32] Patel M., Morrell P.R., Murphy J.J., Skinner A., Maxwell R.S. Gamma radiation induced effects on silica and on silica-polymer interfacial interactions in filled polysiloxane rubber. Polymer Degradation and Stability 2006; (91): 406–413.

[33] Vodák F., Trtík K.V., Sopko O., Kapickova P. Effect of γ-irradiation on strength of concrete for nuclear-safety structures. Cement and Concrete Research 2005; 35: 1447–1451.

[34] Martínez-Barrera G., Brostow, W. Fiber-reinforced polymer concrete: property improvement by gamma irradiation. In: Barrera-Díaz C, Martínez-Barrera G (eds.) Gamma radiation effects on polymeric materials and its applications. Kerala: Research Signpost; 2009, p. 27–44.

[35] Stankovic S.J., Ilic R.D., Jankovic Bojovic K.D., Loncar B. Gamma radiation absorption characteristics of concrete with components of different type materials. Acta Physica Polonica A. 2010; 117 (5): 812–816.

[36] Sukontasukkul P., Jamnam S., Rodsin K., Banthia N. Use of rubberized concrete as a cushion layer in bulletproof fiber reinforced concrete panels. Construction and Building Materials 2013; 41: 801–811.

[37] Najim K.B., Hall M.R. A review of the fresh/hardened properties and applications for plain- (PRC) and self-compacting rubberised concrete (SCRC). Construction and Building Materials 2010; 24: 2043–2051.

[38] Yung W.H., Yung L.C., Hua L.H. A study of the durability properties of waste tire rubber applied to self-compacting concrete. Construction and Building Materials 2013; 41: 665–672.

[39] Ávila-Córdoba L., Martínez-Barrera G., Barrera-Díaz C., Ureña-Núñez F., Loza-Yáñez A. Effects on mechanical properties of recycled PET in cement-based composites. International Journal of Polymer Science 2013; 1: 1–6.

Chapter 12

# Ionizing Radiation Used in Drug Sterilization, Characterization of Radical Intermediates by Electron Spin Resonance (ESR) Analyses

Şeyda Çolak

Additional information is available at the end of the chapter

http://dx.doi.org/10.5772/61052

**Abstract**

In this study, the feasibility of radiation sterilization of drugs/drug raw materials is investigated by using Electron Spin Resonance (ESR) spectroscopy. Experimental data and their theoretical correspondings are presented for Sulfanilamide (SA), Sulfafurazole (SFZ), Sulfatiazole (STZ), Sulfacetamide Sodium (SS), Sulfamethazine (SMH), Butylated Hydroxyanisole (BHA), and Albendazole (ALB). Unirradiated samples exhibited no ESR signal whereas the irradiated samples showed ESR spectra consisting of different number of resonance lines indicating that radiolytic intermediates were produced upon irradiation. Increase in the absorbed dose did not create any pattern change in the ESR spectra of these samples. The results of ESR microwave power studies indicated that saturation is observed to be faster for the studies held below room temperatures. Low radiation yield (G=0.1-0.5) calculated by ESR data for the gamma-irradiated samples showed that these materials can not be used as sensitive dosimetric materials. No significant differences were observed between FT-IR spectra of the unirradiated and irradiated samples and this result is considered to be in agreement with the relatively small G value derived from ESR studies. The decay rates of the ESR peak heights of the samples irradiated at different doses and stored at normal and stability conditions were found to be independent of the irradiation doses. The contributing radical species were determined to decay with different decay characteristics and the decay rates but decaying faster at stability conditions. The discrimination of the samples irradiated at even a low absorbed dose from unirradiated samples was possible for a long storage time after irradiation. Cooling the sample temperature down to room temperature did not create any pattern change in the ESR spectra of irradiated samples except slight reversible increases in the peak heights and at high temperatures irreversible decreases in the peaks heights were observed. Annealing studies indicated that the decay rates of the radical species at high temperatures were higher than the decay rates at low temperatures and the decay activation energies for the radical species were calculated by using Arrhenius plots. Spectrum simulation calculations were also performed and it was concluded that, the molecular ionic fragments and ionic radicals were the main responsible units from the resonance lines of ESR spectra of the gamma-irradiated sulfanomides such as SA, SFZ, STZ, SS, and SMH. Besides these two radical species, some other radical types were also likely produced after irradiation in STZ, SS, and SMH. Besides these two radical species, some other radical types were also likely produced after irradiation in STZ, SS and SMH. As for BHA and ALB, again two other type radical species were believed to produce upon irradiation. Basing on the derived experimental and theoretical data it was concluded that SA, SFZ, STZ, SS, SMH, BHA, and ALB could be safely sterilized by gamma radiation up to permitted drug sterilization radiation doses without causing high amount of molecular damages upon irradiation, and ESR spectroscopy could be used as a potential technique in monitoring the radiosterilization of the drugs, drug raw materials, and drug delivery systems containg present samples as active ingredient.

**Keywords:** ESR, Radiation Sterilization, Sulfanilamide, Sulfafurazole, Sulfatiazole, Sulfacetamide Sodium, Sulfamethazine, Butylated Hydroxyanisole, Albendazole

## 1. Introduction

**Sterilization** is an effective process to eliminate microorganisms such as fungi, bacteria, viruses, spore, etc. from materials. Types of sterilization methods are mainly: *"pressured vapor sterilization"*, *"dry heat sterilization"*, *"ethylene oxide (EtO) sterilization"*, *"aseptic sterilization,"* and "**radiation sterilization**" [1]. **Radiation sterilization** is based on the exposure of the materials to high-energy ionizing radiation such as *gamma radiation, accelerated beam of electrons,* or *X-ray radiation.*

**Sterilization by gamma radiation** has been accepted to be a good alternative and an attractive sterilization method for medicinal products and for pharmaceuticals such as drugs/drug raw materials [2-6]. Sterilization by radiation can overcome some specific difficulties that can occur in the other types of sterilization methods. So radiation sterilization method has a general usage on gaseous, liquid, solid materials, homogeneous, and heterogeneous systems [1, 3, 7-9]. Radiation sterilization is also recognized by all major pharmacopeias for pharmaceutical raw materials and their dosage forms are particularly recommended for thermolabile or chemically reactive samples that cannot be sterilized thermally or chemically. With its high penetration range, sterilization by gamma radiation offers some very important advantages such as the possibility of sterilization of drugs

http://dx.doi.org/10.5772/61052

in their final packages (terminal sterilization), low-cost-effectivity, small temperature rise, low chemical reactivity, and ease of validity as the essential validation process is only based on time variable [3, 5, 7-18]. Besides, radiation sterilization is accepted to be a clean, non-residue-producing technology that is safe for the worker and the community. Irradiated drug/drug raw materials are also completely safe to use immediately after sterilization process [2, 3, 10, 14, 18]. Because of these advantages, sterilization by ionizing radiation, especially gamma radiation, has been successfully applied in many countries [13, 21, 22].

EN 552 and ISO 11137 publications recognize standard for implementing radiation technology on sterilization [19, 20]. For the gamma sterilization method, the reference absorbed dose for terminal sterilization is accepted to be **25 kGy**, where the procedures and precautions employed should yield an SAL of at least $\mathbf{10^{-6}}$.

Nevertheless, besides its advantages, radiosterilization also has some drawbacks. Radiation cannot only eliminates microorganisms included in pharmaceuticals but also can cause a molecular decrease in the amount of active drug by destroying it and, therefore, creating reactive molecular fragments which may result in a toxicological hazard [2, 5, 18, 21-24]. Although the radiolytic products induced upon irradiation are generally in very small quantities [14], the characterization of the radio-induced radicals is very important and necessary, both to determine the feasibility of the radiation treatment and to control it. Therefore, to prove the safety of radiosterilization, the determination of physical and chemical features of the radiolytic products and mechanism of radiolysis should be determined [5, 6, 23, 24]. Thus, it is desirable to establish an effective experimental method to discriminate between irradiated and unirradiated drugs as the regulations of irradiated drugs vary from country to country. Besides, radiation effects on drug molecules cannot be generalized; thus, response to ionizing radiation of each molecule has to be individually studied.

**Electron Spin Resonance Spectroscopy (ESR)**, also called **Electron Paramagnetic Resonance (EPR)**, is a spectroscopy technique used particulary for the determination of the types of radiolytical species in paramagnetic samples that contain unpaired electrons, such as organic and inorganic free radicals or inorganic complexes possessing a transition metal ions. Its high sensitivity, precision, ease and its non-destructive readout are the other important advantages of ESR spectroscopy [24-32]. Thus, this nondestructive analytical technique is used for a variety of applications in biology, medicine, and in material science. Moreover, **ESR** is also a very sensitive method for detection of radical intermediates induced in irradiated drug/drug raw materials (radiation sterilized drug/drug raw materials), which appears to be very well suited for their magnetic characterization. ESR yields, both, qualitative information (i.e., whether or not a sample has been irradiated) and quantitative results (i.e., the dose it received), so by this method it is possible to detect and to distinguish irradiated drugs from unirradiated ones.

ESR is a technique that is based on the absorption of electromagnetic radiation in the microwave frequency region by a paramagnetic sample when it is placed in an external magnetic field [32]. ESR resonance can occur basing on the equation (1) where h is the Planck's constant, $\upsilon$ is microwave frequency, g-value is a constant that is dependent on the nature of the radical type ($g = 2.0023$ for a free electron), $\beta$ is Bohr magneton and H is the applied magnetic field.

$$h\upsilon = g\beta H \quad (1)$$

The results of the ESR studies performed in our laboratory relevant to the structural and thermal properties of the radicals produced in gamma-irradiated **seven** different drugs/drug raw materials will be summarized in the present chapter. The aims of the performed studies were: to investigate the radiation sensitivity of solid drugs/drug raw materials in the dose range of 5-50 kGy through detailed microwave saturation, kinetic and spectroscopic ESR studies and to explore the potential use of ESR technique in monitoring the radiosterilization of the investigated drugs/drug raw materials. Experimental results are presented under two different subsections. The sulfanilamides, which include **Sulfanilamide (SA), Sulfafurazole (SFZ), Sulfatiazole (STZ), Sulfacetamide Sodium (SS), Sulfamethazine (SMH)** and two other drugs/drug raw materials, **Butylated Hydroxyanisole (BHA)** and **Albendazole (ALB)** [16, 33-39].

## 2. Materials and methods

### Samples

The investigated drugs/drug raw materials (Sulfanilamide, Sulfafurazole, Sulfatiazole, Sulfacetamide Sodium, Sulfamethazine, Butylated Hydroxyanisole, Albendazole) were provided from local drug providers and stored at room temperature in a well-closed container protected from light. No further purification was performed and they were used as they were received. Stabilization studies for the samples stored in stability conditions (75% relative humidity; 40°C) were also investigated for some group of drugs.

### Irradiation Process

All irradiations were performed at room temperature (293 K) in dark using a $^{60}$Co gamma cell supplying a dose rate of ~2.5 kGy/hr as an ionizing radiation source at the Sarayköy Establishment of Turkish Atomic Energy Agency in Ankara, Turkey. The dose rate at the sample sites was measured by a Fricke dosimeter and ESR investigations were performed on samples irradiated at different doses (5 kGy-50 kGy).

### ESR Measurements

ESR measurements were carried out using both **Varian 9″-EL X-band** and **Bruker EMX 113 X-Band** ESR spectrometers operating at about 9.5 GHz and equipped with a $TE_{104}$ rectangular double cavity containing a DPPH standard sample (g=2.0036) in the rear resonator which remained untouched throughout the experiments. **The ESR experimental conditions were as follows;** *central field: 350 mT, sweep width: 10 mT; microwave frequency: about 9.85 GHz; microwave power: 0.5-100 mW; modulation frequency: 100 kHz; modulation amplitude: 0.1mT; receiver gain: 5x10$^3$, sweep time: 83.89 s; time constant: 327.68 ms; conversion time: 81.92 ms; temperature: 100-400 K.* Sample temperature inside the microwave cavity was monitored with a digital temperature control system (Bruker ER 4111-VT). The latter provided the opportunity of measuring the temperature with an accuracy of ±0.5 K at the site of the sample. A cooling, heating and subsequent cooling cycle was adopted to monitor evolutions of the free radical signals. The temperature of the samples was first decreased to ~100 K starting from room temperature with an increment of ~20 K, then increased to 400 K, and, finally, was decreased again to room temperature. Variations in the line shape and in the signal intensities with microwave power were also studied in the range 0.005-100 mW for samples irradiated at room temperature and at low temperatures. Signal intensities were calculated from first derivative spectra and compared with that obtained for a standard DPPH sample under the same spectrometer operating conditions. Kinetic studies of the contributing free radicals were also performed at different temperatures. To achieve this goal, the samples were heated to predetermined temperatures (320 K-410 K) and kept at these temperatures for predetermined times (3-200 min); then they were cooled to room temperature and their ESR spectra were recorded. The results were the average of five replicates for each radiation dose. Signal intensity variation data obtained for different annealing temperatures were used to characterize the contributing radiation-induced free radicals.

**FT-IR Studies**

IR spectra of unirradiated and gamma-irradiated samples were also recorded using Nicolet 520 FT-IR spectrometer and a comparison between the principal IR bands of interested drugs/drug raw materials was performed to monitor the radiolytic products silent to ESR spectroscopy.

**Simulation**

Digitized signal intensity data derived from room temperature ESR spectrum of each sample irradiated at different doses were used as input data for spectrum simulation calculations. The simulation calculations based on models of different tentative radical species anticipating from the results of microwave saturation, variable temperature, decay at normal and stability conditions, dose response, and annealing studies were performed to determine the spectroscopic features of the contributing free radicals.

## 3. Experimental results and discussion

Experimental results are presented under two different subsections. The sulfanilamides which include **Sulfanilamide (SA), Sulfafurazole (SFZ), Sulfatiazole (STZ), Sulfacetamide Sodium (SS), Sulfamethazine (SMH)** and two other drugs/drug raw materials, **Butylated Hydroxyanisole (BHA)** and **Albendazole (ALB)** [16, 34-39].

**Sulfonamides [sulfa drugs: sulfanilamide (SA), sulfafurazole (SFZ), sulfatiazole (STZ), sulfacetamide sodium (SS), sulfamethazine (SMH)]** are widely used as antibacterial agents in the treatment of urinary system infections; in meningococcal menengitis prophylaxis; ulcerative colitis; urinary tract infections dysentery; bacterial inflammations of the skin, eye, and the genital area; as well as in veterinary medicine [33]. Sulfonamides were the first substances used to cure and prevent bacterial infections in humans. Investigations on radiosterilization of sulfonamides have been rather devoted to the color change determination, loss of potency, formation of acid and gas in irradiated solid and/or aqueous solution samples [33, 40, 41]. As the sulphonamides are known to undergo decomposition during thermal sterilization, radiation sterilization is a preferrable sterilization technique for the sterilization process of sulfanomide drugs/drug raw materials [18]. **Butylated Hydroxyanisole** (BHA) is an aromatic organic compound with the chemical name 2- and 3-tert-butyl-4-methoxy phenol. BHA is used as a preservative and antioxidant in pharmaceutical preparations and cosmetic formulations containing fats and oils [42], as an antioxidant for some rubber and petroleum products, and is a stabilizer for vitamin A. **Albendazole (ALB)** is an antihelminitic drug. Medicines considered in this group are used to treat ruminants by either roundworms or flatworms. Molecular structures of the investigated drug/drug raw materials are given in Table1.

| | |
|---|---|
| 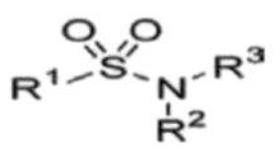 a)Sulfa Drugs | 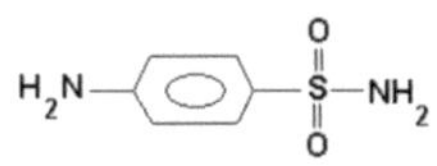 b)SA |
| 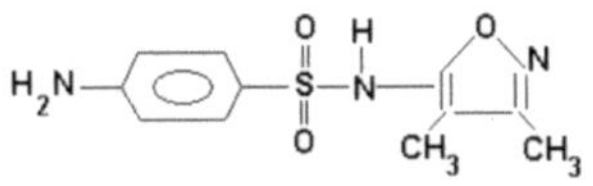 c)SFZ | 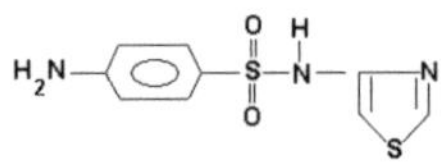 d)STZ |
| 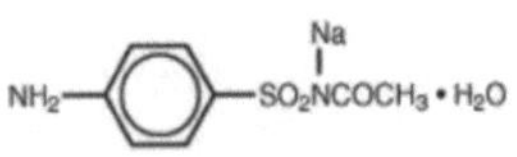 e)SS | 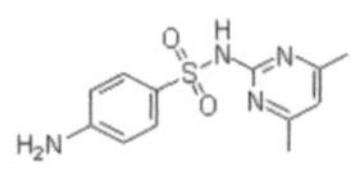 f) SMH |
| 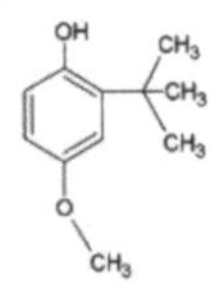 g)BHA | 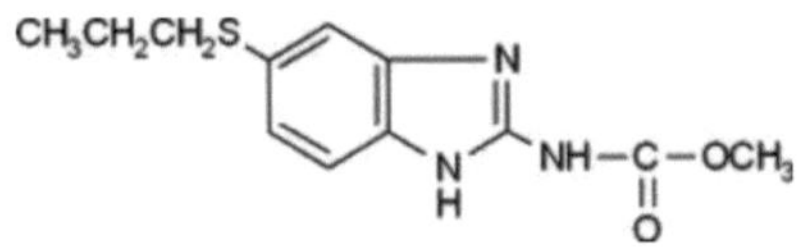 h)ALB |

**Table 1.** Molecular structures of investigated drug/drug raw materials: a) Sulfa drugs, b)SA, c) SFZ, d) STZ, e) SS, f) SMH, g) BHA, h) ALB. (Reproduced from [34], [36], [38], [37], [39] with permissions from Elsevier and Taylor & Francis).

### 3.1. General features of the ESR spectra

Although unirradiated samples exhibited no ESR signal, irradiated samples showed ESR spectrum consisting of different number of resonance lines depending on the sample investigated [16, 34-39]. The presence of ESR signals in irradiated but not in unirradiated samples is the indication that radiolytic intermediates were produced because of the irradiation mechanism. Increase in absorbed dose did not create any pattern change in the room temperature spectra of the samples. Thus, it was concluded that irradiation dose was not an important parameter in the formation of the shape of the ESR spectra of the investigated samples in the adopted radiation dose range. The ESR spectra of the investigated samples are given in Figure 1 with their assigned peak numbers for different gamma-irradiation doses.

Irradiated **sulfanilamide (SA)** samples showed a resolved ESR spectrum with g values of 2.0089, 2.0060, 2.0038 (Figure 1-a). **Sulfafurazole (SFZ)** samples irradiated both at *77 K* and *room temperature* showed a very simple ESR spectrum consisting of three resonance peaks appearing at g values of 2.0081, 2.0067, and, 2.0034, respectively (Figure 1-b). **Sulfatiazole (STZ)** irradiated at room temperature showed a complex ESR spectrum consisting of ten resonance peaks (Figure 1-c). The most intense resonance line appearing in the middle of the spectrum was found to have a g value of 2.0045 and a peak-to-peak width of 5.2 G. **Sulfacetamide-sodium (SS)** irradiated showed also a complex ESR spectrum consisting of seven characteristic resonance peaks (Figure 1-d). The resonance peak appearing in the middle of the spectrum was found to have g value of 2.0045 and peak-to-peak width $\Delta H_{pp}$=5.5 G. Irradiated **Sulfamethazine (SMH)** was observed to exhibit an ESR spectrum consisting of one antisymmetric intense central resonance line and three very weak resonance lines at each side of the central line (Figure 1-e). The g values corresponding to the two peaks of the intense resonance line appearing in the middle of the spectrum and its peak-to-peak width were calculated to be 2.0060 and 2.0028, $\Delta H_{pp}$=0.56 mT, respectively. Irradiated **Butylated Hydroxyanisole (BHA)** samples were observed to present an ESR spectrum consisting of three main resonance peaks (Figure 1-f) with g values of 2.0066, 2.0035, and 2.0007, well-developed at high doses with a hardly observed shoulder at low magnetic fields (g=2.0090). Irradiated **Albendazole (ALB)** samples were observed to present a characteristic ESR spectra consisting of three resonance peaks with g values of 2.0041, 2.0105, and 2.0165 (Figure 1-g).

### 3.2. Microwave power studies

Variations of the signal heights, which were measured with respect to base line and normalized to the receiver gain, masses of the samples, and the intensities of the standart, of these resonance peaks with applied microwave power in the range of 0.5-80 mW, were examined for all the investigated samples. The results of microwave power studies indicated that heights of the assigned peaks increase rather linearly at low microwave powers and saturate homogeneously or inhomogeneously broadened resonance lines at room temperature. Saturation is observed to be faster for the microwave saturation studies which were held below room temperatures. Theoretical functions best fitting to microwave saturation data were calculated assuming *exponentially growing curves* of different characteristics associated with different radical species involved in the formation of experimental ESR spectra of the samples. From the results of these calculations it was concluded that at least *two radical species* of different saturation characteristics were involved for the irradiated samples [16, 34-39].

### 3.3. Dose-response curves and dosimetric features of the samples

Gamma radiation produces damages in the molecular structures of the irradiated samples where the amount of damage will depend on the absorbed dose level of the sample. From the experimental results it was concluded that the discrimination of the irradiated samples at a dose as low as 0.5 kGy, from unirradiated samples was possible even long after irradiation due to the relatively high stabilities of the produced radical species, even if gamma radiation yield of the samples is low. A higher concentration of radicals, generated at the same absorbed dose of radiation, indicates a higher sensitivity of the substance toward the type of radiation used. For the samples, variations of the heigths of the resonance peaks assigned as numbers with absorbed gamma radiation doses were generally found to follow a *linear function* such as I=c+bD or a *growth function* such as $I = I_o (1 - e^{-aD})$ in the dose range of 0-50 kGy (Figure 2). In the given equations, "D" stands for the absorbed dose in kGy, "c" is a constant and "a" is the radiation dose growth parameter constant. The calculated different growth rate parameters indicated that more than one radical species contribute to the formation of each resonance peaks. Functions and parameters best fitting to experimental dose-response curves are calculated and given as solid lines in Figure 2.

In the dosimetric studies, G value (*the number of radical species produced by the absorbed radiation per 100 eV*) was also determined for gamma-irradiated solid sulfonomides. This value is calculated to be fairly small (G=0.5 for SA, G=0.45 for SMH, etc.) compared with those reported for sulfonamide aqueous solutions (3.5-5.1) but stay in the range of the G values reported for solid sulfonamides (0.15-0.6) [40, 41]. The big difference in the G values is believed to originate from hydrated electrons ($e_{aq}$) and hydroxyl radicals (OH) produced in large amount as radiolytical intermediates in irradiated aqueous solutions of sulfonamides. Low radiation yield of gamma-irradiated solid sulfonomides shows that these types of drugs and drug raw material could be suitable candidates for radiosterilization.

### 3.4. Long-term stability of the radiation-induced radicals

Room temperature stabilities of the radicals induced in the irradiated drugs/drug raw materials upon irradiation are as important as the radiosensitivity of these materials. ESR spectra of the samples which were open to air, were recorded in regular time intervals over a long period of time (approximately, 3 months) without changing the position of the sample in the microwave cavity throughout the experiment on normal conditions (*room temperature and normal pressure in the air*). Besides storing at normal conditions, for only the samples of SFZ, STZ, and SS, the decay rates of the samples irradiated at different doses were also investigated for storing at stability conditions (*40°C and 75% relative humidity*). The signal intensity decay data obtained for the samples irradiated were used to get the decay characteristics of the contributing radicals. The decay constants derived from this study for SA, STZ, SS, and BHA are summarized in Figure 3 and theoretical decay results calculated using these parameters are given together with their experimental counterparts (Table 2).

The decay of the peak heights of the samples irradiated at different doses and stored in well-closed container at normal and stability conditions was found to be independent of the irradiation dose. Contributing radicals were determined to decay much faster at stability conditions. The peak heights or spectrum area were observed to experience fast decreases during the beginning of the storage period; after the first days of storage, the decay rate of the induced radicals in the samples upon irradiation was decreased. Model based on the assump-

http://dx.doi.org/10.5772/61052

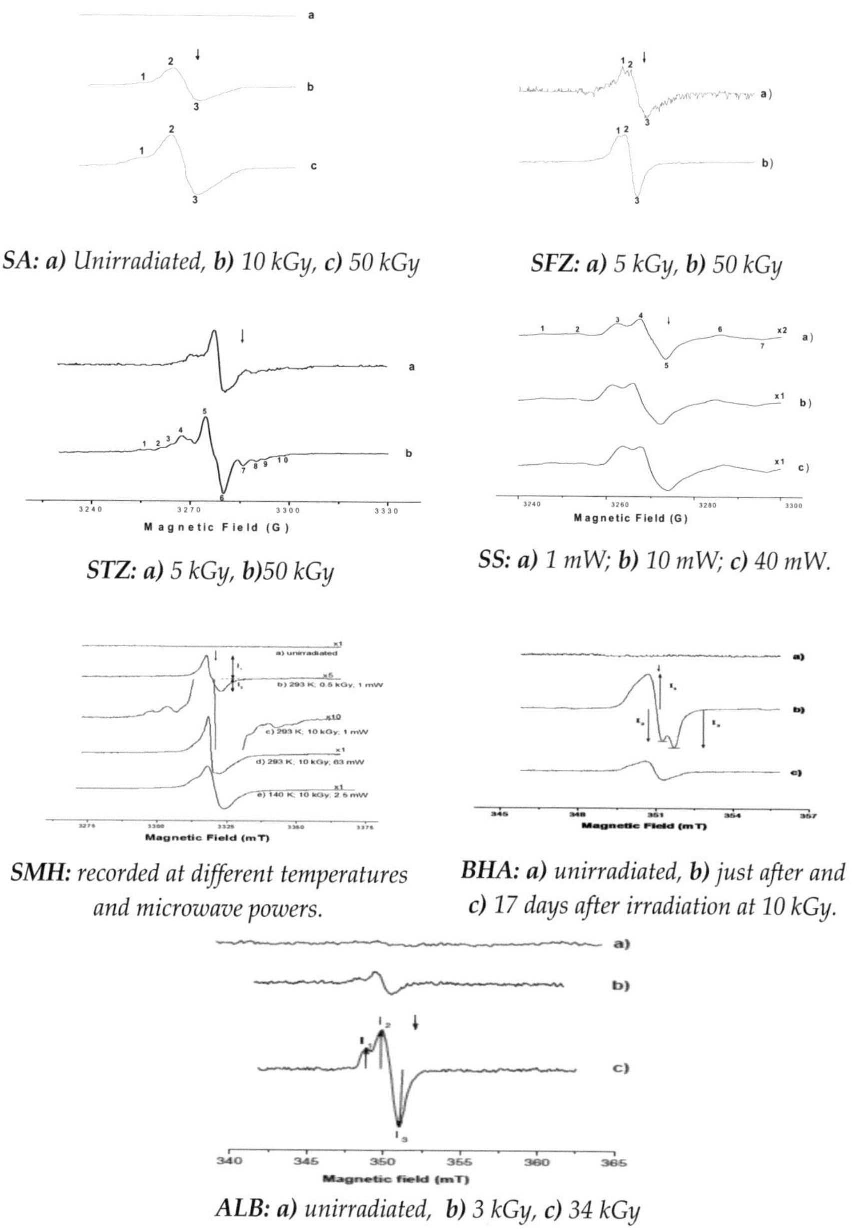


**Figure 1.** Room temperature ESR spectra of SA, SFZ, STZ, SS, SMH, BHA, ALB. Arrow indicates the position of DPPH (standart sample) line. (Reproduced from [34], [36], [16], [38], [37], [39] with permissions from Elsevier and Taylor & Francis).

tions that different numbers of radicals for each sample with different decay kinetics were produced and that they undergo *first-order decay kinetics* was tried to describe the experimental decay data. That is, calculated experimental peak heights were fitted to the equation $I(t)=I_{oA}e^{-kAt}+I_{oB}e^{-kBt}+I_{oC}e^{-kCt}+I_{oD}e^{-kDt}+...$ comprising different number of first-order decay terms with

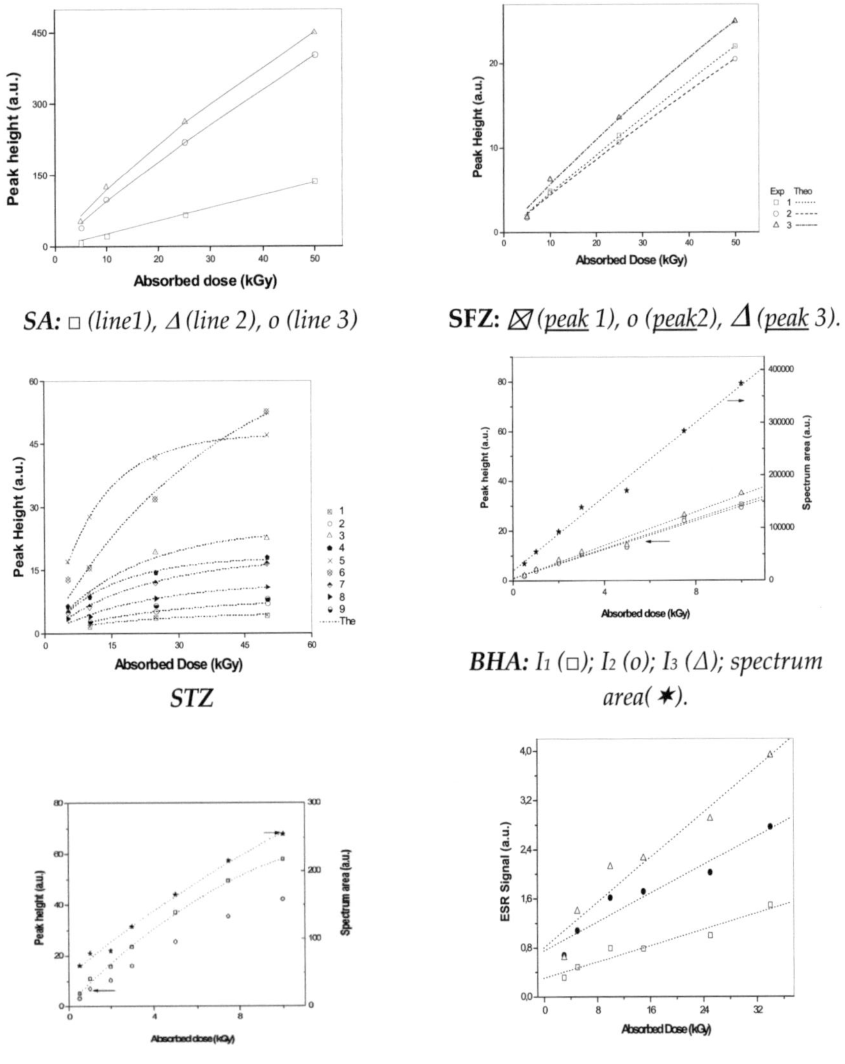


*SA:* □ *(line1), Δ (line 2), o (line 3)*

**SFZ:** ☒ *(peak 1), o (peak2), Δ (peak 3).*

*STZ*

***BHA:*** *I1 (□); I2 (o); I3 (Δ); spectrum area( ✶).*

***SMH:*** *I1 (□); I2 (o); spectrum area ( ✶).*

ALB:. *I1 (□); I2 (•); I3 (Δ).*

**Figure 2.** Dose-response curves for SA, SFZ, STZ, SMH,BHA, ALB. Symbols: experimental; solid line: theoretical curves best fitting to experimental data. (Reproduced from [34], [36], [38], [37], [39] with permissions from Elsevier and Taylor & Francis).

different decay constants and different weights. In the equation, t, $I_{os}$, and $k_s$ stand for the time elapsed after stopping irradiation, the initial signal intensities, and decay constants, respectively, for the contributing radical species in the samples. As the signal height decay data of different resonance lines were calculated to fit to the different exponential functions with different relative weights and decay constants, this result indicated that more than one radical species were induced upon irradiation in the samples which have different decay characteristics.

**For SA samples**, a slight increase in the height of line 1 at the beginning of storage period was the common behavior of the samples irradiated at different doses and stored at room temperature. This likely originates from the transformation of the radicals dominating line 2 and 3 to the radical giving rise to line 1 (Figure 3). **For STZ samples**, the decays over a period of 90 days of the peak heights of the samples stored at normal conditions were calculated best fitting to the sum of four exponentially decaying functions of different weights and of different decay constants. **For SS samples**, the data for the samples stored both at normal and stability conditions were found best fitting to the sum of four exponentially decaying functions with different decay constants. Peak heights of irradiated SS decreased ~90 % in the first 20 days after stopping irradiation when stored in stability conditions. For **BHA sample**, the model based on the presence of two different radical species undergoing first-order decay kinetics describes fairly well the experimental room temperature long-term decay data of this sample. For 25 kGy irradiated **SMH samples**, at the end of storage period of 40 days at normal conditions, a decrease of 10% of signal intensity is observed where it is an indication that the induced radical species upon irradiation in SMH are fairly stable at normal storage conditions. For 34 kGy irradiated **ALB sample** which is stored at normal conditions, at the end of 40 days of storage period the signal intensities of the peak height were decreased to 70% and it was ~85% at the end of 200 days. The discrimination of irradiated Albendazole from unirradiated one was possible even 6 months stored at normal conditions. The decay constants calculated for each sample for the related resonance signal is given in Table 2.

### 3.5. Variable temperature study

The results of variable temperature studies found from the variations of the peak heights with temperature in the range of 290-110 K and 295-400 K are given in Figure 4. The sample was first cooled down starting from room temperature with an decrement of 20 K. Then, the temperature was increased up to high temperatures with the same increment. Spectra were recorded ~5 min after setting the temperature. For the irradiated samples, cooling the sample down to room temperature did not create any pattern change in the spectra except slight reversible increases in the peak heights likely due to classical paramagnetic behavior of the contributing species, as obeying Curie's Law. But irreversible decreases in the intensities at high temperatures were observed for the samples. **For SA sample**, the increase in temperature produced no significant change in the peak heigths of the resonance lines up to 320 K. However, warming the sample above this temperature caused continous decreases in the peak heights of the resonance lines 2 and 3, but it created a slight increase in the height of line 1, up to 375 K, and then a relatively sharp decrease occurred. **For SFZ sample,** the increase of the sample temperature above 320 K caused continuous decreases in the heigths of the resonance peaks up to 360 K, and then a relatively sharp decrease in the heights occurred. **For STZ sample,** relatively sharp decreases especially in the heights of $5^{th}$ and $6^{th}$ peaks were observed for the sample temperature above 320 K. Above ~370 K, ESR spectrum of STZ turned into a singlet resonance line, indicating that most of the radical species were decayed completely at high temperatures, except that or those contributing to the central line (g=2.0045). **For SS sample,** warming the sample above room temperature produced no significant changes in the heights of the observed peaks where this result is accepted to be orginating from the high stabilities

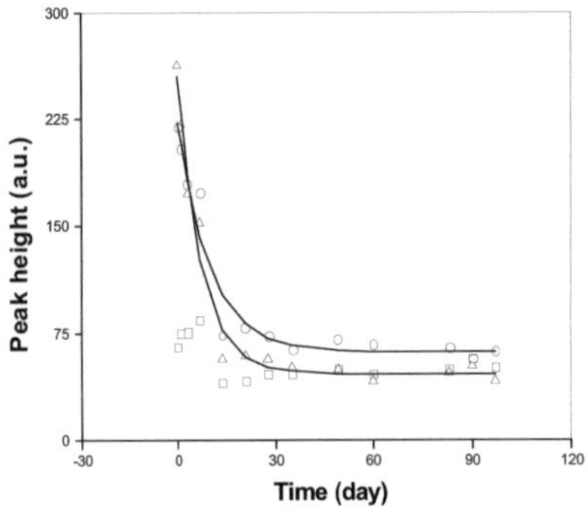


*SA irradiated at a dose of 25 kGy. Symbol:experimental [□ (1), Δ (2), o (3)].*

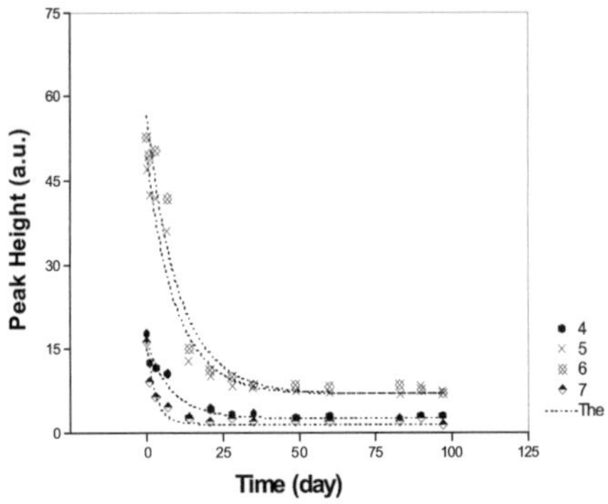


*STZ irradiated at a dose of 50 kGy.*

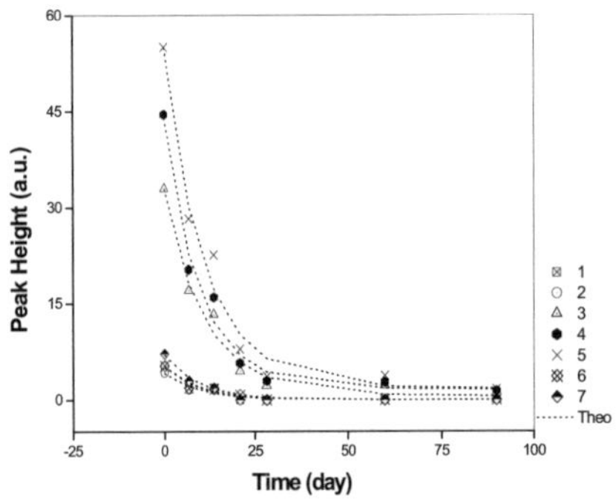


*SS irradiated at a dose of 50 kGy and stored at stability conditions.*

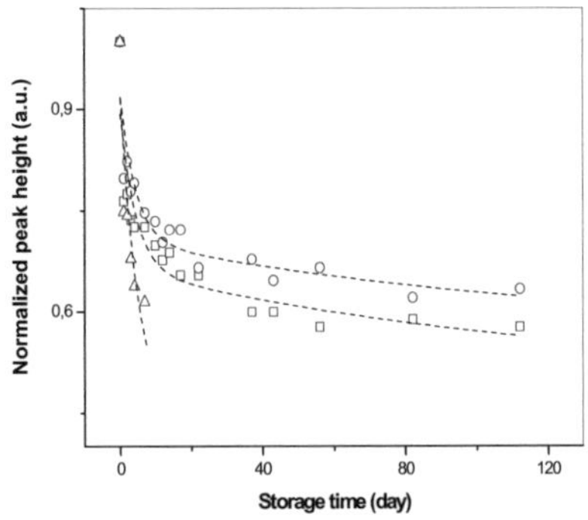


*BHA irradiated at a dose of 34 kGy. $I_1$ (□); $I_2$ (o); $I_3$ (Δ).*

**Figure 3.** Variations of ESR signal intensities of SA, STZ, SS, and BHA samples irradiated at different dose levels and stored over a long period of time. symbols: experimental; solid lines: theoretical. (Reproduced from [34], [36], [16], [38] with permissions from Elsevier and Taylor & Francis)

of the radical species induced in SS, probably because of the existing higher cage effect in its lattice. **For SMH sample**, relatively sharp drops were observed above 370 K. The slight increases of the peak heights in the range of 295-350 K were evaluated as originating from the decay of the radical species giving rise to weak satellite lines taking part at both sides of the central intense resonance line. **For BHA sample**, sharp irreversible decreases in the studied heights, related with important radical decays, were observed above 310 K. **For ALB sample**, irreversible decreases in the heights of all peaks, related with important radical decay, were also observed above room temperature.

### 3.6. Radical decays in annealed samples

Basing on the drastic decreases observed in the peak heights of the samples above room temperature, annealing studies were also performed to determine the kinetic features of the radical species which were responsible from experimental ESR spectra of gamma-irradiated

| | | | | | | | |
|---|---|---|---|---|---|---|---|
| **SA**(*stored at normal conditions*) | | | | | **STZ**(*stored at normal condition*) | | |
| **I** | **Radical Species** | **Relative weight** | **k (day$^{-1}$)** | **r$^2$** | **Radical Species** | **Decay constants(r$^2$) kx10$^5$ (day)$^{-1}$** | |
| 2 | A | 48.11 (±19.65) | 0.00053 (±0.00002) | 0.98 | A | 1483 (±107) | 0.98 |
| | B | 201.64 (±20.49) | 0.14487 (±0.03659) | | B | 8400 (±420) | |
| | | | | | C | 1178 (±95) | |
| | | | | | D | 9800 (±450) | |
| 3 | A | 70.84 (±6.27) | 0.00053 (±0.00002) | 0.94 | | | |
| | B | 155.24 (±13.02) | 0.14487 (±0.03659) | | | | |
| **SS** (*stored at normal and stability conditions*) | | | | | **BHA** (*stored at normal conditions*) | | |
| **Radical Species** | | **Decay constants kx10$^5$ (day$^{-1}$) Normal Stability Condition Condition** | | **r$^2$** | **Radical Species** | **Relative Weights $I_{1\ 3}$** | **k (day)$^{-1}$** |
| A | | 3 9901 | | 0.98 | A | 0.4219 0.3996 0.6062 | 0.008 |
| B | | (±1) (±285) | | 0.99 | | | |
| C | | 2 7856 | | | | | |
| D | | (±1) (±200) | | | | | |
| | | 46 11764 | | | | | |
| | | (±5) (±325) | | | | | |
| | | 167 9677 | | | | | |
| | | (±13) (±210) | | | | | |
| | | | | | B | 0.5181 0.6004 0.3938 ±0.8484 ±0.8731 ±0.8193 | 0.224 |

**Table 2.** Decay constants for contributing radicals calculated for SA, STZ, SS, and BHA. (Reproduced from [34], [36], [16], [38] with permissions from Elsevier and Taylor & Francis).

samples and calculating the activation energies relavant to the radical decay processes. Investigation of the contributing radical species by using ESR signal intensities in annealed samples is very important from the kinetic point of view. The fact that radical decay rates depend on the nature of the matrix containing radicals and annealing is a constant process with local diffusion of radicals and molecules in some softening of defects or irregularities [43]. At room temperature, the decay is very slow and many radical-molecule reactions observed in the liquid state are not observed in the solid state. Irreversible decreases in the intensities at high temperatures would be expected to originate from the decay of the radical species.

Thus, irradiated samples were annealed at different temperatures above room temperature; that is, below their melting temperature range for predetermined times. The decay rates of the radicals at high temperature are found to be higher than the decay rates at low temperatures.

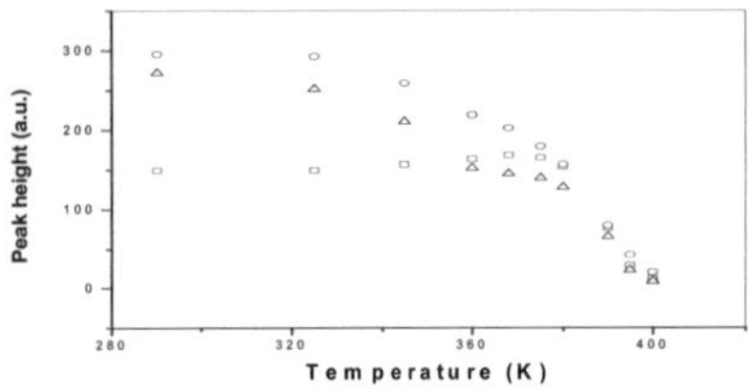


*SA:* □ *(line 1), Δ (line 2), o (line 3).*

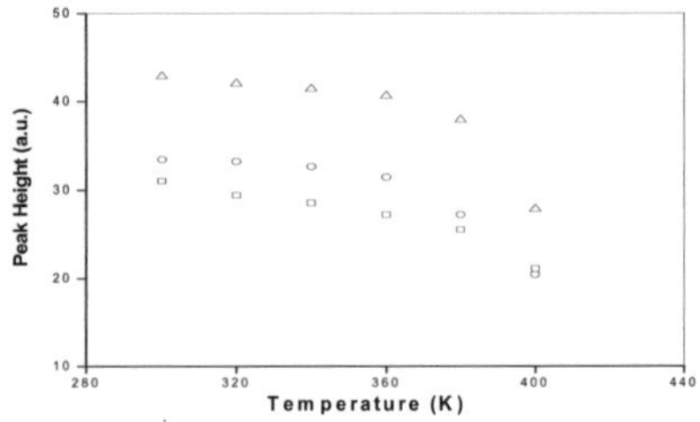


**SFZ:** ☒ *(line 1), o (line 2), Δ (line 3).*

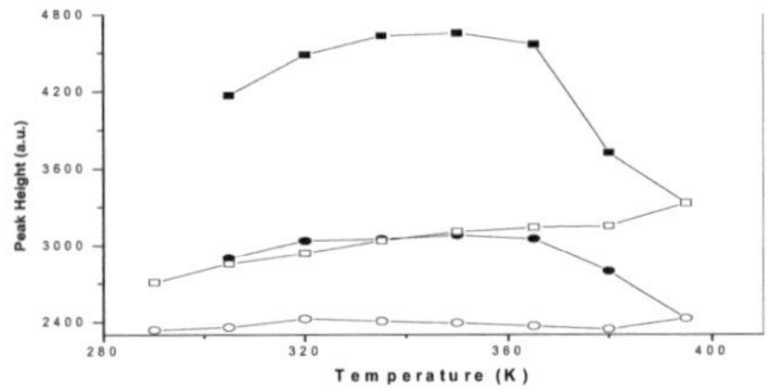


**SS:** □ *(line 1); o (line 2). Bolds: heating; open: cooling.*

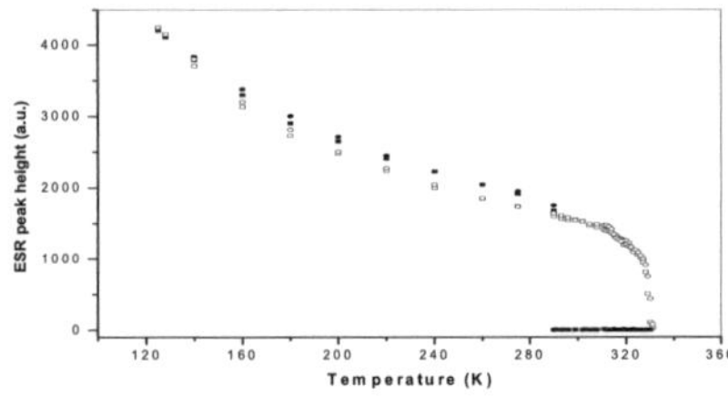


***BHA:*** □ *(line 1); o (line 2). Bolds: cooling; open: heating.*

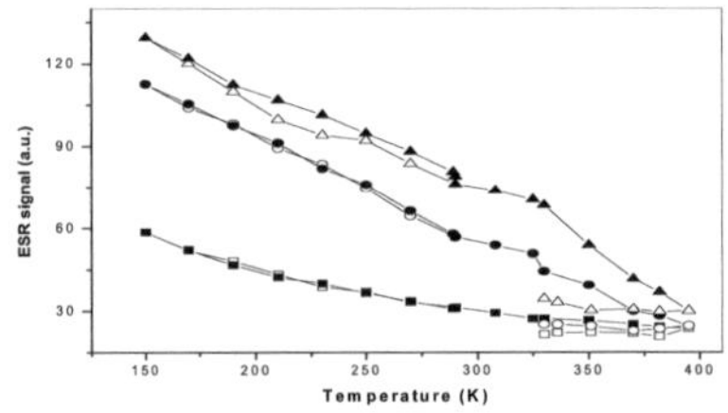


***ALB:*** □ *(line 1); o (line 2); Δ (line 3). Bolds: heating; open: cooling.*

**Figure 4.** Variation of peak heights of SA, SFZ, SS, BHA, ALB with temperature. (Reproduced from [34], [36], [38], [37] with permissions from Elsevier and Taylor & Francis)

Signal intensity decay results obtained for the samples irradiated at different doses and annealed at different temperatures for different times were used to get the decay curves of the resonance lines (Figure 5). Experimental peak height decay data obtained for the samples annealed at different temperatures were used to calculate the decay constants of the contributing species at the annealing temperatures, assuming that radical species induced upon irradiation follow a *first-order kinetics* as in the equation $I(t)=I_{oA}\, e^{-kAt} + I_{oB}\, e^{-kBt} + I_{oC}\, e^{-kCt} + I_{oD}\, e^{-kDt}$ +... where t, $I_{os}$, and $k_s$ stand for the storing times at the initial signal intensities, and decay constants, respectively, for the contributing radical species.

**For SA sample,** the radiation-induced radicals in the sample were observed to be very unstable even at room temperature. The decreasing rate of the intensities of the different peak heights indicated that there are more than one responsible induced radical species in the sample. All peaks suffer drastic intensity decreases above 365 K and the longer the annealing time the higher the intensity decrease. **For SFZ sample,** the signal intensity decay results obtained for

http://dx.doi.org/10.5772/61052

peak 3 of a sample irradiated at a dose 50 kGy and annealed at different temperatures for different times indicated that two radical species with different decay constants were found to fit best the experimental signal intensity decay data. **For STZ sample,** the annealing results relative to the height of peak 5 of a sample irradiated at a dose of 50 kGy are best fitted to a model of four radical species with different decay constants. **For SS sample,** four radical species are found to be responsible for the peak 4 of the SS spectrum, which is irradiated at 50 kGy. **For SMH sample,** the variations of the peak 2 for a sample irradiated at a dose of 10 kGy indicated that a model predicting the presence of three radicals undergoing first-order kinetics were explaining the decay data very well. **For BHA sample**, variations of the peak 2 for a sample irradiated at 10 kGy indicated that side radical species decayed differently at room and at high temperatures. That is, up to the lower limit of melting region (~321 K) the decay is very slow, but in the melting temperature range (321-338 K) the decay rate of the radicals are much faster. **For ALB sample,** the decay rates of $I_2$+$I_3$ peak heights of sample irradiated at 34 kGy, which was annealed, indicated that a model of two radical species could be used best in the fitting process of the experimental decay data recorded. An Arrhenius plot was also constructed to determine the activation energies of the contributing radical species for each irradiated sample. The activation energies are calculated from the slopes of the straight lines of ln(k) = f (1/T) of the Arrhenius plot constructed by using decay constants given in Table 3.

| | | STZ | | | | SS | |
|---|---|---|---|---|---|---|---|
| **Ann. Temp. (K)** | **Type** | **kx10$^5$ (min)$^{-1}$** | **$r^2$** | **Ann.Temp. (K)** | **Type** | **kx10$^5$ (min$^{-1}$)** | **$r^2$** |
| 310 | A | 9 (±1) | 0.96 | 358 | A | 9 (±1) | 0.98 |
| | B | 52 (±7) | | | B | 7 (±1) | |
| | C | 10660 (±325) | | | E | 69 (±4) | |
| | D | 2451 (±95) | | | F | 31 (±3) | |
| 348 | A | 59 (±8) | 0.97 | 393 | A | 1436 (±115) | 0.99 |
| | B | 380 (±30) | | | B | 1234 (±80) | |
| | C | 27000 (±430) | | | E | 2755 (±180) | |
| | D | 11000 (±360) | | | F | 1257 (±105) | |
| 365 | A | 344 (±45) | 0.98 | 413 | A | 4536 (±185) | 0.99 |
| | B | 1230 (±75) | | | B | 3547 (±155) | |
| | C | 46000 (±550) | | | E | 7215 (±220) | |
| | D | 17000 (±280) | | | F | 5390 (±205) | |
| 393 | A | 1575 (±105) | 0.99 | | | | |
| | B | 5500 (±120) | | | | | |
| | C | 68000 (±310) | | | | | |
| | D | 35200 (±180) | | | | | |
| 413 | A | 1675 (±120) | 0.99 | | | | |
| | B | 6462 (±110) | | | | | |

| | | | |
|---|---|---|---|
| | C | 102000(±220) | |
| | D | 56207 (±205) | |

| SMH | | | | ALB | | | |
|---|---|---|---|---|---|---|---|
| Ann. Temp. (K) | Type | kx10$^3$ (min$^{-1}$) | r$^2$ | Ann. Temp. (K) | Type | kx10$^3$ (min$^{-1}$) | r$^2$ |
| 365 | A | 0.9 | 0.99 | 370 | L | 118.65 | 0.98 |
| | B | 3.6 | | | M | 3.17 | |
| | H | 80.0 | | | | | |
| 375 | A | 1.1 | 0.99 | 380 | L | 127.71 | 0.99 |
| | B | 7.9 | | | M | 6.99 | |
| | H | 180.3 | | | | | |
| 385 | A | 3.7 | 0.99 | 390 | L | 172.71 | 0.97 |
| | B | 67.8 | | | M | 7.43 | |
| | H | 221.2 | | | | | |
| 390 | A | 4.5 | 0.99 | 400 | L | 397.14 | 0.95 |
| | B | 118.1 | | | M | 12.16 | |
| | H | 389.2 | | | | | |
| 395 | A | 10.7 | 0.99 | | | | |
| | B | 212.6 | | | | | |
| | H | 440.1 | | | | | |

**Table 3.** Decay constants at different temperatures for the radicals contributing to the ESR spectra of the irradiated STZ, SS, SMH, ALB. (Reproduced from [34], [36], [16], [38] with permissions from Elsevier and Taylor & Francis).

## 3.7. Radical type

Excited molecules are produced both directly and through radical-cations neutralization reactions [44]. They may decompose to radicals by rupture of chemical bonds. However, all species produced after irradiation are expected to undergo immediate germination termination reactions [43] due to cage effect. Consequently the amounts of the species responsible from the ESR spectra would be different depending on the capacity of these species participating to the germination reaction. Excited molecules and, as a result, radicals are localized along the track in region of high local concentration.

### *3.7.1. Proposed radical species for sulfanomides: SA, SFZ, STZ, SS, SMH*

It is believed that the **molecular ionic fragment** (radical A) and **[O=S=O]$^*$ ionic radical** (radical B) are the main responsible units from the resonance lines of ESR spectra of the gamma-irradiated sulfanomides of SA, SFZ, STZ, SS, and SMH. These two radical species were assumed to be produced upon irradiation and giving rise to *isotropic* and *axially symmetric* ESR spectra of the sulfanomide samples. Experimental g values determined in the studies for the

http://dx.doi.org/10.5772/61052

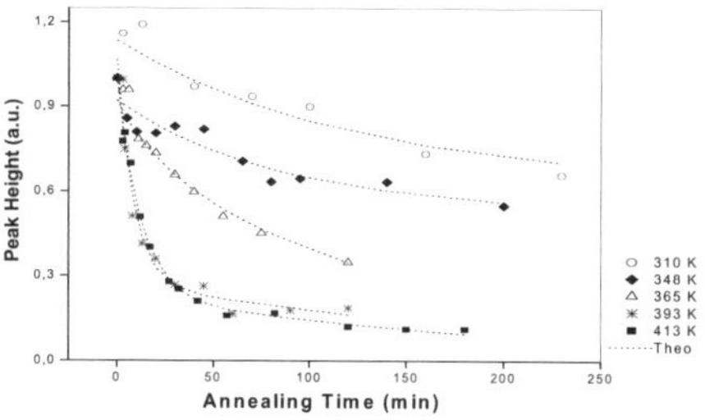


*Peak 5 of 50 kGy irradiated **STZ** and annealed at ○ (310 K), ♦ (348K), Δ (365 K), ж (393 K), ■ (413 K).*

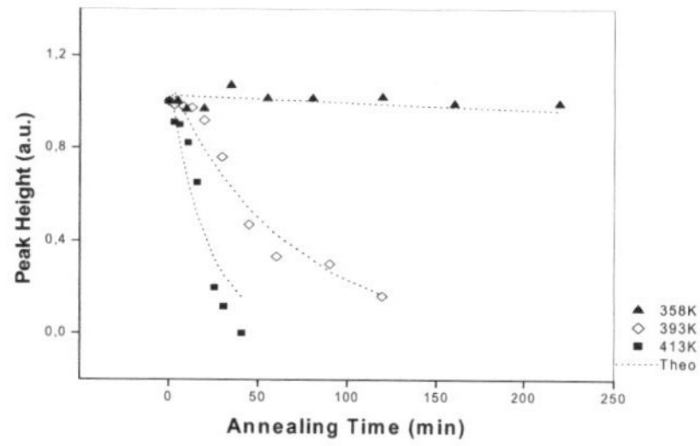


*Peak 4 of 50 kGy irradiated SS and annealed at ▲(358K), ◊(393K), ■Δ (413K).*

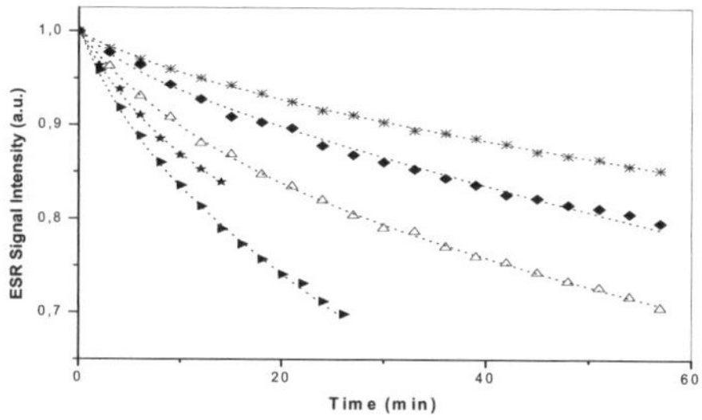


*Peak 2 of irradiated smh with annealing 365K (✵); 375 K (◆); 385 K (Δ); 390 K (★); 395 K (▶).*

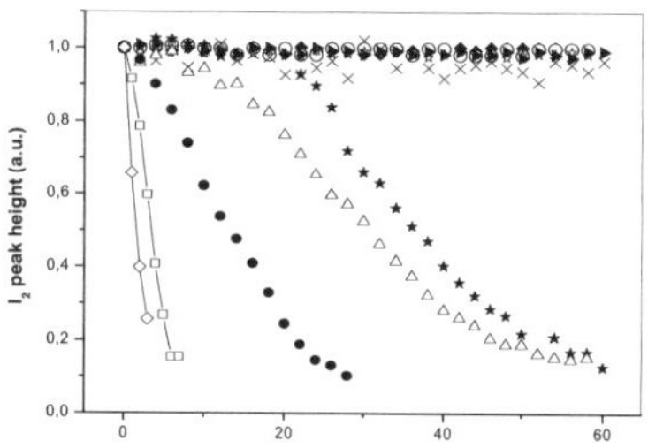


*Peak 2 of BHA with annealing 300 K (◆); 308 K (⊙); 312 K (✰); 318 K (▶); 320 K (x); 326 K (★); 327 K (Δ); 328 K (●); 329 K (□); 330 K (◊).*

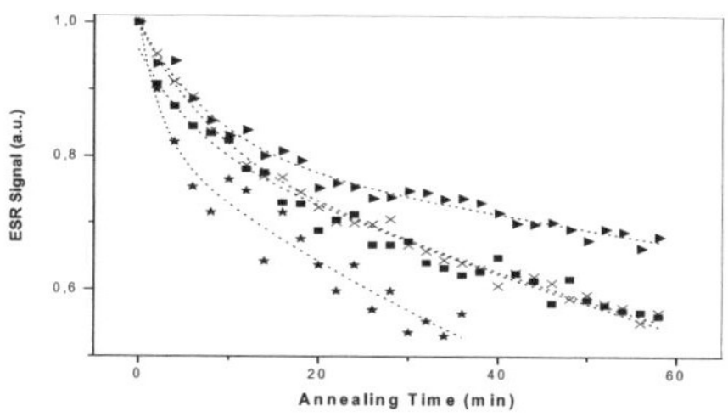


*Variations of the peak height of 34 kGy irradiated ALB with annealing 370 K (▶); 380 K (x); 390 K (■); 400 K (★).*

**Figure 5.** Decay results for the resonance lines of the irradiated samples at different dose levels and annealed at different temperatures for different times. Symbol: experimental, line: theoretical curves. (Reproduced from [36], [16], [38], [37], [39] with permissions from Elsevier and Taylor & Francis).

resonance peaks of the gamma-irradiated samples fall into the expected g value range where sulfur radicals do not exhibit hyperfine structure. The unpaired electron in **$SO_2^-$ ionic radical** (radical A) occupies the antibonding 2b* orbital formed from p orbitals of the S atom. Radicals A and B produced upon irradiation in powder of investigated sulfanomides are randomly oriented and the motion of radical A is restricted in large extent due to the big group attached to it, so that it give rise to powder ESR spectra with principal g values varying between $g_{xx}$=2.0022-2.0031, $g_{yy}$=2.0015-2.0098, and $g_{zz}$=2.0058-2.0066 [45, 46]. As for $SO_2^-$ ionic radical (radical B), its motional freedom is high due to very weak steric effect experienced by this radical and it gives rise to a single resonance line of average spectroscopic g factor varying between 2.0037-2.0059 [44-46].

**For SA and SFZ samples**, only the two radical species (A and B) are assumed to be induced upon irradiation but the spectral parameters of these radical types are different depending on the different lattice specifications of the samples (Table 4). **For STZ sample**, besides these two radical types named A and B, other two types of radical species are also likely produced after hemolytic ruptures of **S-N** (radical C) and **C-S** (radical D) chemical bonds in the irradiated STZ. **For SS sample**, in addition to radical species A and B, (radical E) and (radical F) were found to be involved in the irradiated SS. Radical E is formed after the break of N-Na bond and it has hyperfine splittings as the unpaired electron is localized on nitrogen atom. However, methyl radical is formed after the break of C-C bond and its unpaired electron is expected to be localized on the carbon atom. It has four peaks with 1:3:3:1 relative weight due to the methyl hyrogens. **For SMH sample**, besides the radical species of A and B, a third different radical species (radical H) is also assumed to be responsible from the very weak satellite peaks taking part at both sides of the central intense line which is believed to be produced after hemolytical rupture of **S-N bond** causing the creation of a species having unpaired electron localized rather on nitrogen atom and interacting with an alpha proton. That is, a species of type, which is assigned as **H**.

#### *3.7.2. Proposed Radical Species for BHA and ALB*

**For BHA samples**, species exhibiting single isotropic resonance line is believed to result after removing hydroxyl hydrogen from the molecule. This gives rise to a radical (radical J) with unpaired electron localized on hydroxyl oxygen and not interacting with any hydrogen of the molecule. It is denoted **radical J** in the present work. Second species (radical K) is likely formed by removing one of the neighboring ring hydrogen atoms on ortho or para position conducting to the formation of the radical with unpaired electron localized on one of the ring carbon atoms. Unpaired electron of this species exhibits an axial hyperfine interaction with its nearest neighbor ring hydrogen atom due to planar nature of the ring. Second species is denoted **radical K** in the present work. They are proposed to be produced by removing hydroxal and one of the ring protons from the molecule. Their production is preferential. **For ALB samples, radical L** and **radical M** are believed to be created by dissociations of S-H bonds and one of the C-H bonds at the position five or six on the benzene ring, respectively. Dissociations of S-H bonds create species with unpaired electron localized on sulfur and dissociation of C-H bonds creates species with unpaired electron localized on carbon atom. The neighboring protons to sulfur atoms gives rise to line broadening in species L.

### 3.8. Spectrum simulation calculations and proposed tentative radical species

Simulation calculations were performed to support the idea put forward with the species responsible from the observed experimental resonance peaks of ESR spectra of gamma-irradiated samples and to determine correct spectroscopic parameters of the contributing species. For the simulation calculations, the room temperature experimental signal intensity data obtained from the ESR spectra of the irradiated sample were used as input to perform simulation calculations. A model of different numbers of radical species depending on the

http://dx.doi.org/10.5772/61052

samples was adopted throughout the calculations. Spectral parameter values determined by this technique for contributing radical species and theoretical ESR spectra calculated using the corresponding experimental counterpart are presented together in Table 4. The agreement between experimental and theoretical spectra is fairly good, which indicates that the modelings based on the expected species of different characteristic features explains well the experimental ESR spectra of gamma-irradiated samples.

**SA**

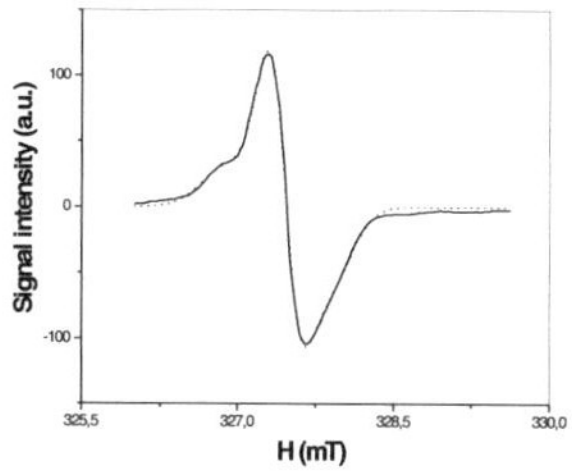


***SA** (solid line: experimental; dashed line: theoretical).*

| Radical Species | Relative Weight | Line width (G) | g factor |
|---|---|---|---|
| A | 7.32 | 2.43 | $g_{II}$=2.0089 $g_{\perp}$=2.0035 |
| B | 226.47 | 1.59 | $g_{av}$=2.0052 |

**SFZ**

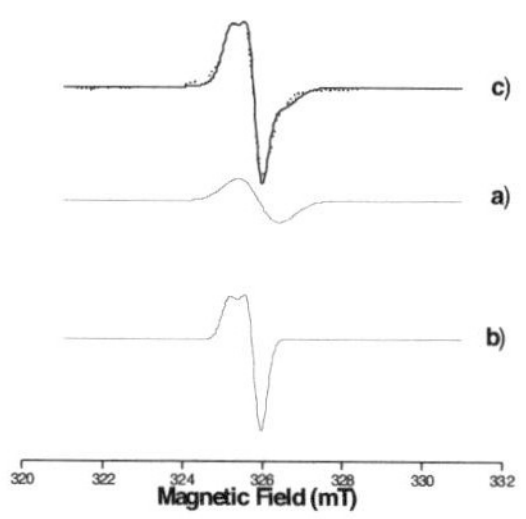


***a)** Experimental and calculated ESR spectra for irradiated **SFZ**, **b)** calculated spectra for radical species **A** and **B**, respectively, **c)** experimental and calculated sum spectra.*

| Radical Species | g factor | Line Width (G) | Relative Weight |
|---|---|---|---|
| A | 2.0042 | 5.10 | 0.52 |
| B | 2.0090 2.0047 | 1.78 | 0.48 |

**STZ**

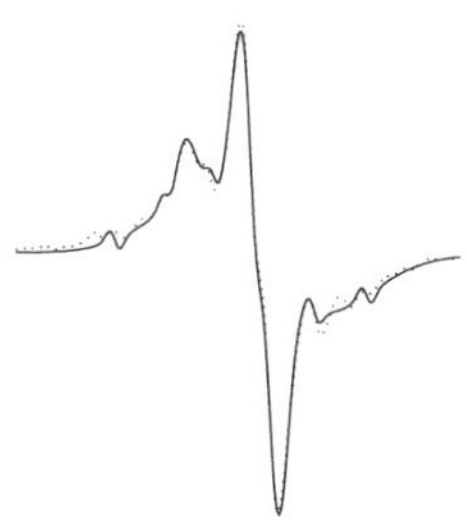

*Experimental and calculated ESR spectra for irradiated **STZ**. Solid line: theoretical, dashed line: experimental*

| Radical Species | g factor | Line Width (G) | $A_N$ AH (G) (G) | Relative Weight |
|---|---|---|---|---|
| A | 2.0051 | 1.5 | - - | 0.06 |
| B | 2.0106 2.0030 | 1.4 | - - | 0.21 |
| C | 2.0059 | 0.8 | 13.7 3.6 | 0.01 |
| D | 2.0039 | 8.0 | - 3.2 | 0.73 |

| | SS | | | |
|---|---|---|---|---|
| | Radical Species | g factor | Line Width (G) | Hyperfine Splitting A (G) |
| 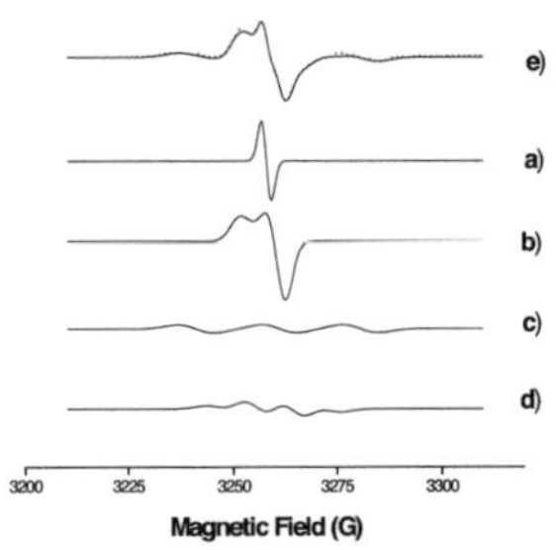 | A | 2.0052 | 1.21 | - |
| | B | 2.0092<br>2.0031 | 2.15 | - |
| | E | 2.0033 | 4.25 | 19.7 |
| *Experimental and calculated ESR spectra for* ***SS*** *irradiated at a dose of 50 kGy.* ***a), b), c), d):*** *calculated spectra for radical species A, B, C and D, respectively.* ***e):*** *experimental and calculated sum spectra.* | F | 2.0040 | 3.50 | 8.6 |



| | SMH | | | | |
|---|---|---|---|---|---|
| | Radical Species | g factor | Line Width (G) | Hyperfin<br>e<br>Splitting<br>$A_N$(mT)<br>$A_H$(mT) | Relative Weight |
| 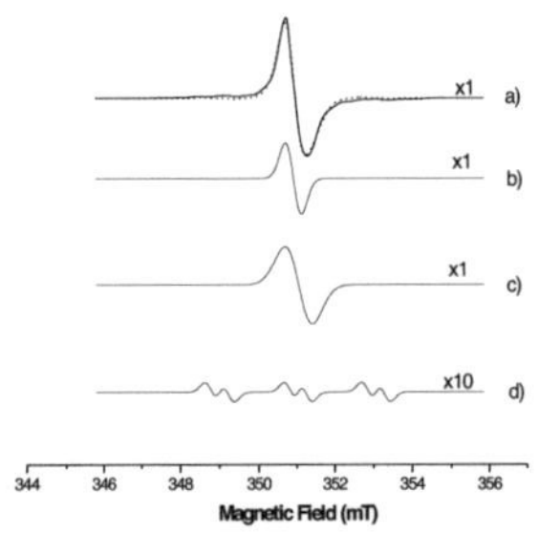 | A | 2.0072 | 0.21 | -<br>- | 0.47 |
| | B | 2.0059<br>2.0075 | 0.35 | -<br>- | 0.51 |
| *Experimental (solid line) and theoretical (dashed line) ESR spectra of irradiated* ***SMH. a)*** *sum spectra,* ***b)*** *species A;* ***c)*** *species B;* ***d)*** *species H* | H | 2.0066 | 0.16 | 2.03<br>0.21 | 0.02 |



| | BHA | | | | |
|---|---|---|---|---|---|
| | Radical Species | g factor | Line Width (G) | Hyperfin<br>e<br>Splitting<br>$A_N$(mT)<br>$A_H$(mT) | Relative Weight |
| 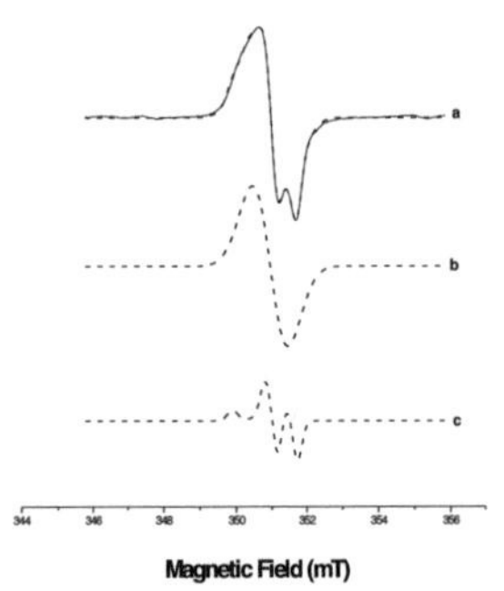 | J | 2.0057 | 0.50 | -<br>- | 0.834 |
| | K | 2.0047<br>2.0067 | 0.14 | 0.1584<br>0.9009 | 0.166 |

http://dx.doi.org/10.5772/61052

*Experimental (solid line) and theoretical (dashed line) ESR spectra of irradiated* ***BHA.a)*** *sum spectra;* ***b)*** *species J;* ***c)*** *species K.*

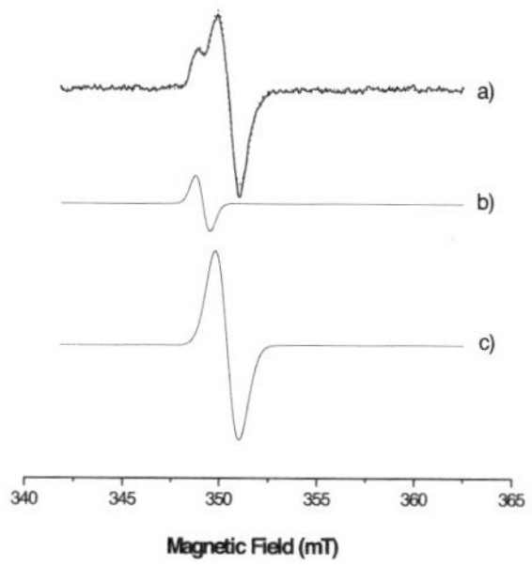


*Experimental (solid line) and theoretical (dashed line) ESR spectra of irradiated* ***ALB. a)*** *sum spectra;* ***b)*** *species L;* ***c)*** *species M.*

| Radical Species | g factor | Line Width (G) | Relative Weight |
|---|---|---|---|
| L | 2.0161 | 3.8 | 0.10 |
| M | 2.0088 | 6.16 | 0.90 |

**Table 4.** Experimental and calculated ESR spectra for irradiated samples and calculated spectral parameters for contributing radical species. (Reproduced from [34], [36], [16], [38], [37], [39] with permissions from Elsevier and Taylor & Francis)

**For SA sample**, a model of **two radical species** was adopted throughout the calculations. Although the line width of radical B is small, it dominates ESR spectrum and radical A contributes mainly to the line centered at g=2.0089. **For SFZ sample,** the line width of species B is small compared with ionic radical A, but both species equally dominate ESR spectrum. **For STZ samples,** simulation calculations were performed assuming the presence of four radical species, exhibiting isotropic (species A, C, and D) and axially symmetric (species B) g tensors. Linewidth and relative weight of the radical D are found to be fairly large compared with other radical species due to unresolved hyperfine splitting. **For SS sample,** the line width of radical E is large compared with other radical species and radicals A and B dominate together ESR spectrum. **For SMH samples**, three radical species were performed in the calculations, exhibiting isotropic (species A and H) and axially symmetric (species B) g tensors. It is seen that radical species A and B dominate the central part of the experimental spectrum; the radical H, with its small weight, gives rise to the appearance of the satellite lines of the very weak intensities. **For BHA samples,** for the simulation calculations two radical species named J and K were assumed and it is concluded that species J is produced more easily and therefore it dominates experimental spectrum. **For ALB samples,** simulation calculations were based on a model predicting the presence of two radical species where species M with its relatively high concentration and big linewidth dominates the ESR spectrum and species L has contributions to the peaks 1 and 2.

### 3.9. FT-IR studies

IR spectroscopy is also used as a complementary technique for ESR spectrocopy in the determination of the radical species induced upon irradiation for the investigated samples. For this purpose, FT-IR spectra of **SFZ**, **STZ**, and **SS** samples both for unirradiated and gamma-irradiated cases were recorded at room temperature. However, no significant differences between the unirradiated and irradiated IR spectra for the radical species created by gamma radiation were observed. Different spectra were also constructed using IR spectra of unirradiated and irradiated samples with using the same masses of the samples but still no meaningful results were obtained. Thus, it was concluded that the amounts of the radiation-induced intermediates were very small where they cannot create any significant detectable changes in the IR bands of the samples. This result is in agreement with that of relatively small G value (approximately, 0.1-0.5) calculated for the investigated samples by using ESR technique.

## 4. Conclusion

Experimental results derived for studied drugs/drug raw materials showed that they were not sensitive (small G value) to high energy radiations. Therefore, they could be sterilized by gamma radiation up to a radiation dose of 50 kGy without causing very much molecular damages upon irradiation. Also, they do not present the features of a good dosimetric material. That is, they cannot be used as an effective dosimetric material in the drug sterilization radiation dose limits. Nevertheless, the detection and discrimination of an unirradiated sample from irradiated samples turned out to be possible even at low radiation doses by ESR spectroscopy. Radical species created upon irradiation were found to decay much faster at stability conditions (40°C and 75% humidity) than at normal conditions as in the case of the samples in liquid forms. This point was considered presenting the possibility of diminishing even getting rid of radiolytical intermediates produced in irradiated samples. Therefore, it was concluded that gamma radiation produces relatively low amounts of radiolytic intermediates in the studies drugs/drug raw materials and that ESR spectroscopy could be used as a potential technique in monitoring the radiosterilization of drugs, drug raw materials, and drug delivery systems containing present samples as active ingredients.

## Acknowledgements

I am grateful to **Prof. Dr. Mustafa Korkmaz** from Hacettepe University, Department of Physics Engineering, for sharing his excellent experience, for very valuable information and for his encouragement.

http://dx.doi.org/10.5772/61052

## Author details

Şeyda Çolak*

Address all correspondence to: seyda@hacettepe.edu.tr

Hacettepe University, Physics Engineering Department, Ankara, Turkey

## References

[1] Silva Aquino K. A. Sterilization by Gamma Irradiation, Gamma Radiation. In: Adrovic F. (ed.), Gamma Radiation. Rijeka: InTech; 2012. P171-206. Available from http://www.intechopen.com/books/gammaradiation/sterilization-by-gamma-irradiation in-chickpea (accessed 21 March 2012).

[2] Jacobs, G.P. A Review of the Effects of Gamma Radiation on Pharmaceutical Materials. J Biomater Applic 1995; 10 59-96.

[3] Reid, B. D. Gamma Processing Technology: An Alternative Technology for Terminal Sterilization of Parenterals. J Pharmaceut Sci Technol 1995; 49 (2) 83-89.

[4] Barbarin, N., Rollmann B. and Tilquin B. Role of Residual Solvents in the Formation of Volatile Compounds after Radiosterilization of Cefotaxime. Int J Pharmaceut 1999; 178 203-212.

[5] Boess, C. and Bögl K. W. Influence of Radiation Treatment on Pharmaceuticals - A Review: Alkaloids, Morphine Derivatives and Antibiotics. Drug Dev Indus Pharm 1996; 22 (6) 495-529.

[6] Maghraby A. M. Ionizing Radiation Induced Radicals. In: Nenoi M. (ed.) Current Topics in Ionizing Radiation Research. Rijeka: InTech; 2012. P649-682. Available from http://www.intechopen.com/books/current-topics-in-ionizing-radiation-research/ionizing-radiation-inducedradicals- in-chickpea (accessed 12 February 2012).

[7] Pourahmad, R. and Pakravan, R. Radiosterilization of Disposable Medical Devices. Rad Phys Chem 1997; 49 285-286.

[8] Jacobs, G. Radiation in the Sterilization of Pharmaceuticals, Sterile Pharmaceutical Manufacturing. Buffalo Groze: Interpharm. Press; 1991.

[9] Duroux, J. L., Basly J. P., Penicaut, B. and Bernard M. ESR Spectroscopy Applied to the Study of Drugs Radiosterilization: Case of Three Nitroimidazoles. Appl Rad Isotopes 1996; 47 (11/12) 1565-1568.

[10] Bhalla H. L., Menon M. R. and Gopal N. G. S. Radiation Sterilization of Polyethylene Glycols. Int J Pharmaceut 1983; 17 351-355.

[11] Wogl, W. Radiation Sterilization of Pharmaceuticals, Chemical Changes and Consequences. Rad Phys Chem 1985; 25 425-435.

[12] Jacobs, G. P., Wills, P. A. Recent Developments in the Radiation Sterilization of Pharmaceuticals. Rad Phys Chem 1988; 31 685-691.

[13] Phillips, G. O. Radiation Technology in Surgery and Pharmaceutical Industry: An Overview of Applications. IAEA Bull 1994; 1 175-185.

[14] Varshney, L. and Patel K. M. Effects of Ionizing Radiations on a Pharmaceutical Compound Chloramphenicol. Rad Phys Chem 1994; 43 (5) 471-480.

[15] Safrany, A. Radiation Processing: Synthesis and Modification of Biomaterials for Medical Use. Nucl Instru Meth Phys Res B 1997; 131 376-381.

[16] Çolak Ş. and Korkmaz M. Investigation of Radiosterilization and Dosimetric Features of Sulfacetamide Sodium. J Pharmaceut Biomed Anal 2004; 36 791-798.

[17] Maksimenko, O., Pavlov, E., Toushov, E., Molin, A., Stukalov, Y., Prudskova, T., Feldman, V., Kreuter, J., Gelperina, S. Radiation Sterilization of Doxorubicin Bound to Poly(butyl cyanoacrylate) Nanoparticles. Int J Pharmaceut 2008; 356 325-332.

[18] Trends in Radiation Sterilization of Health Care Products, International Atomic Energy Agency, Vienna, 2008.

[19] EN 552, Sterilization of Medical Devices: Validation and Routine Control of Sterilization Irradiation, CEN, European Committee for Standardization, Brussels, Belgium; 1994.

[20] ISO 11137, Sterilization of Health Care Products: Requirements for Validation and Routine Control. Radiation Sterilization; International Organization for Standardization, Geneva, Switzerland; 1995.

[21] Gopal, N. G. S., Patel, K. M., Sharma, G., Bhalla, H.L., Wills, P. A., Hilmy, N. Guide for Radiation Sterilization of Pharmaceuticals and Decomposition of Raw Materials. Rad Phys Chem 1988; 32 619-622.

[22] Gibella, M., Crucq, A. S., Tilquin, B., Stocker, P., Lesgards, G. and Raffi, J. ESR Studies of Some Irradiated Pharmaceuticals. Rad Phys Chem 2000; 58 69-76.

[23] Schuler R. H. Three Decades of Spectroscopic Studies of Radiation Produced Intermediates. Rad Phys Chem 1994; 43 417-423.

[24] Miyazaki, T., Arai, J., Kaneko, K., Yamamoto, K., Gibella, M., Tilquin, B. Estimation of Irradiation Dose of Radiosterilized Antibiotics by ESR: Ampicillin. J Pharmaceut Sci 1994; 83 1643-1644.

[25] Bögl K. W. Identification of Irradiated Foods - Methods, Development and Concepts. Appl Rad Isotopes 1989; 40 1203-1210.

[26] Delincée H. Analytical Methods for Irradiated Foods. A Review of Current Literature. IAEA-TECDOC-587, International Atomic Agency, Vienna; 1991.

[27] Desrosiers, M. F., Wilson, G. L., Hunter, C. R., Hutton, D. R. Estimation of the Absorbed Dose in Radiation-Processed Food. Part 1. Test of the EPR Response Function by a Linear Regression Analysis. Appl Rad Isotopes 1991; 42 613-616.

[28] Raffi, J. Delincée, H., Marchioni E., Hasselmann C., Sjöberg A.M., Leonardi M., Kent M., Bögl K.W., Schreiber G., Stevenson H. and Meier W. Final Report on New Methods for the Detection of Irradiated Food, BCR, CEC, Luxembourg, EUR 15 261 EN; 1994.

[29] Ciranni Signoretti, E., Valvo L., Fattibene P., Onori S. and Pantoloni M. Gamma Radiation Induced Effects on Cefuroxime and Cefotaxime, Investigation on Degradation and Syn-Anti Isomerization. Drug Dev Indus Pharmacy 1994; 20 (16) 2493-2508.

[30] Onori S., Pantoloni M., Fattibene P., Ciranni Signoretti E., Valvo L. and Santucci M. ESR Identification of Irradiated Antibiotics: Cephalosporins. Appl Rad Isotopes 1996; 47 (11/12) 1569-1572.

[31] Çolak Ş., Maquille A., Tilquin B. Radiation Sterilization of Ketoprofen: ESR, HPLC, LC-MS, GC-MS Studies. Rad Effects Defects Solids 2006; 161 (1) 75-82.

[32] Hawkins, C. L., Davies, M. J. Detection and Characterisation of Radicals in Biological Materials using EPR Methodology. Biochim Biophys Acta 2014; 1840 708-721.

[33] Olguner Mercanoğlu G., Özer, A. Y., Çolak, Ş., Korkmaz, M., Barbarin, N., Tilquin, B., Özalp, M., Ekizoğlu, M. Radiosterilization of Sulfonamides I: Determination of the Effects of Gamma Irradiation on Solide Sulfonamides. Rad Phys Chem 2004; 69 511-520.

[34] Çolak Ş. and Korkmaz M. Investigation of Structural and Dynamic Features of the Radicals Produced in Gamma Irradiated Sulfanilamide: An ESR Study. Int J Pharmaceut 2003; 267 (1-2) 49-58.

[35] Çolak Ş. and Korkmaz M. Kinetics of the Radicals Induced in Gamma Irradiated Sulfafurazole: An EPR Study. Zeitschrift für Naturforschung 2004; 59a 481-487.

[36] Çolak Ş. and Korkmaz M. Spectroscopic Features of Radiolytical Intermediates Induced in Gamma Irradiated Sulfatiazole: An ESR Study. Int J Pharmaceut 2004; 285 1-11.

[37] Çolak Ş. and Korkmaz M. ESR Response of Gamma Irradiated Sulfamethazine. Rad Effects Defects Solids 2009; 164 (12) 788-799.

[38] Çolak Ş., Korkmaz M., Güneş Ç., Dölek, S. Investigation of Spectroscopic and Kinetic Features of the Radicals Produced in Gamma Irradiated Butylated Hydroxyanisole (BHA). Rad Effects Defects Solids 2009; 164 (2) 101-112.

[39] Çolak Ş. ESR Identification of Gamma Irradiated Albendazole. Rad Effects Defects Solids 2010; 165 (1) 72-82.

[40] Philips, G. O., Power, D. M. and Sewart, M. C. G. Effect of Gama Irradiation on Sodium Sulphacetamide. Rad Res 1971; 46 236-250.

[41] Philips, G. O.; Power, D. M. and Sewart, M. C. G. Effect of $\gamma$-Irradiation on Sulphonamides. Rad Res 1973; 53 204-215.

[42] Osol, A. Remington's Pharmaceutical Sciences. Easton: Mack Publishing; 1980.

[43] Tilquin, B. Composante Radicalaire des Transformations Radio-Initiées dans les Alcanes à 77 K. Thése d'agrégation. UCL, Ciaco-la-Neuve, Belgique;1985.

[44] Huzimura, R. ESR Studies of Radical Ion Centers in Irradiated $CaSO_4$. Jap J Appl Phys 1979; 18 2031-2032.

[45] Bershov, L. V., Martırsyan, V. O., Marfunin, A. S. and Speranskii, A. V. EPR and Structure Models for Radical Ions in Anhydrite Crystals. Fortschrıtte Der Mıneralogie 1975; 52 591-604.

[46] Samoilovich, M. I. and Tsinober, L. I. Characteristics of Radiation Color Centers and Microisomorphism in Crystals. Soviet Phys Crystallography 1970; 14 656-666.